Cowbirds and Other Brood Parasites

Catherine P. Ortega

Cowbirds
and Other Brood Parasites

The University of Arizona Press   *Tucson*

The University of Arizona Press
© 1998 The Arizona Board of Regents
First Printing
All rights reserved
♾ This book is printed on acid-free, archival-quality paper.
Manufactured in the United States of America

03 02 01 00 99 98   6 5 4 3 2 1

Library of Congress Cataloging-in-Publication Data
Ortega, Catherine P.
Cowbirds and other brood parasites / Catherine P. Ortega.
p. cm.
Includes bibliographical references and index.
ISBN 0816515271 (cloth : alk. paper)
1. Cowbirds. 2. Brood parasites. 3. Brood parasitism. 4. Birds,
Protection of. I. Title.
QL696.P2475 O775   1998
598.8'74—ddc21
98-8911
CIP

British Library Cataloguing-in-Publication Data
A catalogue record for this book is available from the British Library.

To Dr. Joseph C. Ortega

who for many years has provided me

with boundless and enduring love,

support, patience, and humor

Contents

Figures

Tables

Preface

In his keynote address at the 1993 North American Research Workshop on the Ecology and Management of Cowbirds held in Austin, Texas, Steve Rothstein expressed his surprise with the number of individuals attending the conference: 170 persons were on the roster of attendants. He remarked that when he began studying cowbirds in the 1960s, he felt as if he were in a lonely business and that no one would read the results of his studies. Herbert Friedmann was his only hero in the field of cowbirds. When I began studying cowbirds in the early 1980s, although much more information was available than when Rothstein started working in the field, I also noted the lack of meaningful long-term studies available in the literature, especially for birds that have such a potentially dramatic effect on the reproductive success of many other species of birds. The 1980s and the 1990s were met with a surge of interest in cowbirds and other brood parasites, and the interest was followed by a wealth of literature which has become almost overwhelming. In BIOSIS as of 1997, there were 690 papers with "cowbird(s)" in the title and 279 papers with "brood parasitism" in the title.

This surge of interest has undoubtedly been kindled by the decline of some migratory songbird populations and the perceived role that cowbirds play in those declines. In general, cowbirds are accused for the declines and are pigeonholed as pests more despicable than predators by those who do not understand them. It is common to read that an author removed all cowbird eggs from host nests, and Friedmann suggested that

> Persons wanting to attract birds to their gardens and have them nest there should destroy all [Brown-headed] Cowbird eggs found, and, if so inclined, destroy Cowbird eggs or young anywhere, but to proclaim the Cowbird an economic outlaw and persecute it accordingly would be a serious mistake. . . . The Cowbird may well be dispensed with locally but . . . on the whole it plays a very real and important part in the general scheme of nature, and consequently should not be outlawed. Because of the harm it does the Cowbird should not be granted protection. Legally it may best be left alone. (1929b:303)

It appears that Friedmann was suggesting that individuals take the law into their own hands, and this undoubtedly occurs across the United States as birdwatchers and perhaps some researchers find parasitized nests. However, Brown-headed Cowbirds are protected under the Migratory Bird Treaty of 1916, and removal of their eggs from host nests is a federal offense. Stephanie Jones of the U.S. Fish and Wildlife Service doubts that most cowbird eggs removed from host nests are removed with the required permit.

For a very brief time in our history, it may have been fashionable to search for virtues in Brown-headed Cowbirds. In 1896, Beal (in Salvage 1897) discovered through a stomach analysis study that Brown-headed Cowbirds eat harmful insects, weed seeds, and waste grain; he believed they should, therefore, be protected because they do more good than harm. Today, Beal might have a different opinion, but his attempt to understand and appreciate the ecological role of cowbirds was, nevertheless, admirable.

Cowbirds are unquestionably a scapegoat for human-caused ecological issues that are difficult to admit and complicated to manage. In addition to being blamed for declines of songbird populations, cowbirds fuel the fires of already contentious grazing issues. Introduction of livestock has allowed cowbirds to spread and return to the same areas year after year, offering no relief to host populations as probably occurred when cowbirds followed migratory herds of Bison. Habitat destruction in both breeding and wintering grounds of hosts has undoubtedly contributed far more to migratory bird declines than cowbirds. Through our pastoral activities and perceived need for lavish lifestyles, we have created the perfect and more or less permanent habitat for cowbirds. Whether or not we admit that cowbirds are scapegoats, we are presently faced with the ethical dilemma of managing the cunning brood parasites who have been part of the natural ecosystem since long before humans arrived in the Americas.

In order to effectively manage cowbirds without further damaging delicate balances in host-parasite relationships, we must understand many aspects of their behavior, their reproductive potential, the regulation of their populations and hosts' populations, and how they respond to different landscape patterns. Although a wealth of information is now available on cowbirds, no collective works have been published since Herbert Friedmann's 1929 book, *The Cowbirds: A Study in the Biology of Social Parasitism*. The profuse and widely scattered information that is now available can be daunting to individuals with an emerging interest in cowbirds or to persons in positions of managing cowbirds, persons who have neither the resources nor time to sort through scattered literature. I therefore perceived a timely need for an introductory book on cowbirds and brood parasitism. This book is written primarily for graduate students, ambitious undergraduates, those in managerial positions in various agencies and nature preserves, ecological consultants, and birdwatchers who would like to gain a more thorough understanding and appreciation of cowbirds.

Acknowledgments

I am indebted to many individuals for information and logistical support provided to me throughout this project. I thank Peter Lowther and an anonymous reviewer for their outstanding evaluations and comments on the first draft of this manuscript. The staff at the John Reed Library at Fort Lewis College, Durango, Colorado, was extremely helpful in fulfilling my relentless requests for interlibrary loans; in particular, I would like to thank Mary Mohs, Jeff Frisbie, and Donna Bidor for their assistance. The following museums provided me with either loans or information: Museum of Vertebrate Zoology, University of California, Santa Barbara; Museum of Vertebrate Zoology, University of California, Berkeley; Museum of Southwestern Biology, University of New Mexico, Albuquerque; National Museum of Natural History, Washington, D.C.; Museum of Natural History, University of Oregon, Eugene; James R. Slater Museum of Natural History, University of Puget Sound, Puget Sound, Wash.; Charles R. Conner Museum, Washington State University, Pullman. I would like to thank the following individuals for providing me with valuable information during the process of writing this book: Ralph Browning, Christy Carello, Jameson Chace, Alexander Cruz, Therese Donovan, Alfred Dufty, Jane Griffith, Enrique Hernandez-Prieto, Mark Holmgren, William Howe, Stephanie Jones, David Leal, Paul Mason, Brian Mauer, Tammie Nakamura, Kenneth Parkes, Bruce Peterjohn, Spencer Sealy, Janice Simpkin, Neal Smith, David Wiedenfeld, and Wayne Werkmeister. I would also like to extend a special thanks to my husband and research partner, Joseph Ortega, for his unfaltering support and enthusiasm throughout this project.

Cowbirds and Other Brood Parasites

Chapter 1

Introduction

The genus name of most cowbirds, *Molothrus*, means "intruder" in Latin. The title is rightly deserved, for these parasites intrude upon other birds by laying eggs in their nests. At this point, the cowbirds' parental duties are essentially done; they take no part in incubation or caring for the nestlings. This peculiar reproductive habit is known as brood parasitism. Among vertebrates, birds are well known for this alternative reproductive strategy, but brood parasitism is also known among several freshwater fishes (Sato 1986; Baba et al. 1990; Baba 1994). In brood parasitism, the hosts incubate parasitic eggs and raise the young as if these were their own, even though raising a parasite is usually detrimental to the host's own reproductive success. More often than not, parasitized hosts rear no offspring of their own.

Avian brood parasites are interesting from evolutionary, ecological, and ethological perspectives and, hence, have long attracted much attention from a variety of biologists. In recent years, nearly every issue of every major ornithological journal has contained at least one paper regarding some aspect of brood parasitism. Wildlife managers have recently become concerned with brood parasites because parasites lower the reproductive fitness of their hosts, and a few hosts have been listed as either endangered or threatened. Even with all the attention, much remains to be learned about brood parasites, and no large synthesis has been published on cowbirds since Friedmann's (1929b) classic book, *The Cowbirds: A Study in the Biology of Social Parasitism*.

Nonobligate Brood Parasitism

Not all avian brood parasites are cowbirds, and not all are obliged to the habit; those not obliged usually lay their eggs in nests of conspecifics as well as laying in their own nests. However, a few of these nonobligate or "facultative" parasites lay their eggs in the nests of other species.

Table 1.1

Reports of Nonobligate Interspecific Parasitism

Species/Source

Black-bellied Whistling-Duck *(Dendrocygna autumnalis)*: Bolen and Cain 1968
Ruddy Duck *(Oxyura jamaicensis)*: Joyner 1975, 1976, 1983; Siegfried 1976
Australian Blue Duck *(Oxyura australis)*: Attiwill et al. 1981
Musk Duck *(Bizura lobata)*: Attiwill et al. 1981
Snow Goose *(Anser caerulescens)*: Höhn 1972
Canada Goose *(Branta canadensis)*: Höhn 1972
Wood Duck *(Aix sponsa)*: Bellrose 1980
Red-crested Pochard *(Netta rufina)*: Amat 1985
Mallard *(Anas platyrhynchos)*: Leach 1994
Northern Shoveler *(Anas clypeata)*: Spencer 1953; Bellrose 1980
Redhead *(Aythya americana)*: Weller 1959; Joyner 1976, 1983; Nudds 1980; Sugden 1980;
    Giroux 1981; Talent et al. 1981; Bouffard 1983; Sayler 1996
Common Eider *(Somateria mollissima)*: Höhn 1972
Common Goldeneye *(Bucephala clangula)*: Mallory and Weatherhead 1993
Barrow's Goldeneye *(Bucephala islandica)*: Bellrose 1980
Hooded Merganser *(Lophodytes cucullatus)*: Eadie and Lumsden 1985
Great Egret *(Casmerodius albus)*: Cannell and Harrington 1984
Black-crowned Night Heron *(Nycticorax nycticorax)*: Cannell and Harrington 1984
Willow Ptarmigan *(Lagopus lagopus)*: Höhn 1972
Virginia Rail *(Rallus limicola)*: pers. obs., Boulder, Colo., 1989
Yellow-billed Cuckoo *(Coccyzus americanus)*: Nolan and Thompson 1975
Black-billed Cuckoo *(Coccyzus erythropthalmus)*: Nolan and Thompson 1975
Mountain Bluebird *(Sialia currucoides)*: Simpkin and Gubanich 1991
Song Thrush *(Turdus philomelos)*: Erard and Armani 1986
Verdin *(Auriparus flaviceps)*: Carter 1987

Nonobligate Brood Parasitism—Interspecific

Few species are nonobligate interspecific parasites. Redheads *(Aythya americana*, Weller 1959; Joyner 1976, 1983; Sugden 1980; Sayler 1996) and Ruddy Ducks *(Oxyura jamaicensis*, Joyner 1975, 1976, 1983; Siegfried 1976) are particularly well known for parasitizing other duck species. Partially interspecific parasitic breeding habits are particularly interesting because this strategy may be considered an evolutionary pathway to obligate brood parasitism. Other species have been reported as nonobligate interspecific parasites (table 1.1), but further records are needed to substantiate parasitism as a regular breeding alternative in many of these species.

In some situations, what appears to be brood parasitism may instead be interspecific squabbles over nest ownership. For synchronously hatching species, incubation is typically not initiated until the clutch is complete;

therefore, the nest may be unguarded for a number of days. Meanwhile, two or more females may lay in the same nest with little to no interactions. Sooner or later, conflict arises, but only one female will eventually incubate the eggs and care for the young. This phenomenon has been well studied in Australia, in Pink Cockatoos (*Cacatua leadbeateri*) and Galahs (*Cacatua [Eolophus] roseicapilla*), by Rowley and Chapman (1986) and Rowley (1991). Pink Cockatoos and Galahs sometimes lay their eggs in the same nest, during which time there is little interaction between the two. Pink Cockatoos almost always win ownership of the nest and consequently incubate Galah eggs and rear Galah chicks along with their own. Pink Cockatoos and Galahs nest in tree cavities, and this misunderstanding of nest ownership may be more likely to occur in cavity-nesting species. Often, tree cavities are the abandoned nest sites of other species, and ownership may be questionable until incubation begins.

Other apparent cases of parasitism could be caused by other means. Avian predators may help themselves to an egg (carrying it whole in the bill), discover another nest, and leave the egg in the other nest. Furthermore, some of these eggs may end up in nests of the predators themselves. Eggs of many species have been found in nests of California Gulls (*Larus californicus*). Generally, this has been attributed to food provisioning rather than interspecific brood parasitism (Sugden 1947; Twomey 1948).

Eadie and Lumsden (1985) suggested that parasitism is not always disadvantageous and may even benefit the host. They found that hatching and posthatching success was higher among clutches of Common Goldeneyes (*Bucephala clangula*) parasitized by Hooded Mergansers (*Lophodytes cucullatus*) than among nonparasitized clutches. They suggested that because the parasitic young they studied suffered higher mortality than host young, parasitic young dilute predation risks. They offered three explanations for why parasitic young may experience higher mortality: (1) the parasitic young may react inappropriately or more slowly to alarm calls by host parents; (2) hosts may give preferential protection to their own young, pushing parasitic young to the periphery of the group; and (3) predators may select odd-looking young or the minority type. Amat (1987) pointed out that predation may not have been the primary cause of mortality among the parasitic Hooded Merganser ducklings. Common Goldeneye and Hooded Merganser ducklings have different diets, and food availability for Hooded Mergansers may have been low in sites used by Common Goldeneyes.

In the host-parasite relationship of Hooded Mergansers parasitized by Common Goldeneyes, Mallory and Weatherhead (1993) found a consistent pattern of parasitic eggs being located on the nest periphery. In all ten of the Hooded Merganser nests (plus two Common Merganser, *Mergus merganser*, nests) parasitized by Common Goldeneyes, more host eggs were situated in the center of the nest. Mallory and Weatherhead's experiments confirmed that hosts moved parasitic eggs to the periphery.

Nonobligate Brood Parasitism—Conspecific

When parasites lay in the nests of their own species, it is known as conspecific nest parasitism (abbreviated CNP) or intraspecific brood parasitism. Conspecific parasites often have no nests of their own (Dhindsa 1983a) or their nests were destroyed during egg laying (Low 1941; Leopold 1951; Rosene 1969; Cooke and Mirsky 1972; Dhindsa 1983b, 1990). Others with nests may seize the opportunity to lay in another nest as well, often a neighbor's nest in a colony (Møller 1987; Brown and Brown 1988, 1989).

Not all conspecific nest parasites lay directly in host nests. Cliff Swallows (*Hirundo pyrrhonota*) lay in their own nests and then may transfer by mouth one of their own eggs to a neighbor's nest. Transferred conspecific eggs were found in about 6% of Cliff Swallow nests (Brown and Brown 1988). This behavior was confirmed by direct observation of the transfer and by the observation that eggs appearing in nests after incubation had already begun hatched simultaneously with the rest of the clutch. Also, previously marked eggs appeared in nests of other individuals. Similarly, Black-billed Magpies (*Pica pica*) and Pinyon Jays (*Gymnorhinus cyanocephalus*) have been observed carrying eggs in their bills from one nest to another (Trost and Webb 1986).

Limitation of nest sites may be one of the most important factors influencing rates of CNP. McLaughlin and Grice (1952) observed that when a population of Wood Ducks (*Aix sponsa*) increased but nest site availability remained constant, CNP also increased. Many conspecific nest parasites use site limited places, such as cavities, abandoned nests, and wetlands. Contrary to this, Rohwer and Freeman (1989) argued that nest site limitation does not explain CNP in precocial birds because several investigators reported parasitism even when unoccupied nest sites were available. They also suggested that nest sites are not in short supply in emergent vegetation. However, other factors—such as spacing, territoriality, food resources, distance to water, and perception of safety—influence nest site selection. Therefore, even when unoccupied nest boxes or apparently suitable vegetation are available, they may not be used. In these situations, CNP may be better than nesting in an unsuitable site or may be a bet-hedging strategy used in addition to nesting in a suboptimal site.

If a bird's nest is destroyed but she has parasitized a neighbor, her chances are increased that at least one of her own offspring will survive. CNP is also advantageous for an unmated female without a territory because it may be her only chance for reproductive success. Even with the advantages of reducing nest mortality risks by spreading eggs among nests, CNP appears to be a relatively rare strategy (MacWhirter 1989).

CNP is known among approximately 162 species, or 1.8%, of all avian species. Some of the more common conspecific nest parasites are numerous species of ducks (Yom Tov 1980), Barn Swallows (*Hirundo rustica*, Møller 1987), European Starlings (*Sturnus vulgaris*, Yom Tov et al. 1974; Evans 1988; Lombardo et al. 1989), and Eastern Bluebirds (*Sialia sialis*, Gowaty 1984).

The rarity of CNP suggests that its advantages may, for some species, be outweighed by the risks or disadvantages. These disadvantages may include (1) aggression toward the parasite, (2) laying eggs in inappropriate places or at inappropriate times, (3) reduced hatching success due to increased clutch size, (4) increased predation risks, (5) host desertion, and (6) egg ejection.

CNP appears to be most common among ducks and other precocial birds. Precocial species are well developed upon hatching and do not require extensive parental care; much of the parasite's dependence on the host is for incubation. Thus, even though hosts may suffer reproductive loss (due to clutch reduction), parasitized ducks generally do not suffer the same degree of reproductive loss as parasitized altricial species. Therefore, selective pressures for defenses against CNP may be less intense for precocial birds than for altricial species. Rohwer and Freeman (1989) suggested that the prevalence of CNP in precocial birds compared with altricial species results from the difference in selection pressure for defenses against CNP. Rohwer and Freeman (1989) also discussed the female-biased philopatric hypothesis that CNP may be tolerated more among waterfowl because neighboring females are more apt to be related than in less female philopatric, altricial species (Greenwood 1980; Rohwer and Anderson 1988). However, within CNP species, such as Snow Geese (*Anser caerulescens*, Lank et al. 1989) and Common Moorhens (*Gallinula chloropus*, Gibbons 1986), individuals may not parasitize relatives.

Many altricial species, particularly passerines, may be determinate egg layers; that is, they lay a predetermined number of eggs regardless of whether eggs are removed from or added to their nests. Parasitism may be less likely in determinate layers because, although spreading risks of nesting by laying an egg in a neighbor's nest may still be advantageous, parasitism does not necessarily result in additional reproductive success; the same number of eggs would be laid whether or not a bird engages in CNP. A possible relationship between determinate and indeterminate egg-laying patterns and the altricial-precocial dichotomy of CNP was investigated by Kennedy (1991). Unfortunately, data are lacking for both the status of determinate-indeterminate laying patterns and whether birds are conspecific nest parasites. Of 104 species in which egg addition or removal experiments had been conducted, only 49 could be classified as to whether or not they were conspecific nest parasites (Kennedy 1991). Among precocial species, 56.8% (n = 44) were indeterminate layers, whereas only 38.3% (n = 60) of altricial species were indeterminate layers (Kennedy 1991). Among conspecific nest parasites, 35.7% (n = 14) were indeterminate layers, whereas 25.0% (n = 12) of birds that are not known to be, or known not to be, conspecific nest parasites were indeterminate layers (Kennedy 1991). While neither test generated statistical significance, with larger sample sizes these relationships could become more clear.

CNP is not as easily detected as interspecific parasitism because host and parasite eggs are usually similar; therefore, the prevalence of CNP may be underestimated. CNP may be suspected in field studies if the following are noted: (1) abnormally large clutches, (2) irregularities in the normal laying sequence, or (3) differences in appearance of eggs within a nest. Suspected CNP may be confirmed by direct observations of egg laying, using a tracing agent (such as tetracycline), allozyme electrophoresis, or DNA fingerprinting.

Biased reporting may also hinder our knowledge of CNP, in part because few reports regarding lack of CNP have been published (Sealy et al. 1989; Lyon et al. 1992; Kempenaers et al. 1995). Unfortunately, publishing negative results such as these is often discouraged in the scientific community. Without more reports on CNP, whether or not parasitism actually occurs, we cannot know whether the lack of information is due to actual lack of CNP or lack of reporting. Detecting the absence of CNP may be simple: number individual eggs as they are laid and look for irregularities in egg laying. This method, however, must be interpreted with caution. Other than an occasional domestic chicken (Romanoff and Romanoff 1949), no birds are known to lay more than one egg per day (Welty 1982). Therefore, while the appearance of two eggs in a single day is suggestive of CNP, interpreting interruption of egg laying is more problematic. For example, if a bird whose clutch size varies from three to five eggs lays three eggs in her nest on three consecutive days and two days later lays a fourth egg, the appearance of the fourth egg may be interpreted as an incidence of CNP.

Interruptions in egg laying have been reported for Wood Ducks (Heusmann et al. 1980), various tit species (*Parus* spp., Dhondt et al. 1983), Willow Ptarmigan (*Lagopus lagopus*, Martin 1984), and Common Moorhens (Gibbons 1986). Interruptions may be a consequence of adverse environmental conditions (MacWhirter 1989) or of poor female condition (Eriksson and Andersson 1982), but interruptions may also indicate that these birds are themselves parasites. For example, a female may lay three eggs in her own nest, lay the fourth in a conspecific's nest, and lay a final fifth egg in her own nest. The above species for which interruption of egg laying have been reported are, in fact, conspecific nest parasites. Deposition of parasitic eggs on the day on either side of host laying, particularly if no host eggs are removed by the parasite, is also difficult to detect. Therefore, if a parasitic egg is laid the day before or after the host laying period, parasitism may go undetected.

For some species, a morphological trait in a nestling unlike that of the parents may indicate CNP. As one can suspect that a brown-eyed child of two blue-eyed human parents is an adopted child or the biological product of only one of the parents, one may also infer that a blue-phased Snow Goose (*Anser caerulescens*) in the nest of two white-phased adults is the result of a CNP or cuckold event.

Appearance of a dissimilar egg may also suggest CNP. Indeed, some investigators have inferred CNP by using this criterion (Yom Tov 1980; MacWhirter 1989). The appearance of odd-looking eggs must be interpreted with caution because some birds, such as House Sparrows (*Passer domesticus*, Lowther 1986, 1988), Eastern Kingbirds (*Tyrannus tyrannus*, Bischoff and Murphy 1993), and Herring Gulls (*Larus argentatus*, Baerends and Hogan-Warburg 1982), show individual variation in egg morphology, perhaps resulting from pigment depletion before clutch completion. Brown and Sherman (1989) suggested that researchers should compare within-clutch variation and between-clutch variation instead of assuming that individuals always lay similar eggs. Also, males apparently sometimes contribute to egg morphology (Romanoff and Romanoff 1949); therefore, if a female copulates with another male, one or a few of her eggs may differ from the rest of the clutch. In other words, it may be difficult to distinguish between a parasitic event and a cuckold event on the basis of dissimilar eggs. However, this criterion may be useful in wholly monogamous species if individual variation is low and variation among clutches is high.

While the appearance of dissimilar eggs and irregularities in laying sequence may be useful in inferring CNP, tracing agents and biochemical analyses aid in detecting parasitism with little doubt. Tracing agents, such as tetracycline (Haramis et al. 1983; Eadie et al. 1987), fat-soluble dyes (Appleby and McRae 1983), and radioisotopes (Dickman et al. 1983) can be injected into or fed to a female and traced to eggs or offspring. These tracing agents, though, can have deleterious effects. Tetracycline, which can be detected under ultraviolet light by a characteristic fluorescence, may inhibit egg laying (Eadie et al. 1987) and may be ineffective with exposure to sunlight (Haramis et al. 1983). Gel electrophoresis is destructive to eggs and requires an unincubated or undeveloped egg (Manwell and Baker 1975; Fleischer 1985; Fleischer and Smith 1992). Furthermore, the detection of parasitic offspring is limited by protein polymorphism within the population and depends on the possession of different genotypes by the (suspected) parasitic parents and host parents. To date, DNA fingerprinting is unquestionably the most powerful method of determining genetic relatedness, as it detects individual identity (Pinxten et al. 1993; Larsson et al. 1995).

Obligate Brood Parasitism

Unlike nonobligate parasites, which can parasitize conspecifics or other species, obligate parasites necessarily parasitize other species. Obligate brood parasites have completely lost the ability to construct nests and incubate eggs. Even with treatments of estrogen and prolactin (hormones responsible for nest construction behavior and incubation), Brown-headed Cowbirds (*Molothrus ater*) failed to produce incubation patches (Selander 1960; Selander and Kuich 1963).

Obligate brood parasitism is thought to have evolved independently at least six times as it is found in five or six (depending on the systematic scheme that is applied) avian families representing four orders. Even though some brood parasites are conspicuous and abundant, only approximately 91 species (1% of all bird species) are known to be obligate brood parasites.

Black-headed Ducks

Obligate brood parasitism occurs in a single duck species, the Black-headed Duck, *Heteronetta atricapilla* (family Anatidae, order Anseriformes), which inhabits South America. Parasitism in Black-headed Ducks was apparently first documented by Rodríguez (1918). Later, Daguerre (1920, 1923), Dabbene (1921), and Wilson (1923) reported on parasitism by Black-headed Ducks. Weller (1967a, 1967b, 1968) and Höhn (1975) further contributed to our knowledge of this duck, and Rees and Hillgarth (1984) published an account of a captive study, but to my knowledge, no more recent studies on parasitism by wild Black-headed Ducks have been published. They are shy and difficult to study; thus, much remains to be discovered about their breeding habits. The list of their 17 known hosts is quite diverse, including ibises, herons, spoonbills, storks, screamers, swans, rails, limpkins, coots, and gulls.

Black-headed Ducks are abundant in some areas, such as Argentina (Weller 1968), and parasitism can be intense for some hosts. Weller (1968) reported parasitism rates by Black-headed Ducks to be 55% ($n = 133$) for Red-fronted Coots (*Fulica rufifrons*) and 83% ($n = 6$) for Rosy-billed Pochards (*Netta peposaca*). Although it is not known whether multiple parasitism (more than one parasitic egg per nest) is by more than one female, it does occur with some regularity, particularly for heavily parasitized hosts (Weller 1968).

Black-headed Ducklings apparently leave the care of their foster parents at an early age, typically in a day or two after hatching, subsequently leading solitary and seclusive lives until they are able to fly. Black-headed Ducklings do not imprint as other ducks do. In an experimental attempt to have Black-headed Ducklings imprint on humans, Weller (1968) found that one very intensely trained duckling displayed only a weak following response, while two less intensely trained ducklings displayed no following response at all. Similarly, Rees and Hillgarth (1984) reported early independence and a weak following response to Rosy-billed Pochard hosts in captivity.

Cuckoos

Roughly 50 of the 138 (36.2%) or so species of cuckoos in the family Cuculidae, subfamilies Cuculinae, Centropinae, Coccyzinae, and Neomorphinae (order Cuculiformes) are brood parasites. Cuckoos are cosmopolitan in their distribution, but most brood parasitic cuckoos are found in the Old World. Only three parasitic cuckoos are found in the New World; these three inhabit Central and South America. The New World parasitic cuckoos belong to the subfamily Neomorphinae, as does the Greater Roadrunner (*Geococcyx californianus*) of North America, and are commonly referred to as ground cuckoos.

Cuckoos range in size from approximately 16 cm (sparrow size) to 60 cm (crow size) and typically have long graduated tails. Although there are a few gaudy exceptions, in general, parasitic cuckoos are dull in plumage compared to their nonparasitic relatives. Caterpillars, lizards, and snakes constitute their primary diet, and many cuckoos preferentially select hairy aposematic caterpillars with no apparent detrimental side effects. Cuckoos can be difficult to study because they are shy and widely dispersed. While some cuckoos have been studied extensively, so little is known about 12 of the "parasitic" cuckoos that their parasitic habits are only presumed (Wyllie 1981). These little-studied cuckoos generally occupy remote areas and/or inaccessible habitats.

One of the most well-known cuckoos, the European Cuckoo (*Cuculus canorus*) of Eurasia and Africa, is the only cuckoo that distinctly calls its name. The "cuckoo" is the male mating song and sounds very much like a typical cuckoo clock. Because of the clarity with which it repeats its name, it has earned the same common name (although spelled in many ways) in most languages throughout its range.

Cuckoos were probably the first brood parasites whose breeding habits were familiar to humans. The first known mention of a brood parasite was found in ancient Vedic literature dating back to around 2000 B.C., where mention was made of the Asian Koel (*Eudynamys scolopacea*), and although nothing was specifically stated of its brood parasitic habits, the name it was given ("anya-vapa," which can be translated as "reared by others") implies that its brood parasitic habits were well known (Friedmann 1964). The brood parasitic habits of cuckoos were also known to Aristotle, and cuckoos have been featured by many writers and artists through the ages. The early Anglo-Saxons coined the word "cuckold," meaning to dupe or fool, from observations of the breeding strategies of the cuckoo. It is interesting to note that in Europe, the parasitic cuckoos appear to be revered (cuckoo clock and all), whereas in North America, the parasitic cowbirds are shunned and relegated to the list of pests.

Honeyguides

First thought to be members of the cuckoo family (Friedmann 1964), honeyguides (family Indicatoridae, order Piciformes) are also brood parasites. Although much remains to be learned about honeyguides, it is thought that most of the 17 species (Clements 1991) are brood parasites. Honeyguides inhabit Africa and Asia, and are well known for their symbiotic relationship with mammals such as humans and Ratels (*Mellivora capensis*).[1] They were named after their behavior of guiding mammals to the vicinity of beehives, where they wait to feed on bits of honeycomb and bee larvae after the mammal has plundered the nest.

Hosts of honeyguides are varied, but they usually parasitize cavity nesters. Honeyguides parasitize some passerines and swifts, but they usually parasitize closely related barbets and woodpeckers. Cavity-nesting kingfishers, rollers, bee-eaters, and hoopoes also frequently serve as hosts (Friedmann 1955). The list of known hosts, however, is probably incomplete, as relatively little is known about honeyguide breeding biology.

Cowbirds

Five of the six cowbirds (family Emberizidae,[2] order Passeriformes) are brood parasites. Cowbirds were named for their habit of following cattle and picking up insects flushed up by the cattle. Before the introduction of livestock into North America, Brown-headed Cowbirds presumably followed Bison (*Bison bison*) herds; they were commonly called Buffalo Birds.

Cowbirds are found only in the Americas. Three species inhabit primarily South America: Shiny Cowbirds (*Molothrus bonariensis*), Screaming Cowbirds (*Molothrus rufoaxillaris*), and nonparasitic Bay-winged Cowbirds (*Molothrus badius*). Shiny Cowbirds were originally restricted to South America, Trinidad, and Tobago, but they have expanded their range northward through the West Indies over the last century and are now found in the United States. Giant Cowbirds (*Scaphidura oryzivora*) range from southeastern Mexico to southern Brazil and northeastern Argentina. Bronzed Cowbirds (*Molothrus aeneus*) inhabit the southern United States through Central America and Colombia, and Brown-headed Cowbirds range throughout most of North America.

Finches

Brood parasitism is also found in two (or more, depending on the source) genera of finches native to Africa (families Ploceidae [and Estrildidae],[3] order Passeriformes). Parasitic finches are also known as weaverbirds, widow birds, combassous, indigobirds, and whydahs. They are gregarious and inhabit savanna grasslands of Africa, feeding primarily on small grass seeds. To the best of my knowledge, brood parasitism in finches was first reported by Roberts in 1907. Most parasitic finches parasitize closely related estrildid finches. The monogeneric Cuckoo Finch *(Anomalospiza imberbis)*, however, chiefly parasitizes grass warblers *(Cisticola* and *Prinia,* Friedmann 1960).

Unlike most brood parasites, males of several parasitic finch species have striking plumage during the breeding season, with bright colors and showy tails. Their captive survival abilities, together with their astounding beauty, make some parasitic finches, such as Pin-tailed Whydahs *(Vidua macroura),* desirable as cage birds. Escapee populations of finches are now established outside the birds' native range, such as in the West Indies. In these areas, they parasitize native species as well as other (nonparasitic) African finches that have, likewise, escaped from captivity. Parasitic finches continue to be sold commercially and can be found in many pet stores throughout the United States.

Even though captivity of parasites carries the risk of creating escapee populations, it offers excellent opportunities to observe breeding behavior. Most aviculturists breed parasitic finches with the assistance of hosts; however, a few have reported varying degrees of breeding activities that are unexpected of obligate brood parasites. Nielsen (1956) reported that a pair of Steel-blue Whydahs[4] *(Vidua chalybeata)* constructed a nest, incubated their eggs, and fed the young. Similarly, Friedmann 1955 reported that a female Steel-blue Whydah incubated the eggs she laid in an old Zebra Finch *(Taeneopygia guttata)* nest. While such reports are rare, these deviations are intriguing and should not be dismissed lightly.

Effects of Brood Parasitism

As a general rule, brood parasites significantly lower the reproductive fitness of their hosts. The degree to which brood parasites harm their hosts varies from very little to total loss of the host young.

Host Siblicide by Parasitic Nestling

Parasites may increase their chance of survival by removing or killing their host siblings. Some cuckoo nestlings, such as European Cuckoos *(Cuculus canorus)* and Horsfield's Bronze-Cuckoos *(Chrysococcyx basalis),* remove host eggs and/or nestlings with a backwards sweep of the wings, holding the egg or nestling in the small of their back and working it up to the edge of the nest. This instinctive ejection reaction begins between 3 and 36 hours after hatching and may wane after four days (Wyllie 1981). For some cuckoo species, the ejection reaction is triggered by a different (cooler) temperature when the brooding host leaves the nest (Payne and Payne 1990). Other cuckoos will eject host siblings even while hosts brood them (Wyllie 1981). Until Jenner (1788) described in detail how nestling cuckoos eject their host siblings, it was presumed that adult cuckoos ejected the host young. Since Jenner's description, the ejection behavior of nestling cuckoos has been described and filmed many times. During the process of ejection behavior, occasionally cuckoo nestlings will inadvertently eject themselves from a nest (Wyllie 1981). In the event that two cuckoos hatch in the same nest, the larger cuckoo will typically succeed in ejecting the smaller one (Wyllie 1981).

Some honeyguides regularly kill their host nestmates by stabbing them with their daggerlike egg tooth. Stabbing and/or ejection of host siblings is known to occur among Scaly-throated Honeyguides (*Indicator variegatus*), Greater Honeyguides (*Indicator indicator*), and Lesser Honeyguides (*Indicator minor*, Friedmann 1955). Young honeyguides also have specialized heel pads, which may provide them with extra leverage and grip when stabbing their nestmates (Friedmann 1955). Like the cuckoos' ejection response, this innate reaction of honeyguides occurs shortly after hatching.

Cowbird nestlings are generally not known to kill their host siblings. However, Dearborn (1996) recently videotaped a Brown-headed Cowbird ejecting an Indigo Bunting (*Passerina cyanea*) host sibling from a nest.

Removal of Host Nestlings by Adult Parasite

Adult Brown-headed Cowbirds sometimes remove nestlings from both parasitized and nonparasitized nests (Du Bois 1956; Tate 1967; Beane and Alford 1990). While removing nestlings in nonparasitized nests may seem pointless, it may stimulate potential hosts to renest.

Removal of Host Eggs by Adult Parasite

Brood parasites can lower host nesting success by removing host eggs. Cuckoos typically remove (and consume) a host egg just prior to laying their own. Greater Honeyguides and cowbirds also often remove host eggs. Egg removal by cowbirds may be more variable than it is in cuckoos. Cowbirds sometimes remove a host egg for every egg they lay (Mayfield 1960; Smith 1981; Finch 1983; Wolf 1987), but in one study, cowbirds removed a host egg from every three parasitized Yellow Warbler (*Dendroica petechia*) nests (Sealy

1992). Larger hosts may not suffer this loss as often as smaller hosts because they generally have larger eggs that may be less easily manipulated by cowbirds. In some years, I found no evidence that cowbirds removed host eggs from parasitized Red-winged Blackbird (*Agelaius phoeniceus*) nests (Ortega 1991). Cowbirds sometimes remove eggs from the nests of uncommonly used hosts, such as American Robins (*Turdus migratorius*, Blincoe 1935), and nests that never become parasitized (Scott et al. 1992). Host eggs are also punctured and left in the nest by some adult parasites (Hofslund 1957).

Damage by Thick Eggshells

A trait common to all obligate parasites (except perhaps Black-headed Ducks) [5] is an unusually thick-shelled egg. Thick shells of parasitic eggs and the damage they can cause has been the focus of many ornithological and oological studies (chapters 2 and 3).

Reduced Clutches

For indeterminate-laying hosts, the addition of a parasitic egg may cause the host to lay fewer eggs (Andersson and Eriksson 1982). For determinate layers, clutch size increased by parasitic eggs may cause lowered incubation efficiency (Biebach 1981) and hatching success (Lerkelund et al. 1993).

Abandonment

Some hosts respond to parasitic eggs by abandoning their nests. Abandonment of parasitized nests may be particularly high in some species, such as Willow Flycatchers (*Empidonax trailii*, Sedgwick and Knopf 1988; Harris 1991). For some hosts, the removal of host eggs may influence abandonment. Sealy (1992) observed that only one of 37 (2.7%) Yellow Warblers whose clutches were reduced from four or five eggs to three eggs abandoned, whereas abandonment occurred in 13 of 42 (31.0%) Yellow Warblers whose clutches were reduced to only two eggs.

Competition for Food

Parasites that do not remove their host siblings are raised in competition with their hosts. However, the parasite is often larger than the host and outcompetes smaller host siblings for food. Birds usually feed the largest nestling or the nestling with the largest gaping mouth and the loudest begging calls. This strategy ensures that in times of food shortages at least one young may survive, and it also ensures that the healthiest (or largest) is the one to survive. Cowbirds outcompete their nestmates so proficiently that they often become the sole occupant of the nest and receive the undivided attention of the host parents (Marvil and Cruz 1989).

Cowbirds have vigorous begging vocalizations (Nice 1939; Ficken 1967; Gochfeld 1979a) that may promote their chances of survival. Compared with their hosts, Shiny Cowbirds, in particular, have exaggerated begging vocalizations accompanied by wing fluttering and climbing on top of other nestlings to present their gaping bills (Gochfeld 1979a). The intensity of these begging vocalizations may be under genetic control as in Gochfeld's (1979a) study. The begging intensity of Shiny Cowbirds remained the same even when Greater Red-breasted Meadowlark[6] (*Sturnella loyca*) hosts were removed.

Indiscriminate aggressive begging vocalizations may not always be advantageous. Predators may locate well-hidden nests through the begging vocalizations of cowbirds (Hudson 1920; Gochfeld 1979a). The disadvantage of indiscriminate begging behavior was also reported by Ficken (1967), who observed a fledgling cowbird, raised by an Eastern Phoebe (*Sayornis phoebe*), begging from an American Crow (*Corvus brachyrhynchos*). The crow initially turned away from the cowbird, but when the cowbird approached the crow twice more, the crow swiftly killed the cowbird and flew off with it.

Brown-headed Cowbirds also have a gape width (the measurement from the corners of the mouth) that is large relative to their weight (Ortega and Cruz 1992a). This, coupled with vigorous begging, may help ensure that they receive more feedings from their host parents (Ortega 1991; Ortega and Cruz 1992a). The large gape width and vigorous begging may be particularly important when competing in nests of hosts larger or older than themselves.

Host Selection Strategies: Generalists and Specialists

Host selection strategies by brood parasitic species have historically been divided into two broad categories: generalists and specialists. Generalists parasitize a wide range of hosts, while specialists use only one or a very few "biological host" species on a regular basis.[7] Cuckoos, some finches, and Giant Cowbirds are difficult to categorize, as they appear to be intermediate between specialists and generalists (Brooker and Brooker 1989a).

Generalist brood parasites use a wide array of hosts, spanning many taxa. For example, the list of species known to be parasitized numbers 214 for Shiny Cowbirds and 226 for Brown-headed Cowbirds. Generalist parasites include Black-headed Ducks, some cuckoos, honeyguides, and three species of cowbird.

Specialist brood parasites are less common than generalists, both in terms of species numbers and abundance within a species. A few cuckoos are specialists (Friedmann 1929a; Soler 1990; Soler and Møller 1990), as are some parasitic finches (Nicolai 1974; Payne 1977). Screaming Cowbirds specifically parasitize the nonparasitic Bay-winged Cowbird.

Parasites, whether specialists or generalists, are necessarily restricted to the ranges of their hosts during the breeding season. However, specialists are restricted to the range of their primary host. Therefore, ranges of specialists are typically much smaller than ranges of generalists. This range restriction, together with dependence on another species for future survival, may place specialist parasites at a high risk for extinction.

A high degree of morphological and behavioral mimicry is common among specialists. Mimicry in parasitic finches is well developed not only in the eggs but also in nestling morphology and nestling begging vocalizations (Nicolai 1974). Mimicry of parasitic finch nestlings suggests that their hosts eject nestlings unlike their own (chapter 3). Mimicry of adult vocalizations is also well developed in finches (Payne 1973a; Nicolai 1974; Payne and Payne 1995) and some cuckoos (Arias-de-Reyna and Hidalgo 1982). It is thought that these songs are learned from the host while in the nest.

Some cuckoos represent a compromise between specialist and generalist parasitism. Some individuals within a cuckoo species specialize on one host, whereas other individuals specialize on another host. The total number of hosts that a parasitic cuckoo species parasitizes, however, may be large (Riddiford 1986). Groups of cuckoos that lay similar eggs and parasitize the same host species are known as gens (pl. gentes), and the ranges of these gentes show the geographical distribution patterns of their hosts.

Cuckoos lay polymorphic eggs, and within a gens, female cuckoos typically lay similar eggs (Brooke and Davies 1988; Brooker and Brooker 1990). Some cuckoo nestlings, such as Great Spotted Cuckoos (*Clamator glandarius*, Mundy 1973), Long-tailed Cuckoos (*Eudynamys taitensis*), and Shining Cuckoos (*Chrysococcyx lucidus*, McLean and Waas 1987) appear to mimic begging calls of their hosts.

Summary

Brood parasitism may be either nonobligate or obligate. Nonobligate parasites can use conspecifics or other species for hosts, whereas obligate brood parasites must necessarily parasitize other species. Recognition of CNP has been a fairly recent event, and the list of known conspecific nest parasites is growing rapidly. Presently, 162 species are known to engage in CNP. Obligate brood parasitism occurs in cuckoos, honeyguides, cowbirds, finches, and a single species of duck, the Black-headed Duck.

Obligate brood parasitism is usually detrimental to the reproductive success of hosts, but the effects of brood parasitism vary depending upon both the parasite and host. Detrimental effects can occur through host siblicide by the parasitic nestling, the adult parasite removing or damaging host eggs or nestlings, damage caused to host eggs by thicker-shelled parasitic eggs, reduced clutch size, competition from parasitic nestlings, and abandonment.

Host selection is divided into categories of generalists and specialists. Specialists parasitize primarily one or two hosts and often display mimicry of their hosts in egg morphology, nestling morphology, nestling vocalizations, and/or adult vocalizations. Generalists parasitize many hosts and are typically more abundant than specialists. Most cuckoos are intermediate between specialists and generalists in host specificity and form gentes or groups of individuals that specialize on a particular host.

Defense Mechanisms

The implementation of defense mechanisms against intruding brood parasites can take place either before or after the nest is parasitized; these mechanisms range from being energetically inexpensive and safe to being energetically expensive and risky. Hosts may reject parasitic eggs (rarely nestlings), or they may deny parasites access to their nests. Rejection of parasitic eggs requires the ability to discriminate between host and parasitic eggs and may take the form of ejection of parasitic eggs from the nest, burial of the parasitized clutch, or nest abandonment. Denial of access to the nest may take the form of aggression toward the parasite at the nest site or refusal to leave the nest.

Rejection of Parasitic Eggs

Species in which nearly all individuals within the species respond to parasitic eggs by rejecting them are referred to as "rejecters." Likewise, "accepters" are species in which nearly all individuals within the species accept parasitic eggs. The accepter-rejecter dichotomy of Brown-headed Cowbird hosts is interesting as it seems that once rejection behavior appears in a population, it becomes fixed rapidly (Rothstein 1975b). Kelly (1987) demonstrated, however, that rejection becomes fixed rapidly only when the rate of parasitism is relatively high. Almost all Brown-headed Cowbird hosts that practice rejection of parasitic eggs do so regularly (Rothstein 1975a).

Other hosts are more complicated to categorize. For example, Black-billed Magpies (Alvarez et al. 1976) and Azure-winged Magpies (Cyanopica cyanea, Arias-de-Reyna and Hidalgo 1982) are both parasitized by specialist cuckoos.[1] Both species are accepters of mimetic eggs but rejecters of nonmimetic eggs. In these cases, they may be termed "nonmimetic egg rejecters" or "nonmimetic egg accepters" (Arias-de-Reyna and Hidalgo 1982).

Egg Acceptance/Rejection Experiments

Experimental parasitism is an important tool for confirming that observed parasitism rates are actual parasitism rates. That is, parasitism rates may be underestimated for some potential hosts because some individuals may eject parasitic eggs before parasitism is discovered by investigators. Therefore, only by experimentally adding parasitic eggs can responses of hosts or potential hosts be determined. Artificial eggs are usually constructed of plaster of Paris (Rothstein 1974, 1975c), wood putty (Cruz and Wiley 1989), plastic filled with water (Ortega and Cruz 1988), or gel coat resin (Davies and Brooke 1988), and are painted various colors with or without designs, depending on the species and experimental purposes. Typically, eggs are left in a nest for five days, after which the bird may be classified as either a rejecter or accepter. The five-day criterion is used because if parasitic eggs are rejected, they are usually rejected within three days. Justification of this five-day criterion, however, should be verified for each test species.

Presence of adult parasites may or may not affect host responses toward parasitic eggs. Investigations of Fieldfare (*Turdus pilaris*) and Brambling (*Fringilla montifringilla*) responses toward European Cuckoo eggs found that the addition of a mounted cuckoo at the nest site had no effect on responses toward cuckoo eggs; all eggs were accepted with or without the mounted cuckoo (Moksnes and Røskaft 1988; Braa et al. 1992). A study of Meadow Pipits (*Anthus pratensis*), however, revealed that these hosts may reject cuckoo eggs more frequently when, in addition to the experimental egg placed in the nest, a model of an adult cuckoo is placed near the nest (Moksnes et al. 1993).

Egg Ejection

Hosts can eject parasitic eggs either by removing the egg whole (grasp ejection) or by spiking the egg and carrying it away impaled on the bill (puncture ejection). Even though ejection of parasitic eggs is an effective antiparasite strategy, ejection is a surprisingly uncommon behavior. Of the 226 species that are host to Brown-headed Cowbirds, relatively few are known to eject parasitic eggs (table 2.1).

Intermediate Responses. In some species, the acceptance/rejection response varies among individuals. If 15–85% of a population responds in one way

Table 2.1

Birds Known to Eject Brown-headed, Bronzed, or Shiny Cowbird Eggs and/or Eggs unlike Their Own

Species/Source

Couch's Kingbird: Carter 1986
Western Kingbird: Rothstein 1975a
Eastern Kingbird: Rothstein 1971b, 1975a; Hamas 1980; Sealy and Bazin 1995
Gray Kingbird: Cruz et al. 1989; Post et al. 1990
Scissor-tailed Flycatcher: Regosin 1994
Fork-tailed Flycatcher: Mason 1986a; Cavalcanti and Pimentel 1988
Great Kiskadee: Mason 1986a
Blue Jay: Rothstein 1971b, 1975a
Scrub Jay: Rothstein 1982b
Red-legged Thrush: Cruz et al. 1989
American Robin: Rothstein 1971b, 1975a, 1982a; Briskie et al. 1992
Gray Catbird: Berger 1951; Nickell 1958; Rothstein 1971b, 1974, 1975a, 1982a
Sage Thrasher: Rothstein 1971a; Rich and Rothstein 1985
Brown Thrasher: Rothstein 1971b, 1975a
Curve-billed Thrasher: Carter 1986
Crissal Thrasher: Finch 1982
Loggerhead Shrike: Rothstein 1982b
Meadowlark spp.:[1] Hergenrader 1962
Common Grackle: Peer and Bolinger 1997
Great-tailed Grackle: Carter 1986
Carib Grackle: Cruz et al. 1995
Greater Antillean Grackle: Cruz et al. 1989
Baltimore Oriole: Smith 1972; Rothstein 1977a; Sealy and Neudorf 1995
Bullock's Oriole: Rothstein 1978b; Rohwer et al. 1989
Altamira Oriole: Carter 1986

Classified as Accepters but Sometimes Eject
Cedar Waxwing: Rothstein 1971b, 1976b
Warbling Vireo: Sealy 1996
Bay-winged Cowbird: Fraga 1986

Classified as Intermediate
Rufous Hornero: Hoy and Ottow 1964; Salvador 1983; Mason 1986a; Mason and Rothstein 1986
Northern Mockingbird: Mason, pers. comm., in Friedmann and Kiff 1985; Cruz et al. 1989; Post et al. 1990
Chalk-browed Mockingbird: Salvador 1984; Fraga 1985; Mason 1986a; Cavalcanti and Pimentel 1988
Brown-and-Yellow Marshbird: Mermoz and Reboreda 1994

[1] The assumption that meadowlarks ejected parasitic eggs was based on finding Brown-headed Cowbird eggs near their nests.

and 15–85% of the population exhibits the opposite response, the behavior may be considered intermediate between acceptance and rejection.

A few Brown-headed Cowbird and Shiny Cowbird hosts have demonstrated intermediate responses. Yellow Warblers regularly abandon parasitized nests (Berger 1951; Clark and Robertson 1981; Burgham and Picman 1989; Briskie et al. 1990), and Northern Mockingbirds (Mimus polyglottus) often eject parasitic eggs (Cruz et al. 1989; Post et al. 1990). Ejection behavior in the mockingbird or thrasher family (Mimidae) is particularly interesting because all members of this family have bills probably large enough to grasp eject cowbird eggs. To date, five have been described as ejectors (table 2.1), two as accepters (Le Conte's Thrasher, Toxostoma lecontei, and California Thrasher, Toxostoma redivivum), and two appear to be intermediate (Northern Mockingbird and Chalk-browed Mockingbird, Mimus saturninus). Le Conte's and California Thrashers have not existed sympatrically with Brown-headed Cowbirds until the twentieth century (Rothstein et al. 1980), and they still do not serve as hosts.

The Village Weaver (Ploceus cucullatus) is a particularly interesting species in which to look at acceptance/rejection behavior because two distinct populations are faced with different selection pressures. Originally from Africa, Village Weavers were introduced into Hispaniola in the eighteenth century at the time of the slave trade (Bond 1936) and are now well established. Village Weavers lay eggs with interindividual variation but little intraindividual variation (Victoria 1972). That is, their eggs differ among clutches but not within a clutch. Village Weaver eggs range from light to dark blue-green, with or without brown spots. Village Weavers in Africa often eject eggs markedly different from their own (Din 1992). Victoria (1972) and Collias (1984) believed the rejection behavior evolved in response to Dideric Cuckoo (Chrysococcyx caprius) parasitism. Until recently, Village Weavers on Hispaniola have been free from interspecific brood parasitism. However, Shiny Cowbirds were first observed on Hispaniola in 1972 (Post and Wiley 1977b).

Cruz and Wiley (1989) took advantage of this "natural experiment" to examine the presumed decline of egg rejection. They found that, overall, only 13.5% of their experimental population ejected artificial conspecific eggs. Village Weavers on Hispaniola, therefore, are accepters, whereas in Africa, they are intermediate between accepters and rejecters (table 2.2). Cruz and Wiley (1989) attributed the nearly total acceptance of foreign eggs on Hispaniola to incomplete decay of a once common response to brood parasitism.

Table 2.2

Rejection of Conspecific and Shiny Cowbird Eggs Introduced into Village
Weaver Nests in a Captive Population from Africa and in the Dominican
Republic

Host Egg Morphology	Village Weaver		Shiny Cowbird	Total
	Plain	Spotted		
Captive population from Africa				
Plain	15/33 (45.5%)	49/67 (73.1%)	—	64/100 (64.0%)
Spotted	44/76 (57.9%)	37/146 (25.3%)	—	81/222 (36.5%)
Total	59/109 (54.1%)	86/213 (40.4%)	—	145/322 (45.0%)
Dominican Republic				
Plain	3/20 (15.0%)	4/30 (13.3%)	4/24 (16.7%)	11/74 (14.9%)
Spotted	4/22 (18.2%)	4/40 (10.0%)	4/34 (11.8%)	12/96 (12.5%)
Total	7/42 (16.7%)	8/70 (11.4%)	8/58 (13.8%)	23/170 (13.5%)

Source: African data from Victoria 1972; Dominican Republic data from Cruz and Wiley 1989a.

Shiny Cowbird parasitism on Village Weavers on Hispaniola was only
1.3% between 1974 and 1977, but by 1982, parasitism had increased to 15.7%
(Cruz and Wiley 1989). Nonparasitized nests of Village Weavers are more
successful than parasitized nests (Cruz and Wiley 1989); therefore, selection
should favor individuals that reject parasitic eggs. One might predict, then,
that rejection in the Hispaniola population might become more prevalent
over time.

Rothstein (1975b) derived an equation to estimate the number of generations that would theoretically be required to change from one allelic
frequency of acceptance to another allelic frequency of acceptance. Using
Rothstein's equation and data from Cruz and Wiley (1989), theoretically, it
would take 134 generations for Village Weavers to evolve from being mostly
accepters of cowbird eggs (13.5% rejection in 1982) to mostly rejecters of
cowbird eggs (13.5% acceptance). The Village Weaver is a particularly good
species to test the validity of Rothstein's (1975b) equation because they are
known to have both egg discrimination and grasp ejection abilities.

The decline of rejection behavior is particularly interesting because it implies that rejection behavior may be costly. However, because egg rejection
experiments were not carried out in earlier times, we do not know whether
the founder population on Hispaniola consisted of accepters or rejecters. It
is possible that individuals introduced to Hispaniola were accepters, and a
decline of rejection behavior never did occur.

Grasp Ejection Indices. Ejection behavior is limited by the ability to remove parasitic eggs, and this may partially explain the relative rarity of ejection behavior. Grasp ejector indices have been created to predict whether species are capable of grasp ejection. Indices are based on bill measurements of known rejecters and accepters, not on actual ejection abilities; therefore, they are not always accurate predictors. Least Flycatchers (*Empidonax minimus*) were classified by Rothstein (1975a) as being probably capable of lifting a Brown-headed Cowbird egg, but when Briskie and Sealy (1987) attempted to fit a cowbird egg between the mandibles of a freshly dead adult Least Flycatcher, they could not do so and concluded that Least Flycatchers are incapable of grasp ejection. Another grasp index predicts that Red-winged Blackbirds are incapable of grasp ejection (Rohwer and Spaw 1988), but Red-winged Blackbirds ejected from their nests objects other than eggs that were as large or larger than parasitic eggs (Ortega and Cruz 1988). I have also observed Red-winged Blackbirds carrying these large objects between their mandibles.

Limitations of the Ejection Response

There are several explanations of why birds accept parasitic eggs, even though some may be capable of grasp ejecting: (1) absence of or low selection pressure; (2) inability to recognize the parasitic egg; (3) inability to remove the parasitic egg; (4) high cost of ejection compared with the cost of acceptance; and (5) parasite hatching success may be relatively low.

Selection Pressure. Selection pressure on hosts to evolve antiparasite defenses are not equivalent in all species; that is, if raising a parasite does not lower a species' reproductive fitness, then ejection behavior is unlikely to become fixed. For example, even though Red-winged Blackbirds are capable of removing Brown-headed Cowbird eggs, the selection pressure may not be intense enough for ejection behavior to become fixed in the population, particularly at a relatively low level of parasitism (Røskaft et al. 1990; Ortega 1991).

Egg Recognition. Hosts may accept parasitic eggs because they do not fully recognize them as parasitic. Eggs have several parameters used in rejection responses: size, color, and maculation. Some species eject foreign eggs differing by only one parameter from their own (Victoria 1972). Other species, such as American Robins and Gray Catbirds (*Dumetella carolinensis*), are less sensitive to fine differences. They will accept eggs differing from their own by one parameter but will eject eggs that differ in more than one parameter (Rothstein 1982a). This built-in tolerance may reduce the likelihood of rejecting their own eggs (Rothstein 1982a). Accepter species, on the other hand, will generally accept any egg-shaped object with a continuously smooth surface, even if the object is colored differently from their own eggs (Ortega and Cruz 1988; Ortega et al. 1993).

Whether rejecter species eject nonmimetic eggs based on actual recognition or discordancy (ejection of eggs that differ from the majority) is an intriguing question. After experimental addition of nonmimetic eggs to the nests of several species, Rensch (1925) believed that ejection was based on discordancy. Rothstein (1975c) disagreed with Rensch's interpretation and believed that the birds in Rensch's study demonstrated true egg recognition.

Egg recognition may be innate or have a learned component. If egg recognition is learned, then tolerance of nonmimetic eggs should be higher early in a bird's experience as the bird is learning the morphology of its own eggs. Rothstein's (1974) experiments suggested that Gray Catbirds learn to recognize their own eggs. American Robins subjected to the same experiments as the Gray Catbirds responded similarly, but their ejection of nonmimetic eggs occurred more quickly when the bird's own eggs outnumbered the nonmimetic eggs, suggesting that discordancy plays at least a minor role (Rothstein 1975c).

High Cost of Ejection. Another explanation of why birds may accept parasitic eggs is the high cost associated with puncture ejection. For birds with bills too small to grasp eject parasitic eggs, puncture ejection and abandonment are the only defenses available once parasitism has occurred. Puncture ejection is a much riskier strategy than grasp ejection. Not only is there a chance that the parasitic egg contents will spill into the nest, increasing the likelihood of predation, but there is also a chance that the host's bill will slide off of the thick shell of the parasitic egg and puncture one of the host's own eggs.

Brown-headed Cowbird eggshells are at least 30% thicker than would be expected by their volume (Spaw and Rohwer 1987), and thickness is highly correlated with strength (Romanoff and Romanoff 1949). Cowbird eggs are also thicker and rounder than other blackbird eggs (Picman 1989), and the shells have a significantly higher proportion of inorganic constituents, particularly calcium (Picman 1989), which contributes to a stronger shell (Romanoff and Romanoff 1949). Hoy and Ottow (1964) suggested that a bird's bill may be deflected by the uniform curvature of cowbird eggs. In a test using a mechanical puncture tester, Picman (1989) determined that Brown-headed Cowbird eggs were more puncture resistant than Red-winged Blackbird and Yellow-headed Blackbird (*Xanthocephalus xanthocephalus*) eggs. He attributed the puncture resistance, in part, to the eggs' roundness, which increases the load bearing capacity of the shell. Spaw and Rohwer (1987) observed that Marsh Wrens (*Cistothorus palustris*) had a more difficult time puncturing Brown-headed Cowbird eggs than eggs of nonparasitic species and proposed that the thicker shell is a specific adaptation to brood parasitism. Blankespoor et al. (1982) found a significantly higher proportion of cracked Red-winged Blackbird eggs than Brown-headed Cowbird eggs among parasitized nests, but Weatherhead (1991) did not find that presence of Brown-headed Cowbird eggs was associated with increased cracking of Red-winged Blackbird eggs. However, disappearance of Yellow Warbler eggs in artificially parasitized nests suggests that Brown-headed Cowbird eggs may have damaged more delicate Yellow Warbler eggs (Weatherhead 1991).

Puncture ejection may be the only ejection strategy for small-billed hosts (Cedar Waxwing, *Bombycilla cedrorum*) and hosts with pendulous nests that make removal of parasitic eggs difficult (orioles, caciques, and oropendolas). Costs of puncture ejection have been clearly demonstrated in several studies. In 61.1% (*n* = 18) of Baltimore Oriole (*Icterus galbula*) nests in which Rothstein (1977a) added puncture-proof artificial eggs, oriole eggs disappeared or were otherwise damaged; he suggested that the damage occurred while the orioles attempted egg ejection. Rohwer et al. (1989) compared host egg damage in Bullock's Oriole (*Icterus bullockii*) nests by experimentally parasitizing oriole nests with Brown-headed Cowbird eggs and Cliff Swallow eggs. Cliff Swallow eggs are only slightly smaller than Brown-headed Cowbird eggs, but they have a thinner shell. Both Cliff Swallow and Brown-headed Cowbird eggs were puncture ejected by the orioles, but damage to host eggs was significantly higher in nests with Brown-headed Cowbird eggs than in nests with Cliff Swallow eggs. Sealy and Neudorf (1995) reported that for every Baltimore Oriole nest they artificially parasitized, 0.4 host eggs were damaged or disappeared. Even slight damage to Bullock's Oriole eggs can reduce hatchability by 55.4% (*n* = 65 eggs) and cost the host 0.26 of its own eggs per ejection; this cost, however, is lower than the cost of raising a Brown-headed Cowbird when the brood is large (Røskaft et al. 1993).

While puncture ejection may ruin other eggs or the entire nest, the option of abandonment is costly and is a worthwhile strategy only with sufficient time to renest. Many passerines can renest rapidly, laying within five to seven days after nest failure (Scott et al. 1987). However, rates of parasitism may increase as the breeding season progresses (Finch 1983; Freeman et al. 1990; Ortega 1991); therefore, the second nest often stands a good chance of being parasitized as well.

Facilitated Incubation. Another possible reason for acceptance of parasitic eggs is that the presence of additional eggs or discordant eggs may facilitate incubation in some species. Some birds place foreign objects, such as smooth stones and pine cones, in their nests and incubate them along with their eggs (Knight and Erickson 1977). Many of these objects are the size and shape of real eggs. Conover (1985) referred to these objects as "pseudoeggs" and suggested that Ring-billed Gulls (*Larus delawarensis*) and California Gulls may not be able to distinguish between their own eggs and the pseudoeggs that they roll into their nests. Conover (1985) found the frequency of pseudoeggs higher in smaller clutches and suggested that they may enhance incubation in smaller clutches. Coulter (1980) often found stones in Western Gull (*Larus occidentalis*) and Common Tern (*Sterna hirundo*) nests with less than a normal clutch size of three eggs and attributed the high incidence of stones in smaller clutches as evidence of the importance of stones as stimuli for incubation behavior. To my knowledge, the possibility of parasitic eggs enhancing incubation has not been investigated, but Rothstein (1986) suggested that Black Phoebes (*Sayornis nigricans*) "prefer" larger Brown-headed Cowbird eggs to their own. The superstimulus value of a larger (parasitic) egg may result in the host spending more time incubating.

Nest and Clutch Abandonment

Nest abandonment can be costly, but it does provide hosts with an alternative to losing an entire clutch and what might be an entire breeding season. If parasitism occurs early in the breeding season, abandonment may offer a host with an opportunity to start a new clutch.

Most hosts that respond to parasitism by abandonment simply abandon the entire nest. Yellow Warblers, however, often construct a new nest floor over a parasitized clutch, abandoning only the clutch. If the nest is parasitized again, the process may be repeated. These multiple layer Yellow Warbler nests are fairly common and may be many layers deep. Burgham and Picman (1989) observed that relining the nest over the parasitized clutch is a significantly more common response by Yellow Warblers than abandonment of the entire nest. They suggested that this behavior is favored because reconstruction may be less conspicuous than building an entire nest; thus, the chances of subsequent parasitism are reduced. Clutch abandonment may be a function of investment. Yellow Warblers abandon or bury their clutches more often with a small number of host eggs (fewer than two) in the nest (Clark and Robertson 1981; Burgham and Picman 1989). Egg burial has been reported for several other Brown-headed Cowbird hosts, but the behavior does not appear as regularly in other species as it does with Yellow Warblers (chapter 7).

Rejection of Parasitic Nestlings

Even though parasitic nestlings may be widely divergent from host siblings, particularly in size (Jensen and Jensen 1969), rejection of parasitic nestlings (either ejection or starvation) is uncommon (Rothstein 1990). Cuckoo and cowbird nestlings are almost always accepted by their hosts. Parasitic finch nestlings are typically accepted, but the fine mimicry displayed by these parasites suggests that nonmimetic nestlings might be rejected. Indeed, Nicolai (1974) found that parasitic finch nestlings with mismatched mouthparts are usually not fed by the host (chapter 3).

Prevention: Denying Access to the Nest

Avoidance of parasitism can be observed as aggression toward intruders at the nest site in some species. Other species, such as Yellow Warblers, may simply sit on the nest, refusing the brood parasite access to the nest.

Nest Cupping

Refusing to leave the nest, or "nest cupping," appears to be a relatively safe method of preventing parasitism—that is, potential hosts are usually not subsequently attacked and supplanted by the parasites (Burgham and Picman 1989; Hobson and Sealy 1989). However, incidences of cowbirds supplanting hosts have occasionally been reported (Prescott 1947). Appropriately, nest-cupping behavior by Yellow Warblers was not observed during a predator defense study (Hobson et al. 1988). Hobson and Sealy (1989) reported that nest attentiveness of Yellow Warblers is significantly higher during the egg-laying stage when parasitism is more likely to have a detrimental effect.

Aggression

Aggression toward the parasite is potentially more dangerous than nest cupping for both host and parasite. For example, I have witnessed on numerous occasions Red-winged Blackbirds decapitate mounted Brown-headed Cowbirds. Similarly, Leathers (1956) reported an American Robin drawing blood from a Brown-headed Cowbird that was watching her nest.

Aggression Experiments. Many investigators of brood parasitism have been intrigued with host aggression toward brood parasites (table 2.3), but quantifying this behavior and standardizing methods have been challenging. Host aggression toward parasites has been observed under natural conditions (Sutton 1928; Hickey 1940; Nice 1943; Prescott 1947; Edwards et al. 1949; Ficken 1961; Selander and LaRue 1961; Robertson and Norman 1976; Slack 1976; Smith et al. 1984; Briskie and Sealy 1989), but these interactions are too few to draw conclusions regarding intensity of host aggression toward parasites. Additionally, birds often display a general level of aggression toward any intruder in their territories—be it parasite, predator, or competitor. Therefore, to supplement observations of naturally occurring interactions between brood parasites and their hosts, and to distinguish antiparasite aggression from territoriality and predator avoidance, experimental simulations have commonly been employed.

Typically, stuffed or freeze-dried parasites and control species, mounted in lifelike positions, have been placed near potential host nests. The primary goal of these experiments is to determine if the host or potential host can recognize parasites as a unique threat; therefore, controls should be species, such as sparrows, that do not represent potential threats. In some more thorough investigations, models of predators (Burgham and Picman 1989; Duckworth 1991) and models of parasites in varying positions were used.

Although observations of experimental simulations were reported earlier (Chance 1940; Edwards et al. 1949), Robertson and Norman (1976) developed quantifiable methods using behavior categories according to the intensity and duration of response. These methods have been followed or modified by others over the last two decades. More recently, they have been criticized, yet few alternatives suitable for most species have been offered.

Rothstein and O'Loglen (unpub. ms.) criticized the methods of numerous studies whose experiments were based on two presumed goals—(1) to determine if the potential host recognizes the parasite as an enemy; and (2) to assess, by the level of aggression intensity, how well the host can prevent parasitism, with the assumption that aggression is an effective defense.

Table 2.3

Species Reported to Display Aggressive Behavior toward Brown-headed
Cowbirds and Shiny Cowbirds

Species/Location/Source

Black-billed Cuckoo: Ontario: Robertson and Norman 1977
Puerto Rican Flycatcher: Puerto Rico: Nakamura 1995
Eastern Wood Pewee: Ontario: Robertson and Norman 1977
Willow Flycatcher: Ontario, Manitoba: Robertson and Norman 1976, 1977
Least Flycatcher: Manitoba: Robertson and Norman 1977; Briskie and Sealy 1989
Eastern Phoebe: Ontario: Robertson and Norman 1976, 1977
Great Crested Flycatcher: Ontario: Robertson and Norman 1977
Western Kingbird: Manitoba: Robertson and Norman 1977
Eastern Kingbird: Ontario, Manitoba: Robertson and Norman 1976, 1977
Gray Kingbird: Puerto Rico: Nakamura 1995
Eastern Bluebird: Ontario: Robertson and Norman 1977
Wood Thrush: Ontario: Robertson and Norman 1976, 1977
American Robin: Ontario, Manitoba: Robertson and Norman 1976, 1977
Brown Thrasher: Manitoba: Robertson and Norman 1977
Cedar Waxwing: Ontario, Manitoba: Robertson and Norman 1976, 1977
Warbling Vireo: Manitoba: Robertson and Norman 1977
Red-eyed Vireo: Ontario: Robertson and Norman 1976, 1977
Black-whiskered Vireo: Puerto Rico and St. Lucia: Nakamura 1995
Yellow Warbler: Iowa: Folkers 1982; Folkers and Lowther 1985.
   Manitoba: Robertson and Norman 1977; Briskie et al. 1992.
    Ontario: Robertson and Norman 1976, 1977; Burgham and Picman 1989.
   Puerto Rico and St. Lucia: Nakamura 1995
Common Yellowthroat: Ontario: Robertson and Norman 1977
Hooded Warbler: Pennsylvania: Mark and Stutchbury 1994
Rose-breasted Grosbeak: Ontario: Robertson and Norman 1977
Indigo Bunting: Ontario: Robertson and Norman 1977
Chipping Sparrow: Ontario: Robertson and Norman 1976, 1977
Clay-colored Sparrow: Manitoba: Robertson and Norman 1977
Field Sparrow: Ontario: Robertson and Norman 1976, 1977
Song Sparrow: British Columbia: Smith et al. 1984.
   Ontario, Manitoba: Robertson and Norman 1976, 1977
Swamp Sparrow: Ontario: Robertson and Norman 1976, 1977
Bobolink: Ontario: Robertson and Norman 1976, 1977
Red-winged Blackbird: Colorado: Ortega 1991; Ortega and Cruz 1991.
   Iowa: Folkers 1982; Folkers and Lowther 1985.
   Ontario, Manitoba: Robertson and Norman 1976, 1977
Yellow-hooded Blackbird: Trinidad: Cruz et al. 1990
Eastern Meadowlark: Ontario: Robertson and Norman 1977
Yellow-headed Blackbird: Manitoba: Robertson and Norman 1977
Brewer's Blackbird: Manitoba: Robertson and Norman 1977
Common Grackle: Illinois: Peer and Bollinger 1997.
   Ontario, Manitoba: Robertson and Norman 1976, 1977
Baltimore Oriole: Manitoba: Neudorf and Sealy 1992.
   Ontario, Manitoba: Robertson and Norman 1976, 1977
Yellow-shouldered Blackbird: Puerto Rico: Nakamura 1995
Greater Antillean Grackle: Puerto Rico: Nakamura 1995
American Goldfinch: Ontario: Robertson and Norman 1976, 1977

However, not all studies of mount experiments state, either implicitly or explicitly, this second goal, which is a very difficult question to answer through mount experiments—or at all. Whether aggression is an effective antiparasite nest defense depends on several factors: (1) if the host or potential host is large enough to effectively intimidate the parasite, (2) if aggression is intense enough to deter a parasite, (3) how vigilant the potential host is throughout the day, and (4) if the host is at the nest site during the critical hours of egg laying by the parasite. For the first two factors, the response of the parasite must be observed, and for the third factor, the potential host would have to be observed throughout each day of the entire egg laying and early incubation periods.

Furthermore, in natural encounters between parasite and host, the parasite may depart in response to an aggressive host, whereas during experiments, the model remains regardless of host activity. The continued presence of the parasite may also cause escalation of aggression by the host (P. Lowther, pers. comm.).

Hobson and Sealy (1989) criticized the methods of Robertson and Norman (1976) for categorizing alarm calling and distraction behaviors as aggressive responses. In their analyses of Yellow Warbler responses to mounted Brown-headed Cowbirds, Hobson and Sealy (1989) did not include vocalizations as aggressive behavior; they considered only contact or close passes as aggression and, consequently, reported little aggression. Aggression, however, is a much broader term, encompassing a variety of behaviors, including vocalizations. Drickamer and Vessey (1982:275) stated that "the intensity of aggression varies widely, from actual physical contact and killing to a subtle threat, such as a direct stare or a raising of the eyebrows to expose the eyelids." As humans, we often describe ourselves as aggressive based solely upon conversation. An aggressive display, be it vocalization or chasing, communicates an animal's mood or intentions, and may inhibit attack by another.

Another factor to consider is that, unlike mounts, live parasites assume different positions, which may elicit either more or less aggression from potential hosts. Brown-headed Cowbirds may appease hosts by bowing their heads and ruffing their neck feathers in what is termed an "invitation to preening display" (chapter 7). In Robertson and Norman's (1976) study, six of eight hosts responded less aggressively toward a Brown-headed Cowbird mounted with head bowed down than in a normal position.

Effectiveness of Aggression. Aggression may be a poor indicator of how well hosts are actually protected from brood parasitism, because cowbirds are persistent, smaller species may be unable to drive away cowbirds, and nest attentiveness varies. Furthermore, cowbirds lay their eggs in the early morning when nests are often unattended; therefore, cowbirds can parasitize nests that are otherwise rigorously defended. Birds that are particularly aggressive toward cowbirds, such as Red-winged Blackbirds, do not altogether escape parasitism. Other studies have reported similar findings with aggressive yet parasitized cuckoo hosts (Edwards et al. 1949; Smith and Hosking 1955; Wyllie 1981). Robertson and Norman (1977) suggested that aggression may be effective in host populations that are dense; in low density populations, aggression may be used as a cue to finding nests.

Aggression Levels. Several studies have suggested that the level of host aggression toward cowbirds is proportional to the intensity of parasitism (Robertson and Norman 1976, 1977; Nakamura 1995). Several hypotheses may explain this. First, aggression at the nest site may alert parasites to the location of an available nest (Ficken 1961; Seppa 1969; Robertson and Norman 1976, 1977; Smith et al. 1984). Uyehara and Narins (1995) observed that noisy Willow Flycatchers were parasitized more frequently by Brown-headed Cowbirds than were more silent individuals. They also reported that Willow Flycatchers chased Brown-headed Cowbirds more frequently when they approached within 2 meters of the nest (10 of 12) than when they approached within 2–10 meters (3 of 16). This is an interesting finding, as it suggests that Willow Flycatchers may respond only when their nest is in imminent peril. It would probably be useful to add a category of distantly perched parasites to aggression experiments.

A second hypothesis is that ejectors may have less reason than accepters to display aggression toward parasites, as egg ejection is a more efficient and less risky defense. It may be less costly to lose one egg (if the parasite removes it) than to face an aggressive encounter with the parasite. Ejector species can also eject the egg if aggression does not work. If aggression, indeed, serves as an effective host defense, one might expect aggression to be more intense among accepter species, because the cost of parasitism is higher and aggression may be the only defense. While both ejector and accepter species have been experimentally tested for levels of aggression toward cowbirds, no clear pattern between the two is readily apparent. Although several ejector species have been the subject of aggression experiments, only Eastern Kingbirds, Gray Catbirds, and American Robins have demonstrated significantly more aggression toward cowbirds than toward controls (Robertson and Norman 1977). Other ejector species may, indeed, be aggressive toward cowbird models, but sample sizes of aggression experiments have been too low to generate statistical significance. Few ejector species have clearly demonstrated a lack of aggression toward cowbirds in mount experiments, but American Robins apparently vary greatly in their responses toward cowbird mounts (Robertson and Norman 1976, 1977; Briskie et al. 1992). American Robins and most other ejector species are much larger than cowbirds and perhaps can deter parasites by their size alone.

A third hypothesis is that aggression may, indeed, function as an antiparasite defense (Robertson and Norman 1976, 1977). Robertson and Norman (1977) reported a positive correlation between the incidence of Brown-headed Cowbird parasitism and aggression within several families and subfamilies of hosts, including Tyrannidae, Muscicapidae (subfamily Turdinae), Emberizidae (subfamilies Parulinae and Icterinae), and Fringillidae. They interpreted this to be the result of increased rates of parasitism selecting against lack of host aggression toward cowbirds. Robertson and Norman (1977) concluded that because communities at both high and low host nest densities were within the same predominant habitat type, and high density areas were parasitized at a relatively low rate or not at all, aggression serves as an antiparasite defense. That is, parasitism is less likely to occur in areas with more attentive adults present.

Change in Aggression Levels over the Nesting Cycle. Burgham and Picman (1989) hypothesized that aggression levels toward Brown-headed Cowbirds should be most intense during nest construction and egg laying, when parasitism is most likely to occur. Their results did not entirely support their hypothesis: they found Yellow Warblers to be most aggressive toward mounted female Brown-headed Cowbirds during egg laying and early incubation, but they were aggressive during all nesting stages. Briskie and Sealy (1989) reported similar results with Least Flycatchers. However, these results are not particularly surprising in light of the fact that Brown-headed Cowbirds have been known to remove host eggs well into incubation, and they have also been known to remove host young from nests (chapter 7).

Aggression Used as Cues. Parasites may selectively choose more aggressive individuals as hosts. Smith (1981) reported that older Song Sparrows (*Melospiza melodia*) are parasitized at twice the rate of yearling Song Sparrows, and that older Song Sparrows are more successful in raising young. Older Song Sparrows responded more aggressively toward cowbird mounts than control mounts, and two-year-old Song Sparrows responded more strongly than they had as yearlings (Smith et al. 1984). These findings support both the hypotheses that cowbirds use aggressive displays to reveal the location of nests and that aggressive displays may be used by cowbirds as cues in finding older, experienced individuals.

Is Aggression toward Parasites Learned or Innate? Whether aggression toward parasites is a learned or an innate response is a difficult question to answer, requiring knowledge of the breeding history of individuals in a population. The experiments of Smith et al. (1984) suggested that aggression is a learned response. Robertson and Norman (1976) also suggested that aggression toward Brown-headed Cowbirds is a learned, rather than innate response, because individuals displaying the highest intensity of aggression toward parasite mounts showed little aggression toward control mounts, whereas individuals displaying little aggression toward parasite mounts also displayed a similar lack of aggression toward control mounts. However, Mark and Stutchbury (1994) found that yearling Hooded Warblers (*Wilsonia citrina*) were equally aggressive as older females toward mounted Brown-headed Cowbirds placed near their nests.

Group Defense. While aggression is risky for both host and parasite, aggression by more than one potential host individual may be particularly risky for parasites and effective in preventing parasitism. Such group defenses, or mobbing, have been observed in Bobolinks (*Dolichonyx oryzivorus*, Robertson and Norman 1976) and Red-winged Blackbirds (Robertson and Norman 1976; Ortega and Cruz 1991). Group defense by Red-winged Blackbirds may be fairly effective in preventing parasitism. During mount experiments, Red-winged Blackbirds emit alarm calls, which often elicit help from neighboring individuals. It is not uncommon for a model cowbird to be attacked by ten or more Red-winged Blackbirds.

Red-winged Blackbirds in marshes typically nest in colonies, whereas Red-winged Blackbirds nesting in upland habitat are more widely scattered. This provides an unusual opportunity to compare the effectiveness of individual versus group defense by way of parasitism rates. The incidence of cowbird parasitism is higher in scattered upland nests than in marshes (Robertson and Norman 1976; Ortega 1991), higher in sparsely populated areas than in densely packed colonies (Robertson and Norman 1977; Linz and Bolin 1982; Freeman et al. 1990), and higher in peripheral portions of marshes than in centrally located areas of marshes, where group defense is apt to be higher (Linz and Bolin 1982; Ortega 1991).

In addition to potential greater group defense, nests in marshes may be more difficult for cowbirds to find. Upland bushes provide perches from which cowbirds can watch the activities of their hosts and thereby select nests to parasitize.

The incidence of Brown-headed Cowbird parasitism is also significantly higher in low nest densities compared with high nest densities among other species, including Willow Flycatchers, Least Flycatchers, Warbling Vireos (*Vireo gilvus*), Yellow Warblers, Clay-colored Sparrows (*Spizella pallida*), and Song Sparrows (Robertson and Norman 1977).

Acting Defenses

Parasitism may also be reduced by other breeding characteristics that act as defenses against brood parasitism but are unlikely to be specific adaptations in response to brood parasitism. Predation is the major source of reproductive failure in most studies, and selection should favor strategies that reduce the risk of predation. Some of these antipredator adaptations may also reduce rates of parasitism.

Nest Concealment

Nest concealment and elusive behavior at the nest site may lower risks of predation and, in some cases, lower the incidence of brood parasitism. Well-hidden nests, however, do not necessarily completely eliminate risks of parasitism.

Certain nest types may reduce parasitism as well as predation. For example, Brown-headed Cowbirds rarely parasitize cavity-nesting species, such as swallows, chickadees, wrens, some flycatchers, and bluebirds, that may otherwise be suitable hosts. Other cavity-nesting species, however, such as Prothonotary Warblers (*Protonotaria citrea*), do not escape parasitism by Brown-headed Cowbirds (Petit 1991). Shiny Cowbirds, by contrast, frequently parasitize cavity-nesting species. Nest deception may elude brood parasites as well as predators. Some waxbills (*Estrilda* spp.) of Africa, and Yellow-rumped Thornbills (*Acanthiza chrysorrhoea*) of Australia, construct an open cup nest on top of a domed nest where they lay their eggs (Collias and Collias 1984). Both species are parasitized, by finches and cuckoos, respectively. The false nest is thought to deceive brood parasites; however, little evidence exists to support this hypothesis (Collias and Collias 1984).

Colonial Nesting

Although predation and food resources may have been the primary factors selecting for colonial nesting, such nesting may also work well against brood parasites in several species, including Red-winged Blackbirds, Yellow-headed Blackbirds, and Tri-colored Blackbirds (*Agelaius tricolor*). Habitat may also influence rates of parasitism. Yellow-headed Blackbirds usually nest in deep water marshes, often in the absence of woody vegetation. Cowbirds may use high perches in searching for nests to parasitize, and lack of woody vegetation may affect the incidence of parasitism. Øien et al. (1996) similarly found that the frequency of European Cuckoo parasitism on Reed Warblers (*Acrocephalus scirpaceus*) was higher near perches, and the density of Eurasian Reed Warbler nests was higher farther from trees, presumably to avoid cuckoo parasitism.

Cooperative Breeding

Poiani and Elgar (1994) expected that cuckoo parasitism would be more common among cooperative breeders than solitary breeders because the additional activity of helpers at the nest would attract the attention of cuckoos. They tested this hypothesis using nine species of cuckoos and 121 hosts or potential hosts in Australia and found no association between host breeding system and incidence of cuckoo parasitism. Alternatively, they suggested that helpers may increase the chance of detecting and repelling intruding cuckoos.

Overnight Roosting

Some birds roost overnight in their nests during the egg-laying period when parasitism is likely to occur (Neudorf and Sealy 1994). Brown-headed Cowbirds laying before dawn (Scott 1991) would, therefore, find their potential hosts sitting on their nests. Neudorf and Sealy (1994) reported a high percentage of overnight roosting in Yellow Warblers after the initiation of egg laying and suggested that this may be one of the best defenses against brood parasitism.

Summary

Two broad mechanisms of defense can be employed against brood parasites: rejection of parasitic egg or nestling and denying access to the nest. Rejection of parasitic nestlings is rare and may occur only in finches. Parasitic eggs may be rejected either through ejection from the nest, abandonment of the whole nest, or abandonment of a clutch through egg burial. Hosts may accept parasitic eggs for a variety of reasons: (1) absence of or low selection pressure, (2) inability to recognize parasitic eggs, (3) inability to remove parasitic eggs, (4) high cost of ejection compared with the cost of acceptance, and (5) parasite hatching success may be relatively low.

Prevention of parasitism may take the form of "nest cupping" or aggression toward the parasite at the nest site. Although these prevention defenses may or may not be effective, they do not altogether prevent parasitism, as parasites may lay their eggs while hosts are away from the nest.

Evolution of Brood Parasitism

Obligate brood parasitism has evolved independently in several taxa of birds, and although the independent evolution of brood parasitism is clear among these phyletic groups, it has not necessarily evolved only once within each taxa. Because of this, there is no reason to expect that various hypotheses on the evolution of brood parasitism are either valid for all taxa or are mutually exclusive.

Breeding Anomalies in Nonparasitic Relatives

Assuming brood parasites are derived from nonparasitic ancestors, one might expect to observe breeding anomalies in closely related species. These anomalies would include irregularities in egg laying and degeneration of nest building, incubation, and care of young. Examination of some close nonparasitic relatives of various parasitic groups reveals that breeding irregularities do, indeed, exist.

In the cuckoo group, flimsy platform nests of nonparasitic Yellow-billed Cuckoos (*Coccyzus americanus*) are made from relatively few twigs, and this simplistic nest construction is thought to hint at degeneration of nest building (Nolan and Thompson 1975). This must be based on the assumption that at some point in their evolutionary history Yellow-billed Cuckoos constructed more elaborate nests. However, cuckoos are more primitive than most passerines that construct intricately woven nests, and the crude nests of Yellow-billed Cuckoos may never have evolved to a more elaborate state. Black-billed Cuckoos (*Coccyzus erythropthalmus*) build cup nests that are somewhat more complicated and stable than nests of Yellow-billed Cuckoos (Harrison 1978), but compared to the fine weaving of many passerines, even the nests of Black-billed Cuckoos appear coarse. Both cuckoos lay eggs at irregular intervals, which is a breeding characteristic shared with European parasitic cuckoos (Nolan and Thompson 1975). Irregular egg laying

results in unusually disparate developmental stages, and failure to synchronize may result in a female producing an egg before she has a nest to lay in.
An unsynchronized egg must be placed somewhere, and the most appropriate place would be in a nest — perhaps of another species. Both Yellow-billed
and Black-billed Cuckoos occasionally engage in interspecific parasitism
(Nolan and Thompson 1975), and Yellow-billed Cuckoos have been reported
as intraspecific parasites (Nolan and Thompson 1975; Fleischer et al. 1985).

Hamilton and Orians (1965) suggested that because nonparasitic North
American cuckoo nestlings are adapted to a coarse omnivorous diet, they
are well suited for receiving various food items from different species. Alternatively, they suggested that if European Cuckoos had difficulty switching
from their diet of noxious and toxic prey items to a diet more suitable for
nestlings, their young would be better raised by foster parents.

The food resources exploited by cuckoos may contribute to abnormal
egg laying patterns. The prey items of North American cuckoos are irregularly available, and these cuckoos may respond to high food abundance
by producing more eggs than they can care for by themselves. Nolan and
Thompson (1975) reported that the incidence of Yellow-billed and Black-
billed Cuckoo parasitism was higher in years of unusual food abundance.

Another relative of cuckoos, the Groove-billed Ani (*Crotophaga sulcirostris*), is a communal nester, which involves more than one female laying
eggs in a single nest; typically, several males and females incubate the eggs.
This breeding system differs from brood parasitism in that most females
that lay eggs in the communal nest also contribute to the care of the young.
Similar to parasitism, however, the dominant female replaces the eggs of
low status females with her own and participates less in incubation than the
lower ranking females (Vehrencamp 1977).

In the cowbird group, nonparasitic Bay-winged Cowbirds show some degeneration of nesting behavior. Occasionally they construct their own nests
but more often use abandoned nests of other species. Nests are acquired by
both males and females; only then are surrounding territories established,
unlike most birds that choose a nest site within an established territory.
Friedmann (1929b) suggested that this secondary interest in territory led to
reduced territorial defense in males.

Breeding anomalies in nonparasitic relatives do not necessarily suggest
that brood parasitism will eventually dominate these species' reproductive
strategies. More important, these anomalies suggest suitable traits for a parasitic mode of reproduction, and these same breeding characteristics may
have been shared by their (now parasitic) relatives at the time of incipient
brood parasitism.

Host Suitability and Parasite Constraints

Certain ecological conditions and breeding characteristics of birds are necessary precursors of brood parasitism, and the evolution of brood parasitism may have been facilitated by some of these traits. Parasites are constrained by the breeding ecology of their hosts, and in order for brood parasitism to appear in a population, suitable hosts must be present. Host suitability includes an incubation period as long or longer than the parasite's, an overlapping breeding season, acceptance of parasitic eggs, an appropriate diet, and perhaps low aggression toward intruding parasites.

Incubation Period

The number of days required to incubate parasitic eggs is critical; if parasitic eggs took longer to incubate than host eggs, the parasite would be at a competitive disadvantage, and brood parasitism would not likely become established. Even for species that eliminate host competition, such as some cuckoos and honeyguides, a longer incubation period would not allow for successful parasitism as later-hatched parasites would be much smaller and unable to eliminate their host siblings. Therefore, the parasite's incubation period was probably as short as or shorter than the host incubation period at the time of incipient brood parasitism.

Cowbirds have an incubation period of 9–11 days, which is as short or shorter than those of their hosts. Many blackbirds also have short incubation periods compared with other passerines. For example, Red-winged Blackbirds have an incubation period of 8–13 days, with a mode of 10 (Ortega 1991). The short incubation period of cowbirds is, therefore, probably not a specific adaptation to brood parasitism but a phylogenetic condition or trait that is suitable for brood parasitism. Both parasitic and nonparasitic cuckoos also have short incubation periods, which may have been an adaptation for rapidly exploiting unpredictable food resources (Hamilton and Orians 1965).

Overlapping Breeding Seasons

The breeding seasons of host and parasite must overlap. Species and individuals that breed outside the breeding season of local parasites can escape parasitism altogether. In temperate North America, the breeding seasons of most passerines and cowbirds overlap in late spring and early summer. In tropical South America, where passerines breed throughout the year, Shiny Cowbirds have a longer breeding season than Brown-headed Cowbirds.

However, Wiley (1988) reported that Shiny Cowbird parasitism did not extend beyond the breeding season of the highest quality hosts.

Acceptance of the Parasite's Egg and Young

Brood parasitism is probably the primary selective agent for egg and nestling discrimination. In the absence of brood parasitism, no eggs are added to nests, and eggs remain in the owner's nest unless they are removed by predators. Under these conditions, there may be little or no recognition of one's own eggs. Presumably, prior to the evolution of brood parasitism, there were no selection forces operating against acceptance of clutch sizes larger than the bird's own clutch. Therefore, at the time of incipient brood parasitism, parasitic eggs in the nests of hosts were probably not selected against. Brood parasitic eggs and/or increased clutch size may, in fact, have been preferred as a magnification or superstimulus promoting incubation.

Presently, for many species there is probably an upper limit to the superstimulus value of larger eggs as suggested by the size of parasitic eggs. Cuckoos and cowbirds both lay small eggs for their sizes. The eggs are also relatively small compared with those of their nonparasitic relatives. However, they are typically much larger than the eggs of many hosts and probably represent a compromise between the advantage of larger eggs producing larger young and the discriminative abilities of hosts. Conformity in size of parasite and host eggs suggests that at some point, more important hosts may have discriminated against larger parasitic eggs.

The ludicrous sight of a cuckoo nestling being fed by a warbler standing on top of the cuckoo's head in order to reach its gaping mouth raises the question of why the host does not abandon the huge nestling. It is so obvious to us that the nestling is not her own. Do hosts also recognize this? We will never know the answer to this age-old question. Nevertheless, it is interesting to speculate about why hosts expend such effort to care for parasites in the event they do recognize parasites as not being their own. Host acceptance of parasitic nestlings so distinctly different from their own may be partially explained by instinctual feeding behaviors that simply will not be denied. Occasionally, feeding instincts even transcend species outside the context of brood parasitism. Welty (1982) cited an example of a Northern Cardinal (*Cardinalis cardinalis*) feeding goldfish; the cardinal may have lost his own brood but retained an overwhelming urge to feed.

As with parasitic eggs, presumably before the appearance of brood parasites there was no selection against caring for nestlings of another species in the nest. The largest, noisiest mouths are the ones that typically are fed in a brood, and for most hosts, there is no apparent adaptive limit to this process (a few parasitic finch hosts, however, apparently discriminate against parasitic nestlings with "mismatched" mouth parts). As with the willing incubation of larger parasitic eggs, larger, louder parasitic nestlings may be preferentially fed because they provide a supernormal stimulus, eliciting maximum effort by foster parents. Dawkins and Krebs (1979) drew the following analogy to explain why hosts are duped into feeding parasitic nestlings: like a drug addict who knows that a drug is killing him, but who cannot stop taking it because the drug is manipulating him, the host cannot stop feeding the parasite because it cannot resist the supernormal stimulus of the parasite.

For some hosts, there may be a disadvantage in discriminating against parasitic nestlings as the behavioral repertoire that would be necessary for this discrimination might also weaken their responsiveness to their own nestlings, and this would incur a much higher cost than the occasional losses incurred by brood parasitism.

Not all brood parasites are larger than their hosts. Although cowbirds are larger than many of their hosts, they are smaller than some, most notably other blackbirds. When compared with other (male) Yellow-headed and (male) Red-winged Blackbirds, however, Brown-headed Cowbirds have a gape width (mouth opening) that is large relative to their size (Ortega 1991). The wide gape may stimulate foster parents to provide relatively more food to cowbird nestlings. The same gape width differential is observed between smaller female and larger male blackbirds (Ortega and Cruz 1992a).

Appropriate Diet

Many brood parasites, including cowbirds, require a high-protein insectivorous diet, which is easily accommodated as many passerines feed their nestlings a similar diet. Brood parasites that require an insectivorous diet do not survive well in the nests of birds that feed their young a diet primarily of seed. For example, Brown-headed Cowbirds apparently have low success in the nests of American Goldfinches (*Carduelis tristis*). In one study, of 13 cowbirds that hatched in American Goldfinch nests, 12 died within four days of hatching; the other was poorly developed and died at 12 days (Middleton 1991).

Unlike other brood parasites, the diet of parasitic finches consists primarily of small grass seeds. They parasitize hosts that feed their young mostly seeds, sometimes with small amounts of insects (Friedmann 1960), as this is apparently an appropriate diet for the parasitic finches.

Host Size

Host size may be an important constraint on parasites, but defining the upper limit is speculative. Some brood parasites do not regularly parasitize larger species that are capable of rearing their own offspring in the face of parasitism. For example, Yellow-headed Blackbirds are larger than Brown-headed Cowbirds; they accept cowbird eggs and successfully raise cowbirds, yet they are rarely parasitized (Ortega and Cruz 1991; but see Dufty 1994). Although the answer to lack of parasitism may lie in their dense colonial nesting, size could also play an important role. Yellow-headed Blackbirds are not particularly aggressive toward cowbirds in some mount experiments. However, their larger size may be intimidating to smaller cowbirds, and they could pose a threat of serious injury to cowbirds caught in the act of egg laying. The high nesting density together with the larger size could make parasitism of Yellow-headed Blackbirds a daunting experience for Brown-headed Cowbirds.

Some cuckoos have an apparent upper size limit of host suitability that is surprisingly small. Species larger than thrushes are rarely parasitized by some cuckoos. Fieldfares, roughly 25 cm in length, are not parasitized by European Cuckoos (Moksnes and Røskaft 1988) 33 cm in length. As with Yellow-headed Blackbirds, Fieldfares do not display aggression toward parasites in mount experiments, and they accept eggs unlike their own (Moksnes and Røskaft 1988). However, Fieldfare eggs may be too large for cuckoo nestlings to eject, Fieldfares may provide an inadequate diet for cuckoos, and predation rates of Fieldfares are higher than other hosts (Moksnes and Røskaft 1987, 1988). Experimental parasitism might provide insight to these questions.

Association with Members of Own Species

Brood parasites seem to know just who they are at an early age. Brown-headed Cowbirds join flocks of their own kind after they gain independence from their foster parents. This species recognition is most likely under genetic control and is a necessary precondition of brood parasitism.

Brood Parasites as Observers

Brood parasites are observers of their surroundings and the breeding activities of their hosts. Instead of spending their time busily constructing nests, incubating, and taking care of young, brood parasites spend their time sitting and watching the nesting activities of their hosts. Many cuckoos have a hawklike hunting strategy that requires the same sit and watch approach, with long periods of motionless waiting for suitable prey items. This contrasts with the foraging strategy of many other arboreal birds that constantly flit from branch to branch in search of their food. Whether this sit and watch strategy evolved as an adaptation for hunting or for reproductive activities, of course, is speculative. However, Yellow-billed and Black-billed Cuckoos also demonstrate this hunting strategy, which suggests that this behavior may have been in place prior to brood parasitism.

Hawk Mimicry

The resemblance of many cuckoos to small hawks has sometimes been viewed as a specific adaptation for brood parasitism. In form, plumage, flight pattern, and threat displays, many cuckoos resemble accipiters and small falcons. Kuroda (1966) further suggested that the yellow eyelids of some cuckoos mimic the yellow iris common among hawks. The accipiter-like form of a long tail together with short rounded wings is characteristic of nonparasitic cuckoos as well. Kuroda (1966) noted that the hawklike pattern of plumage occurs in most parasitic cuckoos but not in nonparasitic cuckoos. This resemblance could reduce aggression if a host perceives the hawklike appearance as a life threat and flees the nest, but aggression may be increased if a host mobs the cuckoo as it might mob a hawk. In either case, the nest is exposed, allowing the female cuckoo momentary access to the nest.

Craib (1994) believed that the hawklike appearance is, indeed, mimicry because potential hosts mob cuckoos as they would a hawk. However, potential hosts may also mob cuckoos because they recognize them as cuckoos. Smith and Hosking (1955) and Duckworth (1991) found that in mount experiments some birds could distinguish between Sparrowhawks (*Accipiter nisus*) and cuckoos. Aggressive host response toward cuckoos waned as the breeding cycle progressed, when cuckoos were less of a threat, whereas they continued to respond strongly toward Sparrowhawks and Eurasian Jays (*Garrulus glandarius*), other nest predators (Duckworth 1991). Nevertheless, a live female might effectively defend herself from host attack by displaying the hawklike pattern under her tail and wings (Kuroda 1966).

Instead of being an adaptation for brood parasitism, the hawklike resemblance could be an adaptation for reducing foraging competition. Accipiters feed on small songbirds that feed on caterpillars and other invertebrates. The resemblance to accipiters may function to eliminate songbirds from optimal foraging grounds. The similar appearance of nonparasitic cuckoos suggests that the resemblance is not a specific adaptation to brood parasitism but is a precondition that also facilitates host distraction.

Loss of Nest-building Behavior

Hamilton and Orians (1965) suggested that brood parasitism may be more likely to evolve among species in which males build nests. In this situation, females are already accustomed to laying in nests constructed by others, and the site of a nest may stimulate egg laying. This may have contributed to the brood parasitic mode of finches. In many closely related nonparasitic finches, males share in nest construction or totally construct nests; others take over abandoned nests. Many of these nonparasitic finches nest colonially, and their breeding activities are highly synchronous following rains. Coloniality and synchrony allow abundant opportunities for parasitism, and accidental placement of eggs might be common. Furthermore, in some finch species, males may construct more than one nest, which could function as a predator confusion strategy (Collias and Collias 1984). Other finches use nests to sleep in, with four or five birds per nest (Collias and Collias 1984), adding to the confusion over which nest to lay eggs in. Hamilton and Orians (1965) suggested that because some of these nests are so elaborate and take so much time to construct, the time-saving aspects of brood parasitism may have been particularly important.

Selection for Parasitism

Presuming that brood parasites were derived from nonparasitic ancestors, in order for brood parasitism to evolve and become established in a species, certain ecological conditions must have conferred selective advantage to those individuals with parasitic tendencies. We will never know exactly what these ecological conditions were—whether environmental conditions changed or whether empty niches were filled. However, it is useful to explore some of the more common selective forces that shaped avian breeding ecology. Predation and resource limitation are among the most common selection pressures and may have contributed to the evolution of brood parasitism.

Limitation of Nest Sites

Limitation of nest sites may have led to the degeneration of nesting instincts and nesting activities, and degeneration of nesting activities is common to all parasitic taxa. For cavity-nesting species, appropriate nest sites may be one of the most important factors limiting population. This may be especially pronounced for those that do not excavate their own nests but instead either evict the rightful owners or take over abandoned cavities. For example, Jones and Leopold (1967) reported that the population of Wood Ducks soared after the provisioning of nest boxes.

In some cases, such as nest competition between Galahs and Pink Cockatoos, nest ownership may be in question and eggs may be deposited by both species but reared by only one (chapter 1). In other cases, the unquestionable nest owner may be forced from its nest by another species, and the eggs left behind may be incubated by the evictor. Species that are regularly evicted are less defensive of their nests and may eventually experience further degeneration of nesting behavior if they are continually forced to abandon their nests, but they may still enjoy reproductive success through care of their young by their evictors. Brood parasitism in honeyguides and Screaming Cowbirds may have evolved by this mode. Honeyguides frequently parasitize related cavity-nesting woodpeckers, and honeyguides may have formerly been cavity nesters themselves (Hamilton and Orians 1965). Bay-winged Cowbirds usually construct their nests in cavities or take over other cavity nests; Screaming Cowbirds may have also used cavities prior to brood parasitism.

The fact that CNP is so common among ducks and cavity nesters is suggestive that nest site shortage may have been a factor in selection for brood parasitism; however, some investigators have reported that CNP occurs even when there is an excess of nest sites (Andersson and Eriksson 1982).

Exploitation and Limitation of Abandoned Nests

Limitation of nest sites or nesting materials may have led to the use of abandoned nests and the subsequent loss of nest-building abilities in numerous species. Incipient brood parasitism is suspected in some of these species. White-throat Munias (*Lonchura malabarica*) will sometimes construct their own nests, but they also habitually use abandoned nests of three species of weaverbirds (*Ploceus* spp.). Dhindsa (1983a) reported CNP in White-throated Munias and attributed it to a shortage of weaverbird nests. Dhindsa and Sandhu (1988) investigated the possibility of incipient brood parasitism of White-throated Munias on Baya Weaverbirds (*Ploceus philippinus*). White-throated Munias regularly enter occupied nests of Baya Weaverbirds, but no White-throated Munia eggs have been found in Baya Weaverbirds' nests. Baya Weaverbirds eject White-throated Munia eggs from their nests; therefore, the question of parasitism remains open because White-throated Munia eggs may be ejected before investigators can discover them (Dhindsa and Sandhu 1988).

Destruction of Nests

Predation of eggs and nestlings is often the major cause of nest mortality. When a nest has been depredated before the last egg is laid, a female must deposit her egg somewhere or reabsorb it. Another nest is the most appropriate place in which to lay that egg; among colonial species, another nest is most likely to belong to a conspecific, while species with large territories might have to lay in nests of other species. If a female successfully lays her egg in the nest of another and renests as well, her reproductive success may be higher than if her first nest was successful. Many individuals face the dilemma of where to lay an egg after destruction of a nest, yet not all species lay their eggs in the nests of other birds. Hamilton and Orians (1965) suggested that the propensity to actively seek out other species' nests must be greater than occasional dumping; more eggs would have to be laid in the nests of other species than in one's own nest.

In a test of the Hamilton-Orians (1965) hypothesis that interspecific brood parasitism evolved from laying in conspecific's nests after nest destruction, Rothstein (1993) removed Red-winged Blackbird nests after the first egg was laid. Removal of nine nests resulted in no extra eggs being laid in neighbors' nests, but in three cases, eggs were found the next day floating in water near the original nest. Whether the eggs were laid on the vegetation or in a neighbor's nest and subsequently ejected was not known, but the latter is unlikely as Red-winged Blackbirds generally accept conspecific eggs (Ortega and Cruz 1988). Similarly, I found only one suspected case of CNP among 650 Red-winged Blackbird nests studied, and Harms et al. (1991) found CNP in only 34 of 7,805 (0.4%) Red-winged Blackbird nests. Rothstein (1993) expressed surprise that he could not elicit CNP in Red-winged Blackbirds given that many conspecific nest parasites are colonial. However, as with many other passerines, predation is high among Red-winged Blackbird nests, and being as well studied as they have been, if CNP occurred, it would likely have been discovered. Rothstein suggested that nest removal experiments be conducted on other species that are not known to be conspecific parasites to determine if CNP can be induced. Yezerinac and Dufour (1994) argued that the Hamilton-Orians hypothesis could not be effectively tested on a single species because experiments provide information only on what is currently maintaining a population. Additionally, nest destruction is already found to be the major source of nest loss in nearly all studies of passerines, making removal experiments unnecessary. A comparative approach considering predation pressure and frequency of brood parasitism across taxa may be a more appropriate test of the Hamilton-Orians hypothesis (Yezerinac and Dufour 1994).

Foraging Conditions

Foraging conditions and foraging modes may have favored parasitism in some groups. Many brood parasites are nomadic, following food resources. Cuckoos wander around assessing local foraging conditions, becoming numerous during local population outbreaks of caterpillars. Honeyguides also wander in search of beehives, and cowbirds presumably used to follow roaming herds of Bison. These three groups of brood parasites, therefore, are nomadic, and becoming sedentary for the time required to set up a territory, construct a nest, incubate eggs, and raise young would be problematic. We will never know whether these foraging modes were in place at the time of incipient brood parasitism; perhaps these groups of brood parasites opportunistically exploited their freedom from nesting activities and filled a previously empty ecological role after the onset of brood parasitism.

Mistaken or Accidental Parasitism

Brood parasites sometimes lay their eggs in the nests of unsuitable hosts that do not provide an adequate diet, that are too large and outcompete the parasitic nestling, or that are too small (e.g., hummingbirds). Although it may be useless in most cases, this kind of "accidental" parasitism may not be extraordinarily costly. When the parasite cannot find a nest at the appropriate time to parasitize, it is better to take the chance in the nest of an unsuitable or new host than to unequivocally lose an egg on the ground. Some new hosts may have been discovered by accidental parasitism.

Endocrine Basis

Failure to synchronize nesting stages may have contributed to the loss of broodiness, and this endocrine imbalance may have facilitated brood parasitism in some groups. Estrogen is at least partially responsible for the control of nest building (Lehrman 1958; Warren and Hinde 1959), and brood patch development apparently requires both estrogen and prolactin (Bailey 1952; Höhn 1962). Brood parasites are probably not deficient in estrogen, as production of eggs is estrogen dependent, nor are they deficient in prolactin secretions (Höhn 1962). However, even with treatments of estrogen and prolactin, female Brown-headed Cowbirds failed to produce brood patches (Selander 1960; Selander and Kuich 1963). Höhn (1962) reported that prolactin secretions in Brown-headed Cowbirds are similar to those in Red-winged Blackbirds and concluded that failure to develop a brood patch is due not to lack of prolactin secretion but to failure of skin to respond to the hormone. Höhn (1962) suggested that obligate brood parasitism could have evolved from a mutation of skin insensitivity to hormones, resulting in the loss of brood patch and incubation behavior.

The Evolutionary Arms Race

An adaptation in one group of organisms may modify the selection pressures on another group. If the other group evolves counteradaptations and these counteradaptations are reciprocated, then a "runaway escalation" or "arms race" may result (Dawkins and Krebs 1979). Evolutionary arms races are either symmetric or asymmetric. In an asymmetric arms race, one group enjoys a distinct advantage. The coevolution between brood parasites and their hosts is an example of an asymmetric arms race, with the brood parasites having the selective advantage over their hosts (Dawkins and Krebs 1979).

Selection favors parasites that most efficiently exploit their hosts, while it also favors hosts that avoid parasitism; however, selection pressures may be unequal. The asymmetric selection pressures in a predator-prey arms race may be referred to as the "life-dinner principle," and is applicable to parasite-host arms races as well (Dawkins and Krebs 1979). The "life-dinner principle" suggests that an inherent imbalance exists between the predator and prey. "The rabbit runs faster than the fox, because the rabbit is running for his life while the fox is only running for his dinner" (Dawkins and Krebs 1979:493). As selection pressures are high for prey to run from predators, there may be greater selection pressure on parasites to deceive their hosts than there is on hosts to recognize and thwart occasional parasitism (Dawkins and Krebs 1979). The parasitic young has to dupe or manipulate its host parent successfully or it will not survive, whereas host parents may raise some of their own young along with the parasite and also may be able to breed successfully at a later time. At a low to moderate rate of parasitism, many individuals will not even face parasitism and those that do may be capable of raising offspring in subsequent clutches or years (Moksnes et al. 1990).

Brood parasites may be ahead or behind in the arms race depending on the adaptive defenses employed by hosts. Presently, rejecters are ahead of parasites in the evolutionary arms race, and accepters are behind. Brood parasites have been able to counter most host adaptations to some degree (at least with many hosts), and this is why they have been so successful.

Egg Rejection and Counteradaptations

Ejection behavior may have originated from selection pressures for sanitary nest conditions and predator avoidance. Birds usually carry away from their nests fecal sacs, eggshell fragments, and dead nestlings. The motor patterns of this nest sanitation behavior are virtually identical to the motor patterns used in egg ejection and may be a precursor of egg ejection ability (Rothstein 1975a). Even eggshell fragments glued onto otherwise intact artificial eggs elicits ejection in some birds (Kemal and Rothstein 1988). The fact that virtually all accepter species have the basic motor patterns needed for egg ejection suggests that for some species there may be either a physical inability to eject parasitic eggs or a cost associated with the ejection response that exceeds the cost of acceptance. Rejection costs include mistaken rejection and damage to host eggs.

Egg Mimesis. Rejection of nonmimetic eggs implies that in the coevolutionary history of host and parasite, selection pressures favored egg mimesis by some parasites. Several experiments have revealed that hosts of parasites that lay mimetic eggs reject eggs unlike their own (Higuchi 1989; Moksnes 1992). Once a parasite has evolved an egg type that mimics host eggs nearly perfectly (or perfectly enough to be accepted), the evolutionary process is not necessarily complete. Selective pressures may be sufficient to favor changes in host eggs so they are recognizably divergent from parasitic eggs. This evolutionary change may take two different pathways: the host may evolve polymorphic eggs (variation within the species), which would make it difficult or impossible for parasites to match, or the host may evolve a single new egg type.

Egg polymorphism in some species, such as weavers, may have evolved in response to brood parasitism by cuckoos (Swynnerton 1918; Victoria 1972; Cruz and Wiley 1989). Alternatively, polymorphic eggs may have evolved in response to CNP. Eggs that are widely divergent from a potential parasite's eggs, such as the blue eggs of American Robins and blue-green eggs of Gray Catbirds, are thought to have evolved in response to the selective pressures of brood parasitism (Fretwell 1973). Rothstein (1974) suggested that simultaneous selective pressures for rejection behavior and a new host egg type facilitates egg discrimination and that these selective pressures favor both egg rejection and learned egg recognition.

Davies and Brooke (1989b) observed that the distinctive egg markings of Meadow Pipits and White Wagtails (*Motacilla alba*) did not differ between populations in Iceland and Britain, which may suggest that egg patterns in these two species did not evolve in response to brood parasitism as brood parasites do not exist in Iceland. However, possibly the Icelandic founder population was derived from parasitized populations elsewhere and retained the egg morphology but not the egg discrimination abilities.

A change in host egg morphology may be countered with a change in egg morphology of the parasite to mimic that of the host. The host may then evolve a new egg type, and this could theoretically go on, ad infinitum, in an unresolvable cycle, eventually leading to evolution of specialists or gentes.

The degree of apparent egg mimicry in European Cuckoos varies depending on the degree of rejection by the host. Cuckoos that parasitize nondiscriminating hosts generally do not lay mimetic eggs, whereas cuckoo eggs closely resemble eggs of discriminating hosts. Experiments with mimetic and nonmimetic eggs revealed that Meadow Pipits rejected white nonmimetic eggs at about the same rate (8.3%, $n = 24$) as mimetic eggs (5.3%, $n = 19$, Moksnes et al. 1990); therefore, it would appear that little selection pressure exists for cuckoos to evolve an egg type mimetic of Meadow Pipits. Other cuckoos, such as Horsfield's-bronze Cuckoos, lay eggs mimetic of their hosts', yet their hosts show no discrimination of nonmimetic eggs (Brooker and Brooker 1989b). It is possible that egg mimicry evolved during a time when the hosts may have been rejecters of nonmimetic eggs.

Two additional hypotheses may explain evolution of mimesis in cuckoo eggs. Harrison (1968) suggested that egg mimesis may have evolved in response to selective pressures of predation and that discordant eggs may, somehow, attract predators. However, I know of no evidence that directly supports this hypothesis. The egg morphology of generalist parasites, such as cowbirds, is typically distinct from the eggs of their hosts, but differential predation on parasitized and nonparasitized nests will not necessarily reveal whether the discordant egg was used as a cue for discovering the nest. The question of differential predation is difficult to answer as predation is confounded by other factors, such as habitat, temporal distribution, age or experience of the host, and brood parasites using some of the same cues to find nests as predators use. Conceding that such analysis may be inappropriate, Davies and Brooke (1988), nevertheless, suggested that predation does not select for egg mimesis in cuckoos because they found no difference in predation rates between nests in which they added nonmimetic cuckoo egg models and nests which contained no nonmimetic cuckoo eggs or cuckoo egg models.

Brooker and Brooker (1989b) and Brooker et al. (1990) suggested that egg mimicry may be selected for because it reduces the likelihood of a second cuckoo recognizing a competitor's egg and subsequently removing it. Individual cuckoos are known to lay only one egg per host nest. Nestling cuckoos eliminate all nest mates upon hatching and, therefore, the laying of a second egg might be wasteful. That is, the second egg would eventually be ejected by the first-hatched nestling. However, multiple parasitism does occur occasionally (Gosnell 1932; Brooker and Brooker 1989b). If a second-laying cuckoo can recognize the first cuckoo's egg, she may remove it. Even with egg mimicry, many second-laying cuckoos remove the egg of the first-laying cuckoo (Brooker and Brooker 1989b), suggesting that perhaps even more eggs would be removed by a second cuckoo if they were nonmimetic. This evidence suggests that, in addition to selective pressures afforded by host discrimination, competition from other cuckoos may also select for egg mimicry.

Maintenance of Gentes. How are gentes maintained? That is, how do cuckoos know which host nests to parasitize and who to mate with? Undoubtedly, the effects of imprinting exert some influence on host selection in cuckoos. However, conceding that experiments and captive conditions may have been deficient, Brooke and Davies (1991) failed to show host preference through imprinting by captive cuckoos. Imprinting normally occurs early in life, at which time young cuckoos learn the appropriate host. As an adult, a female will return to parasitize the species she was raised by, and her eggs will match the host (presuming she came from stock that produced eggs mimetic of the host). Her offspring, in turn, will subsequently return to parasitize the same host species. This succession seems simple enough, but how does she know what male to mate with? Likewise, how does the male cuckoo know which females might be most likely to be successful with his genetic contribution?

Male domestic fowl do contribute to eggshell color, and some evidence suggests that it is a male sex-linked characteristic (Romanoff and Romanoff 1949). Inheritance of eggshell color is not well understood in brood parasites (Brooke and Davies 1991). If males do not contribute to the eggshell color, then maintenance of cuckoo gentes can be more easily understood. If, however, male contribution does influence eggshell color, males must be faithful to their host species or habitat in order to maintain the integrity of gentes. Male cuckoos defend territories, and their territories may reflect a host preference. That is, the territories they defend theoretically would have a high density of the host they were raised by. For example, if a male was raised by a Reed Warbler, he would return to reed beds and mate with a female who also returned to the hosts by which she was raised, and she would parasitize the nests of Reed Warblers. Obviously, this system of gentes maintenance would not be perfect. There is habitat overlap in some major hosts (Brooke and Davies 1988). Indeed, the matching of host eggs by cuckoos is not perfect, but mismatching does not necessarily suggest that males contribute to egg color. Females may also lay in inappropriate host nests when appropriate hosts are limited, regardless of the mechanisms of gentes maintenance. In a model of population dynamics of a cuckoo-host association, Takasu et al. (1993) assumed that egg patterns and colors are inherited from mothers and are not influenced by the mating partner.

Interpretation of Mimesis. Problems often arise with the interpretation of acceptance or rejection of mimetic eggs. Presence of only mimetic cuckoo eggs in nests of their hosts may superficially suggest that cuckoos lay only in nests of hosts with eggs similar to their own (Payne 1967b). This assumption, however, should be verified through experimental parasitism (Rothstein 1971b). If cuckoo hosts eject nonmimetic eggs from their nests but accept mimetic eggs, one would expect to observe more mimetic eggs in nests than would be expected by random placement. Even results from experimental parasitism need to be interpreted cautiously, as hosts may (Korsnes et al., unpublished in Moksnes et al. 1990) or may not (Moksnes and Røskaft 1988) reject cuckoo eggs more frequently when a mounted cuckoo is placed near the nest.

Intensive studies of individual brood parasites may reveal just how species specific host choice is. For example, Chance (1922) reported that an individual European Cuckoo laid 58 of 61 of her eggs in Meadow Pipit nests.

Egg Size. Compared with their nonparasitic relatives, cuckoos and cowbirds lay small eggs relative to their size. However, cuckoo eggs and cowbird eggs are typically much larger than those of their hosts, and the size probably represents a compromise among various selective pressures, including the advantage of larger eggs, discriminative abilities of hosts, and selective advantage of laying eggs rapidly. This compromise may be especially important in cuckoos, as larger cuckoos hatch from larger eggs. If a host nest contains two cuckoo eggs—one larger than the other—from two different females, and if they hatch at the same time, the larger cuckoo will have an advantage toward becoming the sole occupant of the nest. Conformity in size of parasite eggs to host eggs may suggest that, at some point in the coevolutionary history between host and parasite, some hosts discriminated against larger parasite eggs or that larger eggs were not incubated efficiently by smaller hosts.

Ankney and Johnson (1985) suggested that the high rate of egg production by Brown-headed Cowbirds may result in smaller egg size; however, they found that, compared to European Starlings, Brown-headed Cowbirds do not sacrifice egg quality for a larger quantity of eggs produced. Alternatively, the relatively small size of parasitic eggs may also have been selected for because small eggs might be laid more rapidly than larger eggs.

Thick Eggshells of Parasites. Evolution of thick eggshells may prevent smaller hosts, with smaller bills, from ejecting parasitic eggs. For these hosts, thick eggshells of parasites may or may not be an evolutionary dead end. Generally, hosts cannot puncture thick-shelled parasitic eggs without damaging their own eggs (chapter 2).

Why have hosts not evolved a thicker eggshell to counter that of the parasite? Thick eggshells might reduce damage caused by puncture ejection, yet no hosts, other than Bay-winged Cowbirds, are known to have the extraordinarily thick eggshells of parasites (Rothstein 1972a). Friedmann (1929b) observed that the eggshells of Bay-winged Cowbirds are more brittle than other cowbird eggs, but Rahn et al. (1988) listed egg and eggshell measure-

Table 3.1
Eggshell Thickness and Weight for Cowbirds

Species	Eggshell Thickness	Weight
Bay-winged Cowbird	0.127 mm	4.10 g
Screaming Cowbird	0.134 mm	3.64 g
Shiny Cowbird	0.135 mm	4.30 g
Bronzed Cowbird	0.125 mm	4.15 g
Brown-headed Cowbird	0.115 mm	3.12 g
Giant Cowbird	0.170 mm	11.20 g

Source: Schönwetter 1983.

ments for all cowbirds and showed that eggshells of Bay-winged Cowbirds are as thick relative to shell mass as those of parasitic cowbirds (table 3.1) and considerably thicker than eggshells of other icterines. This is a significant observation and may shed some doubt on the characteristic as a specific adaptation to brood parasitism. They also reported that eggshells of parasitic species are 39% thicker than eggshells of nonparasitic birds and calculated the strength of eggs from parasitic species to be 92% greater than nonparasitic species. Hoy and Ottow (1964) also noted the thickness of Bay-winged Cowbird eggshells and suggested that eggshell thickness is probably not an adaptation for brood parasitism but rather an old phylogenetic trait.

The fact that Bay-winged Cowbirds have thick eggshells does not necessarily preclude the possibility that thick eggshells evolved as an adaptation to brood parasitism because thick eggshells may have evolved independently in parasitic cowbirds and Bay-winged Cowbirds. In the cowbird group, the ability to produce thick eggshells was probably present in the gene pool before the evolution of brood parasitism. Therefore, just as parasitic cowbirds could have evolved thick eggshells as a specific adaptation for brood parasitism, Bay-winged Cowbirds could have evolved thick eggshells in response to pressures exerted by lowered reproductive success in parasitized nests. The answer to why other hosts have not evolved thick eggshells may also lie in potentially lower success of eggs with thick eggshells.

Compared with some hosts, a smaller proportion of Brown-headed Cowbird eggs may hatch (Phillips 1951; Hofslund 1957; Walkinshaw 1961). Lower hatching success may partially be due to insufficient incubation, a relatively larger proportion of infertile eggs, and compromised eggshell pore size. Parasitic eggs are not always laid at appropriate times, and if an egg is laid after incubation is well under way, insufficient incubation may interfere with hatching success. Eggshell thickness, together with pore size, dictates the amount of gas exchange, which determines survival of the developing embryo. Eggshell thickness represents a compromise: the eggshell must be thin enough for the chick to pip, yet thick enough to support the egg contents and weight of the incubating bird.

In a lab and field study, Carello (1995) did not detect a significant difference in diffusive gas exchange between Brown-headed Cowbird eggs and Red-winged Blackbird eggs. She suggested that the larger pore size of Brown-headed Cowbird eggs must compensate for the thick eggshell. Larger pores may cause excessive loss of CO_2 and water vapor and allow bacteria and fungi to invade the egg (Carey 1986), particularly if the egg becomes wet (Romanoff and Romanoff 1949). Although other barriers, such as the shell membrane and albumen, may also prevent microorganisms from entering the yolk, increased pore size may carry risks of lethal contamination by invasive microorganisms.

A parasite that lays 25–40 eggs in a season can afford the risk of lowered hatching success that thick eggshells may confer. A host that lays four eggs cannot take such a risk, and this may be why hosts have not evolved a thicker eggshell in response to the thick eggshells of parasites. For the same reason, the thick eggshells of parasites may signal a one-sided victory, favoring the parasite, in the evolutionary arms race.

Why would specialist parasites, such as some cuckoos, have thick eggshells when egg mimesis works well enough? Perhaps in the history of a parasite-host relationship, egg mimesis did not always work well enough. Takasu et al. (1993) suggested that the cost of ejection for hosts might be totally responsible for the existence of cuckoo gentes. If there were no cost associated with ejection, then ejection behavior might become fixed in the host population, driving the cuckoo to extinction.

Removal of Host Eggs. Many hypotheses have been proposed as to why parasites remove host eggs when laying their own.

1. *Host deception.* Rejection of parasitic eggs may, in part, be a response to "feeling" an inappropriate number of eggs in the clutch. Many brood parasites remove a host egg from nests in which they lay their own egg, countering any response to numerical stimulation to the brood patch.
2. *Food for the parasite.* Many brood parasites that practice host egg removal also eat host eggs. Consumption of host eggs seems more likely to be an opportunistic foraging mode rather than a primary reason for removing an egg, as egg consumption does not occur on a consistent basis (Scott et al. 1992).

3. *Reduces competition.* Removing host eggs also reduces sibling competition in some cases; however, even cuckoos that evict their nest mates upon hatching also remove host eggs.
4. *Reveals incubation status.* Livesey (in Sealy 1992) proposed that cuckoos test whether hosts are at an appropriate stage of incubation by removing eggs.

There is no reason to expect any one hypothesis to be applicable to all parasites. For honeyguides and cuckoos that kill their host siblings, the most parsimonious explanation for parasites removing host eggs is to satisfy the host that no change has taken place in the nest (Hamilton and Orians 1965). Removal of host eggs, even for honeyguides and cuckoos, may be important, as it reduces sibling competition in situations where the parasite does not hatch first.

Evidence that Rejection Evolved in Response to Brood Parasitism

Comparison of responses to experimental parasitism of sympatric and allopatric host populations can provide evidence that rejection of nonmimetic eggs evolved specifically in response to brood parasitism. Unfortunately, there are very few allopatric areas left in which to test host responses, particularly for cowbird hosts.

Manitoba, Canada, provides a suitable area to test the responses of hosts that are sympatric and allopatric with Brown-headed Cowbirds. Brown-headed Cowbirds breed in the southern half of Manitoba but not the northern portion. Briskie et al. (1992) placed real cowbird eggs in nests of American Robins and Yellow Warblers in Churchill (northern Manitoba) and in Delta Marsh (southern Manitoba). As they predicted, all ($n = 25$) American Robins studied at Delta Marsh ejected the cowbird eggs, and 66.7% ($n = 18$) of American Robins studied at Churchill rejected the cowbird eggs. Similarly, Yellow Warblers tolerated cowbird eggs more at Churchill than at Delta Marsh. To rule out the possibility that rejection behavior evolved in response to CNP, Briskie also experimentally parasitized American Robin and Yellow Warbler nests with real eggs of conspecifics; none were rejected by either species. Yellow Warblers also responded more intensely toward mounted Brown-headed Cowbirds at Delta Marsh than at Churchill. American Robins, however, did not respond aggressively toward mounted Brown-headed Cowbirds in either location.

In a similar experiment, Davies and Brooke (1989a) provided evidence that egg discrimination by two hosts evolved in response to parasitism by European Cuckoos. With nonmimetic model cuckoo eggs of another gens, Davies and Brooke experimentally parasitized Meadow Pipits and White Wagtails in Britain, where they are parasitized by European Cuckoos, and in Iceland, where they do not coexist with brood parasites. Both Meadow Pipits and White Wagtails in Iceland demonstrated less egg discrimination than their counterparts in Great Britain. Host acceptance of eggs unlike their own is a necessary precondition for brood parasitism to appear in birds. Brown et al. (1990) conducted similar experiments with closely related *Acrocephalus* warblers, one of which is a favored host of the European Cuckoo in Europe and one of which is not known as a host of cuckoos in Australia. They predicted that warblers in Australia would not reject eggs unlike their own. Indeed, the Australian warblers showed significantly less rejection than their parasitized counterparts in Europe.

It is also useful to compare sympatric populations that are parasitized and not parasitized. I found that in Colorado, Yellow-headed Blackbirds were not parasitized even though Red-winged Blackbirds nesting in the same marshes were parasitized. In this population, Yellow-headed Blackbirds accepted 87.5% (*n* = 32) of artificial cowbird eggs and 100% (*n* = 23) of real cowbird eggs placed in their nests (Ortega and Cruz 1988). Dufty (1994), however, found that in Idaho, where parasitism was 17–21%, Yellow-headed Blackbirds ejected 33.3% (*n* = 15 nests) of Red-winged Blackbird eggs placed in their nests.

Evolutionary Lag or Inability? In a review of experimental parasitism on many North American cowbird hosts, Rothstein (1982b) suggested that the simplest explanation for acceptance of parasitic eggs is evolutionary lag; that is, genetic variants for egg ejection have not yet appeared in these species. Presumably, once ejection behavior appeared in a population, it would spread rapidly when selection pressures for antiparasite defenses were high. Rohwer and Spaw (1988) noted that most rejecters have large bills, whereas most accepter species have small bills and may be incapable of grasp ejection. Rohwer and Spaw suggested in an alternative hypothesis that the thick shells of cowbird eggs force acceptance on small hosts; in other words, most accepters are incapable of removing cowbird eggs without damage to their own eggs.[1] Rohwer and Spaw argued that for evolutionary lag to explain the acceptance of parasitic eggs, there should not be an association between bill size and acceptance. However, there is no reason to assume that one or the other hypothesis is operational for all hosts or that they are mutually exclusive. For some host species, such as many small warblers and flycatchers, the inability to remove parasitic eggs without cost to their own eggs may explain widespread acceptance. For other species that have the physical ability to remove parasitic eggs, lack of ejection might be best explained by evolutionary lag.

Reasons other than lack of genetic variance in the population may also explain evolutionary lag of ejection responses. When parasitism is low to moderate, a large proportion of accepter individuals experience reproductive success. For species that are able to raise their own young in parasitized nests, selective pressure may not be high enough to allow ejection behavior to become fixed. Also, ejection behavior may not become fixed in host populations that learn to distinguish their eggs from parasitic eggs after the first year of breeding (Lotem et al. 1992).

Responses of and to Nestling Rejection

Most hosts, other than estrildid finches, do not reject parasitic young no matter how divergent their morphology and behavior are from their own. Estrildid finches are noted for having remarkably unusual mouths. The palates of various colors (white, red, yellow, blue), depending on the species, are marked with a conspicuous and complex design of symmetrical spots (usually 3–5). Additionally, reflective globules at the corners of the gape vary in shape (from simple thickening to "pearls") and color (white, blue, yellow, violet). These mouth structures are assumed to function as directive stimuli for parents. The gape papillae reflect light and illuminate the pattern of the mouth in the semidarkness of domed nests. Apparently, none of the 125 or so species of estrildid finches have identical markings (Nicolai 1974).

Whether the remarkable similarity between host and parasite in these complex mouth parts is a product of mimetic evolution or similar phylogeny is debatable. According to Nicolai (1974), mouth parts of parasitic finches mimic their specific hosts so perfectly that they are nearly indistinguishable from their foster siblings, and parents discriminate against nestlings with mismatched mouth parts by not feeding them. Parasitic nestlings, therefore, must look like their species-specific host in order to be accepted and provided for. In addition to mouth part mimicry, plumage and vocalizations of young parasitic finches mimic their hosts. Nicolai (1974) argued that this fine-tuned mimicry and rejection of deviants implies that selective pressures have been intense for both the host to evolve complex patterns and discrimination against deviants and for the parasite to mimic the host.

Friedmann (1960) suggested that similarities of eggs, nestling mouth markings, and nestling plumage are a consequence of common descent, not mimetic evolution. Parasitic finches may not be as species specific as Nicolai (1974) implied; Pin-tailed Whydahs are known to parasitize at least 18 species (Friedmann 1960). Other *Vidua* species are not confined to one host, and some occur in locations where hosts with corresponding mouthparts do not even exist (Friedmann 1960). Furthermore, mouthparts of the Paradise Whydah *(Vidua paradisaea)* do not correspond with those of its main hosts, pytilias *(Pytilia* spp.). Friedmann (1960) pointed out that only if each *Vidua* species were confined to one host could these characteristics have evolved as specific adaptations. The evidence that Friedmann provided in his argument against mimetic evolution, however, are ecological factors that may have changed. For example, the Pin-tailed Whydah may have been more specialized earlier in its history, and viduines that presently occur in locations without hosts with corresponding mouthparts may have expanded their ranges in recent times.

That finches would refuse to feed their own young with less than "perfect" mouthparts is a difficult concept. However, in the semidarkness of their domed nests, nestlings without reflection globules may be overlooked by parents. This does not necessarily explain refusal to feed nestlings with mismatched nonreflective buccal markings. The reflective structures and buccal markings, however, may be linked characteristics and both may be constrained by yet another characteristic (Gould and Lewontin 1979). In other words, refusal of a parent to feed a nestling with mismatched mouthparts may have nothing to do with the mouthparts, per se, but with some other characteristic. Cross-fostering experiments would be a tremendous contribution to our understanding of whether similarities of mouthparts are due to mimetic evolution or common descent.

Aggression toward the Parasite and Counteradaptations

Parasites can reduce aggression by hosts in several ways. Detection may be reduced by dull or cryptic colors; many female brood parasites are dull in plumage compared with their nonparasitic counterparts. Even some cuckoos that may be considered brilliant, such as the emerald green *Chrysococcyx*, are cryptic amid dense vegetation (Reed 1968). Furtive behavior that is common to most brood parasites may also help to reduce detection. Intensity of aggression may be reduced by other behaviors that distract or appease the host, such as the "head-down" display used by cowbirds. Hawk mimicry of cuckoos may also inhibit aggression. Additionally, parasites are generally larger than their hosts, and they may be undeterred by host aggression.

Duration of host aggression is reduced by rapid egg laying. Brown-headed Cowbirds often lay eggs in less than 30 seconds, with an average laying time of 41.0 seconds (Sealy et al. 1995). This is considerably shorter than most passerines (20.7–103.7 minutes), including related icterines (21.5–53.4 minutes, Sealy et al. 1995). Cuckoos lay eggs in less than 10 seconds (Seel 1973; Gaston 1976; Brooker et al. 1988; Soler 1990). Rapid laying suggests that aggression toward intruders may be an effective defense, and minimal time spent at the nest reduces contact and altercations with hosts.

Parasitism as a Risky Business

As with other types of parasites, populations of brood parasites are directly tied to populations of their hosts. Brood parasites, however, rely on their hosts only for their reproductive success, not for their entire existence. When brood parasites kill the offspring of their hosts, they kill their own potential hosts and the potential hosts of their own offspring, and this is a risky business. An individual brood parasite does not have to attain the dynamic relationship with its individual host other parasites do, but the population of brood parasites must be in balance with the population of hosts if the parasite and host are to coexist.

Presumably, because brood parasites have coexisted with their hosts for long periods of time, a fine population balance is possible. This dynamic balance may be similar to predator-prey oscillations. When host availability is high, the brood parasite population increases. The resulting increase in brood parasites may, in turn, decrease host availability for a period of time. Decreases in host populations will then result in a decrease of brood parasites.

When brood parasites expand their range where hosts and parasites have not coexisted, both face a new situation with new selection forces. In areas of recent range expansion of the parasite, it is apparent that parasite-host populations are not in dynamic equilibrium, as evidenced from declines in host communities and increases in parasite populations. It is often assumed that the cause of this unbalanced population growth of parasites is due to hosts being ill-adapted to cope with brood parasitism; that is, they are behind in the evolutionary arms race and have not yet evolved antiparasite defenses. However, little evidence has been provided to support this claim for hosts of cowbirds. Few cowbird hosts display effective antiparasite defenses, and some species that have probably never been parasitized possess behaviors which we generally assume to be antiparasite defenses (chapter 9). As with other types of parasites, brood parasites may kill their hosts if the hosts are weakened by disease. In this case the "disease" is loss of breeding habitat, increased predation (caused by the forest edge syndrome and introduction of predators) and loss of habitat on wintering grounds (chapter 9).

Systematic History and Evolution of Cowbirds

As with many other birds, cowbird systematics has a history marked by numerous changes. Presently, five of the six cowbirds are placed in the genus *Molothrus*, and the Giant Cowbird is in the monophyletic genus *Scaphidura*.

The cowbird complex is a particularly valuable group in which to study phylogeny and evolution of brood parasitism because it is a small group, comprising only six species. Within this small group are diverse reproductive strategies, ranging from a non–brood parasite, to a brood parasite that specializes on one specific host, to generalist brood parasites. These reproductive strategies may represent various stages in the evolution of brood parasitism, and while a well-substantiated phylogeny may not resolve the sequence of events leading to interspecific brood parasitism, it may allow us to eliminate some alternatives. The polyphyletic origin of brood parasitism within a taxon may seem unlikely; however, there is no reason, a priori, to assume that evolution of a reproductive strategy such as brood parasitism, specialized as it is, can occur only once within a taxon. Furthermore, some conditions and behaviors shared among several members, such as association with large roaming mammals, may lead to the independent development of brood parasitism.

Friedmann (1929b) suggested phylogeny of the six cowbirds based on assumptions that certain attributes of geographic distribution, behavior, and morphology are either primitive or recent. He suggested that Bay-winged Cowbirds are the most primitive for several reasons: (1) Their range is within South America, which is considered by most ornithologists to be the origin of the genus *Molothrus* and blackbirds in general (Orians 1985), and it is more likely that ancestral species inhabit the geographic areas of origin. (2) Their courtship is virtually nonexistent, whereas the other molothrine species have courtship behavior ranging from rather simple (Screaming Cowbird) to more complex with bowing and fluffing. (3) Bay-winged

Cowbirds have a more simple song than the other cowbirds. And (4), Male Bay-winged Cowbirds have no black plumage, whereas males of the other cowbird species do. Black is thought to be a more advanced characteristic state as compared with browns. Friedmann's (1929b:346) best estimate of cowbird phylogeny was the following:

3. *M. aeneus* ——— 4. *Scaphidura*
1. *M. badius* ——— 2. *M. rufoaxillaris*
3. *M. bonariensis* ——— 4. *M. ater*

Some doubt has been cast upon Friedmann's (1929b) classification and whether Bay-winged Cowbirds are indeed cowbirds at all (Lanyon 1992).

Beecher (1951) used skull structure, jaw musculature, and feeding functions to interpret phylogenetic relationships in blackbirds. He suggested that the molothrine cowbirds arose from the finch family and were the most primitive of blackbirds, giving rise to three lines: agelaiine, quiscaline, and cacique. Beecher (1951:428) stated that "*Tangavius* is clearly a cowbird but seems to be transitional to the Giant Cowbird *Psomocolax* which appears to be near the direct line leading to the caciques and oropendolas." More recently, Lanyon (1992) conducted a phylogenetic study of the cowbird complex using DNA sequences of the mitochondrial cytochrome-b gene. Cladistic analyses indicated that brood parasitism evolved a single time within the blackbird group, and in a single most parsimonious tree, *S. oryzivora* is a subset of the *Molothrus* genus. The mtDNA data Lanyon used neither supported nor refuted placement of *M. badius* adjacent to or within the parasitic cowbird assemblage. Lanyon (1992) noted a trend from host specialist to host generalist, suggesting that the generalist strategy is a derived condition.

Freeman and Zink (1995), using restriction enzyme cleavage sites in mtDNA, suggested that brood parasitic cowbirds are not monophyletic, *M. rufoaxillaris* was not placed as a sister clade, and brood parasitism evolved at least twice within the icterines. Freeman and Zink's analyses also are contrary to Beecher's (1951) suggestion that molothrine cowbirds are the most primitive of blackbirds. Freeman and Zink (1995) attributed the differences between their study and Lanyon's to the phylogenetic signal in the sequence data being stronger than restriction site data, and they suggested further testing. When Freeman and Zink restricted their data set to only the taxa used in Lanyon's study and combined their data with Lanyon's, a single most-parsimonious tree was identical in topology to Lanyon's.

Summary

Obligate brood parasitism has evolved independently in several avian taxa; therefore, various hypotheses on evolution of brood parasitism will not necessarily be shared by all taxa. What may be shared by all taxa are certain traits and constraints, such as incubation periods as short or shorter than the hosts', overlapping breeding season, host acceptance of the parasitic egg, and an appropriate diet for parasitic nestlings.

Selective pressures, such as limitation of nest sites, predation, and foraging conditions, may have differed depending on the taxa. For example, limitation of nest sites for cavity nesting species is common, and presuming that honeyguides were once cavity nesters, limitation of such sites may have facilitated brood parasitism in honeyguides. Similarly, limitation of abandoned nests may have promoted brood parasitism in finches and cowbirds. Bay-winged Cowbirds and some estrildid finches use abandoned nests of other species, and these, at times, may be in short supply. Several species of brood parasites (cowbirds, cuckoos, and honeyguides) have unusual foraging modes. These foraging modes may require a somewhat nomadic lifestyle, but it is unknown whether foraging modes were partially responsible for the evolution of brood parasitism or whether brood parasites were able to fill empty niches once they were freed from their nesting responsibilities.

Selection favors brood parasites that successfully exploit their hosts, while selection favors hosts that avoid parasitism. This may result in an escalating series of adaptations and counteradaptations or "arms race." Adaptations and counteradaptations include (1) egg rejection by the host with resulting mimesis, compromised egg size, and thick-shelled eggs by the brood parasite, new egg morphology by the host, and further mimesis by the parasite; (2) nestling rejection by the host with resulting nestling mimesis by the parasite, and (3) aggression toward parasites by the host and resulting dull plumage, hawk mimicry, rapid egg laying, laying when host is away, distraction displays, and stoic determination by the parasite.

Evolution of the cowbird complex is particularly interesting because they are classified as a small group and have diverse reproductive strategies, ranging from a non–brood parasite to a brood parasite that specializes on one specific host to generalist brood parasites. In light of the unquestionable value of studying the phylogeny of cowbirds, it is remarkable how few studies have been conducted, and there is disagreement among the few studies that have been conducted.

Chapter 4

Bay-winged and Screaming Cowbirds

Great advances were made in our knowledge of the remarkable Bay-winged and Screaming Cowbirds in the 1970s and 1980s with Rosendo Fraga's and Paul Mason's intensive long-term research, and some of the curious observations of earlier researchers take on refreshing new meaning in light of more recent discoveries.

Description and Subspecies

Bay-winged Cowbirds

Bay-winged Cowbirds are sexually monomorphic in plumage. The crown, nape, back, and upper-tail coverts are mousey brown; the chest and belly are a slightly lighter mousey brown than the dorsal plumage. The chin and throat are a lighter mousey brown than the belly. Under-tail coverts are also mousey brown, but may have a tinge of rufous (see Bianchi 1971, for variations of rufous on the head and back). The tail is dark brown. Secondaries are dark brown and heavily edged in beige or rufous on the leading edge. The proximal three-quarters of the primaries are rufous, and the distal quarter is dark brown. Their bills are black, and their feet are very dark brown. They have a striking and handsome black lore, or spot, between the eye and bill, which extends in a thin line under the eye. The black lore is a characteristic shared by most male icterines with otherwise brightly colored heads and is thought to enhance visual perception during gape foraging by reducing glare (Orians 1985). Bay-winged Cowbirds are approximately 17–19 cm long and have short rounded wings. Although Fraga (1992) reported females to be significantly lighter in weight than males by 6.2%, the difference is not as pronounced as in other cowbird species (males: 46.6 ± 2.4 SD g, $n = 19$; females: 43.7 ± 2.5 SD g, $n = 17$, Fraga 1992).

Historically, Bay-winged Cowbirds were divided into two species: Bay-winged Cowbird (*M. badius*) and Pale Cowbird (*M. fringillarius*, Friedmann

Table 4.1
Common Names of Bay-winged and Screaming Cowbirds

Bay-Winged Cowbirds	
	Músico
English	Mulato or Mulata
Bay-winged Cowbird	Havia Estéro
Bay-wing or Baywing	Mulato Músico
Brown Finch	El Bayo
	Aza de Telha
Native	Guitarerro (Pereyra 1938; Bianchi 1971)
Murajú or Murajó	
Charé	**Screaming Cowbirds**
Burraca	
Urraquita	*English*
	Screaming Cowbird
Spanish	
Tordo	*Native*
Tordo Colorado	Urrucum (Bolivia)
Tordo Bayo	
Tordo Mulato	*Spanish*
Tordo Argentino	Guitarerro
Tordo de las Cienagas	Chopi [1]
Tordo Músico	Tordo de Pico Corto
Ala-canela	Violinista (Pereyra 1938)

Source: Friedmann 1929b.
[1] Chopi is a misleading name for Screaming Cowbirds because Chopi Blackbirds (*Gnorimopsar chopi*) are also entirely black.

1929b). More recently, three subspecies of *M. badius* have generally been recognized: *M. b. badius* (type locality, Paraguay), *M. b. bolivianus* (type locality, Yungas, Bolivia), and *M. b. fringillarius* (type locality, Minas Gerais, Brazil). *M. b. fringillarius* is paler than the other two forms, and *M. b. bolivianus* may be distinguished from *M. b. badius* by its larger size (wing length of *M. b. bolivianus*, 100 mm; wing length of *M. b. badius*, 90 mm, Friedmann 1929b). As with most birds, Bay-winged Cowbirds are known by numerous common names (table 4.1). Many Spanish common names are forms of Tordo, which is equivalent to *cowbird*.

Screaming Cowbirds

Screaming Cowbirds are also sexually monomorphic in plumage. They are shiny black with a purple sheen all over and have rufous axillars. Their plumage is more loose and silky than other cowbirds. Hudson (1920) described their plumage as having a strong musky odor. The degree of sexual size dimorphism may vary—some (Lowther 1975) have reported them to be not very dimorphic, while others (Fraga 1979; Hudson 1920) have reported males to be significantly heavier than females. Males in Buenos Aires weigh roughly 63 g, whereas females weigh approximately 50 g (Fraga 1979). Screaming Cowbirds are about 19 cm long. The wing length of males varies from 105 mm to 124 mm, but no geographic patterns in the variation are apparent (Navas 1970). In the field they look similar to Shiny Cowbirds and may be overlooked or mistaken for Shiny Cowbirds (Ridgely and Tudor 1989). As one former scientific name (*brevirostris*) and one Spanish vernacular name (Tordo de Pico Corto, see table 4.1) imply, the bill is short relative to other cowbirds. The type locality of Screaming Cowbirds is Maldonado, Uruguay. To my knowledge, no subspecies have been described.

Distribution and Abundance

M. b. badius and *M. b. bolivianus* occupy adjoining regions of central and southern South America, from Bolivia and southern Brazil to central Argentina (fig. 4.1). *M. b. fringillarius* is apparently scarce and occupies a small area of northeast Brazil, disjunct from the other two subspecies (Ridgely and Tudor 1989). Dunning (1982), however, shows a continuous range for Bay-winged Cowbirds extending from Argentina through northeast Brazil. *M. b. badius* and *M. b. bolivianus* are generally common through most of their range (Ridgely and Tudor 1989). Although they are found at high altitudes, their distribution does not extend west of the Andes (Olrog 1963; Fraga 1986). Bay-winged Cowbirds were introduced into Chile at some time prior to Friedmann's (1929b) book; they may have established a population there for at least a time, but Olrog (1963) did not mention Bay-winged Cowbirds inhabiting Chile. Ridgely and Tudor (1989) reported them to be rare but not resident in Chile. Sight records are reported occasionally, but only one specimen has apparently been collected (Johnson 1967). It is unclear whether the high Andes present a barrier to Bay-winged Cowbird occupation of Chile. All species whose nests are most commonly used by Bay-winged Cowbirds are also absent from Chile, and most of these have ranges very similar to the Bay-winged Cowbirds'.

The range of Screaming Cowbirds is similar to that of Bay-winged Cowbirds, but they probably do not overlap with the *M. b. fringillarius* subspecies in Brazil (fig. 4.2). Screaming Cowbirds have recently extended their range through Paraná, Brazil, where Bay-winged Cowbirds apparently do not exist

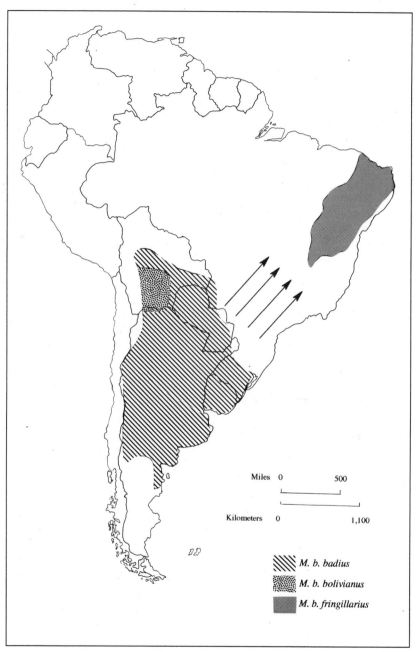

Fig. 4.1. Range of Bay-winged Cowbirds. Arrows indicate probable range expansion.

Miles 0   500

Kilometers 0   1,100

M. b. badius
M. b. bolivianus
M. b. fringillarius

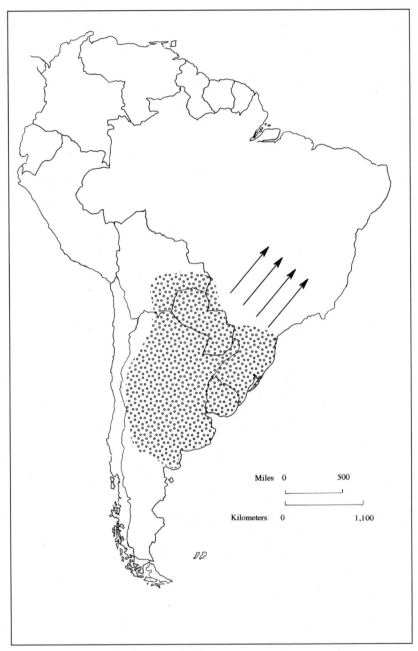

Fig. 4.2. Range of Screaming Cowbirds. Arrows indicate probable range expansion.

Miles 0   500

Kilometers 0   1,100

(Ridgely and Tudor 1989). Hoy and Ottow (1964) described Screaming Cowbirds as being very numerous in northwestern Argentina. Neither cowbird is fully migratory, but both have recently extended their range eastward in Brazil through areas that were formerly forested (Willis and Oniki 1985; Fraga 1986), and Screaming Cowbirds are spreading northward in Brazil and perhaps elsewhere (Ridgely and Tudor 1989). In Argentina, Fraga (1986) observed that Bay-winged Cowbirds outnumber Screaming Cowbirds by a 3.7:1 ratio.

Natural History of Bay-winged Cowbirds

Feeding

Bay-winged Cowbirds are omnivorous ground foragers, consuming mostly insects, grains, and seed. They forage around stables and stockyards but less frequently than other cowbirds (Friedmann 1929b). During the breeding season, individuals nesting close to the forest edge often forage outside their territories in open treeless grasslands (Fraga 1991). They can also be observed foraging in mixed flocks with Shiny and Screaming Cowbirds.

Habitat

Bay-winged Cowbirds are common in numerous habitats, including open xeric and mesic woodlands, grasslands, and more modified agricultural, pastoral, and suburban landscapes. They are common around houses, and undoubtedly human alteration of forested habitat and the planting of trees in the pampas—previously devoid of trees—has facilitated the range expansion of Bay-winged Cowbirds.

Mating System

Monogamy appears to be the dominant mating system in Bay-winged Cowbirds (Friedmann 1929b; Fraga 1991; but see Fraga 1972, for a case of successive polyandry). Pair bonds may not extend beyond a breeding season; Fraga (1991) observed that no pair bond lasted more than two consecutive seasons. Age of first breeding is not well understood; however, Fraga (1991) reported that two males, banded as nestlings, did not breed until they were three years old.

Courtship

Friedmann (1929b) reported that Bay-winged Cowbirds have no courtship at all and that after gathering in small flocks for the winter, they quietly pair off in the spring without displays. Fraga (1991) observed that most copulations (96%, $n = 46$) of individuals occur in their own territories.

Territories

Generally, Bay-winged Cowbirds are noncolonial nesters. Early in his research, Fraga (1972) described Bay-winged Cowbirds as semicolonial in some areas, but later (Fraga 1991) suggested that this may be fairly unusual. Sizes of territories vary greatly (Friedmann 1929b). Fraga (1991) estimated an average territory size of 0.19 ha (0.09–0.3 ha) and a distance to the next closest nest of 43–50 m (Fraga 1991). Factors affecting territory size have not been well studied, but food resources may play a more important role than available nests as nest boxes apparently do not affect the nesting density of Bay-winged Cowbirds (Fraga 1991). In his earlier work, Fraga (1972) suggested that high concentrations of Bay-winged Cowbirds may have been correlated with high densities of birds whose nests are used by Bay-winged Cowbirds. Site tenacity among Bay-winged Cowbirds appears to be strong, particularly among males. Fraga (1991) found individuals returned to the same nesting sites (up to six consecutive seasons), and 30% of nests were built at traditional sites.

Among many birds, males establish territories without assistance from females before nest sites are selected, and territories are presumed to be established on the basis of resource quality; that is, the nest site is usually selected secondarily to the territory. However, as Bay-winged Cowbirds are already paired off before nesting sites are selected, females may be equally, if not more, responsible than are males for establishing territories. Therefore, nest sites are selected first, and territories are established secondarily by extending outward as needed, depending on the food resources (Friedmann 1929b). Usually, males sing from the nesting tree, and territories are defended by breeding parents and their helpers. Small groups of visiting Bay-winged Cowbirds may be expelled, especially from the immediate vicinity of nests.

Song and Calls

Bay-winged Cowbirds sing throughout the year (Hudson 1920; Friedmann 1929b). Females sing, but males sing more frequently. Some authors have described the song as long and melodious, with loud notes that can be heard from a long distance (Hudson 1874). Friedmann (1929b) discussed three types of call notes. The "chuck" call is a harsh hoarse utterance and probably functions as a contact call. Another chuck call may function as an alarm (Fraga 1986) and is similar to the contact chuck call, but sharper and higher in pitch. Friedmann (1929b) observed this call being used while Bay-winged Cowbirds were usurping nests of other birds. Bendire (1895) and Hudson (1920), however, both described a trilling alarm call. The third type of call is a whistle that may correspond to the flight whistle of Brown-headed Cowbirds. It is infrequently used but is apparently used equally by both sexes.

Breeding Season

Bay-winged Cowbirds breed from October through mid-March, with some variation depending on year and location. They are among the latest breeders in Argentina (Friedmann 1929b). The late timing of the breeding season may be influenced by a higher availability of preferred nests and less competition for the same nests by other species. By breeding late, Bay-winged Cowbirds also escape most of the Shiny Cowbird breeding season. Addition of nest boxes in Fraga's (1988) study had no effect on the breeding season.

Nests

Bay-winged Cowbirds vary in nest building and nest acquisition behavior. Some individuals build their own nests, while others use abandoned nests of other species (Hudson 1874; Hoy and Ottow 1964). Still others take over active nests of other species, sometimes fighting ferociously to obtain ownership of a nest (Hudson 1874; Bendire 1895; Friedmann 1929b). Friedmann (1929b) observed that the majority of Bay-winged Cowbird nests are usurped from other species. Hoy and Ottow (1964) never observed fighting for nests, and Fraga (1988) rarely observed this.

Bay-winged Cowbirds often use the same nests or nest sites for more than one season, particularly if they are successful. For example, the same male used a hornero nest for five of six consecutive years, and another pair used the same nest box for five consecutive years; both produced fledglings

Table 4.2

Species Whose Nests Are Commonly Used by Bay-winged Cowbirds

Species/Source

Green-barred Woodpecker *(Chrysoptilus melanochlorus)*: Hoy and Ottow 1964
Golden-breasted Woodpecker *(Chrysoptilus melanolaimus)*: Orians et al. 1977
Rufous Hornero (Ovenbird) *(Furnarius rufus)*: Miller 1917; Friedmann 1929b; Hoy and Ottow
    1964; Mason 1985 several species of Spinetails *(Synallaxis* spp.): Friedmann 1929b
Short-billed Canastero *(Asthenes baeri)*: Hoy and Ottow 1964
Red-fronted Thornbird *(Phacellodomus rufifrons)*: Miller 1917; Friedmann 1929b; Hoy and
    Ottow 1964
Little Thornbird *(Phacellodomus sibilatrix)*: Fraga 1986, 1988
Leñatero (Fire-wood Gatherer or Woodhewer) *(Anumbius annumbi)*: Hudson 1874; Bendire
    1895; Daguerre 1924; Friedmann 1929b; Fraga 1972, 1988
Brown Cacholote *(Pseudoseisura lophotes)*: Orians et al. 1977; Fraga 1988
Great Kiskadee *(Pitangus sulphuratus)*: Fraga 1972, 1988; Mason 1985
Chalk-browed Mockingbird *(Mimus saturninus)*: Miller 1917
Brown-and-Yellow Marshbird *(Pseudoleistes virescens)*: Hudson 1920; Friedmann 1929b

every season (Fraga 1988). In Fraga's (1988) study, successful nests in boxes were more likely to be reused in subsequent years than unsuccessful nests in boxes.

Nest Types. Bay-winged Cowbirds use a variety of nests, including open cupped nests, but they seem to prefer covered structures: domed nests, cavities, boxes, or holes in adobe walls and cacti. Fraga (1986) reported that Bay-winged Cowbirds even accept tin cans for nest sites. Nests are typically 2–8 m from the ground (Friedmann 1929b) and are found in a variety of site types (Fraga 1988). Bay-winged Cowbirds use nests of several species (table 4.2).

Nest Construction. Bay-winged Cowbirds will often add to a nest they take over or put finishing touches on a nest found in an unfinished state. When Bay-winged Cowbirds construct their own nests, the nest may be rather neat and elaborate (Hudson 1874) or a loosely woven open cup made from dried grasses and roots (Hoy and Ottow 1964). Nest lining materials vary but include fine grasses, horsehair, and feathers. When Bay-winged Cowbirds use nest boxes, they construct a nest of grass stems (Fraga 1992). Bendire (1895) and Hudson (1920) believed that Bay-winged Cowbirds modify domed nests by poking a hole in the side to allow light to penetrate the inside and provide an easier entrance. Fraga (1988) found that Bay-winged Cowbirds made entrance gaps in most new Leñatero (*Anumbius annumbi*) nests but did not similarly modify nests of other species.

Friedmann (1929b) suggested that males contribute more to nest building than females, but he based his judgement on only two males with nesting materials. Of 14 banded pairs of Bay-winged Cowbirds in Buenos Aires, Argentina, Fraga (1992) observed only two males help in a major way with nest construction in nest boxes, and males were rarely observed helping with nest construction or modification at traditional sites. Most nesting materials are gathered within 30 m of the nest site (Fraga 1992).

Nest Defense. Males and females both contribute to nest defense. Females may roost on nests during the night before eggs are laid and on the site (nest box) before the nest is even constructed. Nocturnal roosting behavior may incur high costs at times; Fraga (1992) reported that all cases of predation occurred nocturnally. Fraga (1992) found that males unequivocally chase away intruding heterospecific birds from the vicinity of nests more often than do females; most heterospecific visitors were Screaming Cowbirds, but numerous visits were made by Chimangos (*Milvago chimango*) and Roadside Hawks (*Buteo magnirostris*).

Significance of Degenerative Nesting Behavior. Nest-building instincts vary among individual Bay-winged Cowbirds, but it is probably fair to describe the nest-building instinct as weak for the species in general. Friedmann (1929b) suggested that because the nest-building instinct is so weak, Bay-winged Cowbirds would rather fight to take ownership of an already constructed nest than build one for themselves. However, some amount of alteration or addition of materials is common, even for nests that are taken by force. Friedmann (1929b:16) observed that Bay-winged Cowbirds take an unusually long time to construct a nest and appear to look very "casual and disjointed as though the birds felt like quitting once in a while."

Eggs and Egg Laying

As with many passerines, the number of days between completion of nest construction and egg laying varies greatly. Fraga (1992) reported a range of 3–12 days (*n* = 21), and Friedmann (1929b) noted that usually nests are guarded for a couple weeks before eggs are laid. Females may replace clutches lost to brood parasitism one to two times within the season (Fraga 1992), but second broods are unusual (Friedmann 1929b; Fraga 1991).

Eggs are laid one per day (Friedmann 1929b). Clutches range from one to six eggs, with a mode of four (Hoy and Ottow 1964) or five (Friedmann 1929b). Fraga (1986) found an average clutch size of 3.75 ± 0.5 SD eggs. Mean measurements of Bay-winged Cowbird eggs are 23.7 mm × 17.7 mm, n = 226, weighing 4.0 g, n = 113 (Fraga 1983). Hoy and Ottow (1964) reported eggs to weigh less, at 2.7 g (2.3–3.0 g, n = 43). Bay-winged Cowbird eggs vary in both background color and markings. Background colors may be white, grayish white, pinkish white, bluish, greenish, or bluish green; the color of markings may be reddish brown, light red, pinkish gray, or light brown (Hoy and Ottow 1964). Geographic variation is not readily apparent (Friedmann 1929b).

Early in our knowledge of South American cowbirds, a few reports existed of Bay-winged Cowbirds occasionally laying eggs in nests of other species (Hudson 1874). After discovering the parasitic mode of reproduction in Screaming Cowbirds, Hudson (1874) believed that all such previous reports could be attributed to Screaming Cowbirds. Some scattered reports of CNP in Bay-winged Cowbirds also exist (see Friedmann 1929b); Hoy and Ottow (1964) suspected CNP to occur more often than supposed, but their sample size was too small to generate definitive conclusions (2 of 12 nests). Early observers were unaware of helping behavior in Bay-winged Cowbirds and may have become confused by the presence of additional birds (helpers) in the vicinity of nests.

Incubation

Incubation is carried out solely by females (Fraga 1992), which is apparently invariable among blackbirds (Orians 1985). Only females have a brood patch (Fraga 1986), and typically eggs are not incubated until the clutch is complete (Friedmann 1929b). The usual incubation period for Bay-winged Cowbirds is 13 days (Friedmann 1929b; Fraga 1986). Although females are the sole incubators, males spend a great deal of time in or near the nest and occasionally feed incubating females (Fraga 1992). Prior to egg laying and during the first half of incubation, males spend significantly more time in or within 10 m of the nest than females; Fraga (1992) suggested that females may spend less time at the nest than males during the first half of incubation in order to replenish their reserves after clutch formation. The pattern of female incubation and male guarding is similar to other blackbirds.

Parasitism on Bay-winged Cowbirds

Hudson (1874) believed that Bay-winged Cowbirds were parasitized exclusively by Screaming Cowbirds. Even though he recognized polymorphism among Shiny Cowbird eggs, he believed that it was not possible to mistake Shiny Cowbird eggs for Bay-winged Cowbird eggs. We now know that in addition to being parasitized by Screaming Cowbirds, Bay-winged Cowbirds are also parasitized by Shiny Cowbirds, sometimes simultaneously by both species (Fraga 1978, 1986; Mason 1980; de la Peña 1983). Bay-winged Cowbirds are parasitized by Shiny Cowbirds at a low to moderate rate (22.3% of 85 nests, Fraga 1986) and are not the primary host for Shiny Cowbirds; other hosts in the same areas are parasitized at a much higher frequency (more than 50%).

Shiny Cowbird parasitism usually lowers the reproductive success of Bay-winged Cowbirds, due primarily to egg puncturing by the Shiny Cowbird. Fraga (1986) observed the clutch size of Bay-winged Cowbirds to be smaller in nests parasitized by Shiny Cowbirds (2.83 ± 0.98 SD eggs) than non-parasitized nests (3.75 ± 0.50 SD eggs). While Shiny Cowbirds lower the reproductive success of Bay-winged Cowbirds, their own success in nests of Bay-winged Cowbirds is extremely low, at least in some areas. Fraga (1986) reported that most Shiny Cowbird eggs (85.0%, $n = 20$) failed to hatch.

Parasitism by Screaming Cowbirds is often intense, and multiple parasitism is common in some regions. Fraga (1986) reported that 88.2% of 85 Bay-winged Cowbird nests were parasitized by Screaming Cowbirds at one of his study sites; at another study site, 100% of 17 nests were parasitized. Bay-winged Cowbird nests parasitized by Screaming Cowbirds have a lower mean clutch size (3.13 eggs) than nonparasitized nests (3.75 eggs, Fraga 1986). Fraga (1986) found that 39.1% of Bay-winged Cowbird eggs disappeared or were punctured and estimated host productivity loss of 37.9%, which is higher than loss from predation. Negative effects of Screaming Cowbird parasitism occur primarily during the egg stage as nestling Screaming Cowbirds apparently do not affect the growth of Bay-winged Cowbirds. Relatively few Bay-winged Cowbird nestlings are lost to brood reduction as a result of parasitism (Fraga 1986).

Egg Ejection

It is difficult to classify Bay-winged Cowbirds as accepters or rejecters as they may accept or reject parasitic eggs depending on their investment in

Table 4.3

Acceptance, Ejection, and Desertion of Bay-winged Cowbird Nests
According to Number of Eggs of Host and Parasite, from Buenos Aires
Province

Species	Number of Screaming Cowbird Eggs	Accepted	Ejected	Abandoned
0–1 Bay-winged Cowbird eggs				
Parasite	0–3	3	4	5
Parasite	≥4	0	6	1
2–4 Bay-winged Cowbird eggs				
Parasite	0–3	23	0	9
Parasite	≥4	5	3	5

Source: Fraga 1986.

their nest. The general pattern is that they accept parasitic eggs only after
they have laid their own eggs. If the nest becomes a dump nest, the host may
abandon the nest whether or not she has laid her own eggs (Hoy and Ottow
1964; Fraga 1986). In the process of ejecting parasitic eggs, Bay-winged Cowbirds often eject their own eggs. Ejected eggs may be found strewn about the
entrance or outside the nest cup where incubation takes place (Fraga 1986).
Fraga (1986) found the maximum number of eggs (host plus parasite) that
Bay-winged Cowbirds would incubate was eight.

Clutch composition may influence acceptance behavior. Fraga (1986) observed that acceptance increases with investment, whether the nest is singly
or multiply parasitized (table 4.3). When a Bay-winged Cowbird ejects all
eggs, she will often lay another clutch in four to five days, after adding new
nest lining. Parasitic eggs in all parasitized nests (including rejected nests)
outnumber Bay-winged Cowbird eggs in a ratio of 1.2:1. Among incubated
clutches only, Bay-winged Cowbird eggs predominate over parasitic eggs in
a ratio of 1.5:1, illuminating the effect of egg ejection. Hoy and Ottow (1964)
reported no morphological differences between accepted and rejected eggs.

Through experimental addition of artificial parasitic eggs, Fraga (1986)
observed that all prematurely "laid" eggs were ejected. No eggs added after
the onset of their own egg laying were ejected by Bay-winged Cowbirds;
however, 20% of these experimentally parasitized nests were abandoned.
These 20% were also naturally parasitized by Screaming Cowbirds, and in
some nests, the host's eggs were punctured. This is somewhat different than

Fraga's (1986) observation of naturally parasitized nests (table 4.3). Some of the experimental eggs he used were made of plaster of Paris; if Bay-winged Cowbirds puncture eject eggs, even if only some individuals eject by this mode, then the ejection rate of artificially parasitized nests might be less than naturally parasitized nests. However, he also used real eggs of both Screaming and Shiny Cowbirds and obtained the same results. Fraga suggested that repeated visits to nests by the parasites may also stimulate Bay-winged Cowbirds to reject parasitized clutches. This hypothesis could be tested by placing a mounted Screaming Cowbird next to half of the nests with experimentally added eggs.

Nesting Success

The main causes of nest failure among Bay-winged Cowbirds are heavy infestations of ectoparasites, brood parasitism, and predation. The most negative effects of brood parasitism occur during the egg laying and incubation stages. Fraga (1986) reported that 81 of 220 (36.8%) observed Bay-winged Cowbird eggs hatched in parasitized nests, and 4 of 15 (26.7%) hatched in nonparasitized nests; however, actual hatching success per egg laid in parasitized nests is probably much lower. Success of Bay-winged Cowbirds during the nestling stage is much higher (100% for nonparasitized nests and 76.5% for parasitized nests, Fraga 1986). Fraga (1986) observed the largest number of nestlings successfully fledged from an individual Bay-winged Cowbird nest was five (three Bay-winged Cowbirds and two Screaming Cowbirds), and this nest was attended by five adults. Predation rates were remarkably low in Fraga's (1986) study. Only 8.9% ($n = 79$) of nests were preyed upon during the egg stage; he attributed low predation to relatively safe sites in nest boxes or cavities and to helpers mobbing potential predators near nest sites.

Young

Bay-winged Cowbirds are altricial, hatching naked or with a tiny amount of down, and with eyes closed. Their eyes begin to open by the time they are two to three days old. Pin feathers begin to emerge when they are five days (Friedmann 1929b) or seven to eight days old (Fraga 1986). Upon hatching, Bay-winged Cowbirds weigh approximately 3.6 ± 0.4 SD g (Fraga 1986) and measure roughly 42 mm in length (Friedmann 1929b); their predicted asymptote weight is 36.7 g (Fraga 1986). They leave the nest between 12 and

16 days (mean of 14 days, Fraga 1986). Asynchronous fledging is common (40%, Fraga 1986) and does not result in neglected individuals in the nest. Fledglings are able to run and climb, but not fly, by 13 days. By the time young fledge, their plumage resembles adult plumage, with the exception of having no black lores (Friedmann 1929b).

Bay-winged Cowbirds give a peep call at a frequency range of 4–8 kHz (Fraga 1986). At about eight to nine days, they begin to emit a stereotypic call that probably functions as a contact call, in the range of 2–9 kHz, with two to nine similar elements (Fraga 1986). Begging vocalizations were described by Fraga (1986) as a loud, fast, nasal rattle with repeated series of up to nine elements. Begging vocalizations continue in the postfledging period until independence.

Immediately after fledging, Bay-winged Cowbirds depend exclusively on their parents (and/or helpers) for food. Fledglings may remain in the vicinity of the nest for a few days (Friedmann 1929b). After fledging, juveniles may be escorted by their parents and up to eight potential helpers. Food provisioning by parents and helpers discontinues at approximately 30–35 days postfledging (Fraga 1991). Despite escorting and provisioning by numerous adults, juvenile mortality is high. Fraga (1991) estimated juvenile mortality to be 26% during the first 48 hours postfledging and 49% by nutritional independence (35 days postfledging).

Parental Care

The pattern of contribution by both sexes is similar to many other icterines in that males defend the nest while females incubate. Both contribute to feeding of young, but males may contribute more compared to other male blackbirds. Fraga (1992) suggested that contribution by males is substantial, with males and females contributing equally to food provisioning. As with other cooperative breeding species, this pattern is not altered with helpers at the nest. Parents may induce nestlings to fledge between 12 and 13 days by bringing food items to the nestlings and moving backwards from the nest (Fraga 1986). After leaving the nest, fledglings are often surrounded by what Fraga (1972) described as noisy and excited birds, including parents and helpers. Nestlings and fledglings are fed arthropods, primarily grasshoppers, mantids, stinkbugs, and caterpillars (Fraga 1986).

Brood Reduction

Brood reduction, or partial mortality of the brood in response to food shortage, is difficult to ascertain because there are other causes of nestling mortality: partial predation of the brood, disease, metabolic malfunctions, and ectoparasitism. If evidence of other causes is lacking, and if only the late-hatched nestlings disappear, one can fairly safely assume that brood reduction has occurred. Brood reduction is not uncommon among Bay-winged Cowbird nests, especially for Screaming Cowbirds, because they are usually among the last to hatch (Fraga 1986). Brood reduction for Bay-winged Cowbirds, however, is usually not caused by presence of Screaming Cowbirds in the nest (Fraga 1986).

Helpers at the Nest

Many early investigators found Bay-winged Cowbirds to be very sociable near their nests, but they did not make the connection with helping behavior (Hudson 1920; Hoy and Ottow 1964). After studying a banded population of Bay-winged Cowbirds, Fraga (1972) was the first to report that they behave as truly cooperative breeders. Orians et al. (1977) also found helping behavior among Bay-winged Cowbirds.

Helpers were observed at 19 of 20 nests, and the number of helpers at each nest varied from one to four (Fraga 1991). Even in rare cases in which helpers were not observed at the nest, upon fledging, extra birds showed up and shared in the duties of feeding and protecting fledglings (Fraga 1972). Bay-winged Cowbird helpers perform at least eight behaviors that are beneficial to breeding pairs: feed nestlings, feed fledglings, feed breeding female, allopreen nestlings, detect and chase predators, and detect and chase intruding brood parasites. Helpers do not participate in care throughout all phases of breeding. Of 14 banded pairs of Bay-winged Cowbirds, Fraga (1992) found no helpers during the prehatching period, but he (1991) noted that numerous extra unmarked individuals were chased from nests by the breeding parents during the prehatching period. During the nestling phase, helpers were present at 12 of the same 14 nests. Helpers might make the difference between success and failure at nests in which one parent disappears. Fraga (1992) found that of two nests in which one parent disappeared, the one with a helper succeeded, whereas the one without a helper failed due to a combination of starvation and heavy mite infestation.

Bay-winged Cowbirds share numerous life history traits with other cooperative breeding birds: they are sedentary birds displaying philopatry, delayed breeding, and longevity. Most second-year birds that remained at a study site were observed helping (Fraga 1991). Individuals with breeding status may help only after their own breeding attempts fail (Fraga 1991). Although the sex of all helpers was not known, Fraga (1991) suggested that helpers were predominantly males. He rarely observed copulations during the nestling period, but all copulations that occurred during this time were with helpers instead of the territorial male. Interestingly, in two cases ($n = 4$), the female subsequently mated with the helper.

Some advantages of helping behavior in birds include: (1) survival chances are increased on familiar ground due to experience in foraging and vigilance; (2) helpers may inherit high quality territories; (3) helpers gain valuable experience, which may be particularly important for species that regularly experience brood parasitism; and (4) helpers benefit indirectly by helping closely related kin. In most species, helpers are related to the breeding pair; often they are offspring from the previous year (Stacey and Koenig 1990). Fraga's (1991) data suggest that helpers may, indeed, often be related; of 12 individuals marked as nestlings, nine helped one or both of their parents in subsequent years (Fraga 1991).

Cooperative breeding in birds may be explained by narrow habitat requirements, nest shortages, scarce food resources, and a high rate of mortality among early dispersers. Fraga (1991) ruled out that nest shortage was operational and suggested that further studies need to be conducted on other populations before habitat requirements and scarce food resources could definitively be ruled out. Group defense against brood parasites may be yet another explanation for cooperative breeding among Bay-winged Cowbirds; group defense may be more effective than defense by one or two parents against intruding brood parasites. Fraga (1991) pointed out that the disjunct population of Bay-winged Cowbirds in northeast Brazil (fig. 4.1) is in an area uninhabited by Screaming Cowbirds, but helping still occurs in that population. However, the northeast population is not free from brood parasites; selection pressure by Screaming Cowbirds may be replaced by Shiny Cowbirds in this region.

Social Organization

Gregarious foraging and roosting flocks are formed outside the breeding season. Fraga (1991) observed that individual flocks are not cohesive and are generally larger during roosting than during foraging. As with many other icterines, Bay-winged Cowbirds engage in collective agonistic displays throughout the year, with threatening head-up (or bill-up) displays, singing by many or all individuals in the flock, and sometimes fighting or supplanting. Many adults hold pieces of vegetation (leaves, bark, twigs, etc.) in their bills during group agonistic displays, which Fraga (1991) termed "leaf gathering displays." He noted that most leaf gathering displays end with regrouping into subsets. Yearlings often stand on the periphery or turn their heads away from the adults. Leaf gathering displays may take place in trees or on the ground and are variable with regards to the number of individuals (8–26) and the length of interactions (4–21 minutes); however, the length of the display increases with the number of participating individuals. The context of leaf gathering displays varies, but the most common is interception and subsequent expulsion of small groups escorting fledglings that travelled too close (within 60 m) to a nest (Fraga 1991). Fraga (1991) has even observed family members siding together in leaf gathering displays.

Movements

Bay-winged Cowbirds are nonmigratory; they wander in small flocks, but never move far. Fraga (1991) found that throughout the winter, marked individuals commuted to and from communal roosting sites; observations of these winter roosts suggested an influx of individuals. Based on roughly 3000 observations of 150 banded individuals, Fraga (1986) believed that Bay-winged Cowbirds rarely move more than a few kilometers.

Sex Ratio

Bay-winged Cowbird populations are slightly male biased (1.75:1, Fraga 1972), but not nearly to the extent of some other cowbirds.

Plumage and Molt

The natal down of Bay-winged Cowbirds is mousey gray. The juvenile plumage, acquired by a complete postnatal molt, is grayish buffy brown above and olive sandy brown below, with rufous in the primaries and secondaries. The juvenile plumage is described in detail by Friedmann (1929b).

Beyond the postnatal molt, Bay-winged Cowbirds apparently do not have age-related plumage. After molting into their first winter plumage, young are indistinguishable from adults. Bay-winged Cowbirds molt once a year (a

complete postnuptial molt); nuptial plumage, which does not differ markedly from the winter plumage, is acquired by wear (Friedmann 1929b).

Allopreening

Allopreening is common among Bay-winged Cowbirds. Individuals solicit
preening not only from conspecifics but, at least in some contexts, also from
other species. Selander (1964) observed that in a captive situation, a male
Bay-winged Cowbird habitually solicited preening from a Chestnut-capped
Blackbird (*Agelaius ruficapilus*) and was apparently successful. The manner
of solicitation is rather peculiar as it appears to simultaneously elicit a presumably voluntary service in a crouched down position—which might be
considered a submissive behavior—and control the subsequent allopreening event with other mildly threatening postures, such as bill up.

The motivation and function of allopreening remains unclear in some
contexts. The function of adults allopreening nestlings is probably to remove ectoparasites. Selander (1964) described solicitation directed only to
the neck area, with feathers held erect and head cocked away from the
preening individual. Although this behavior has sometimes been interpreted
as an appeasement gesture in parasitic cowbirds (Selander 1964; Selander
and La Rue 1961; see also chapter 7), the behavior in Bay-winged Cowbirds
may have other functions. Some possibilities are to rid themselves of ectoparasites and to establish and maintain social relationships. Interspecific
preening solicitation, as Selander (1964) observed in his aviary, may be less
common among wild birds, and his captive Bay-winged Cowbird may have
elicited preening from heterospecifics only in the absence of conspecifics.
Only further field observations might reveal the existence and function of
this behavior in wild populations.

Natural History of Screaming Cowbirds

Hudson (1874) was the first ornithologist to discover and report the parasitic
habits of Screaming Cowbirds. He was jubilant with his discovery, enthusi-
astically reporting that he felt as if he had discovered a new planet. Early
in his ornithological career, he had searched many times for the nests of
Screaming Cowbirds, but obviously failed. Prior to his discovery, he believed
that Bay-winged Cowbirds occasionally engaged in interspecific parasitism
because he found what he believed to be Bay-winged Cowbird young being

fed by a Brown-and-Yellow Marshbird *(Pseudoleistes virescens)*; he later assumed that these were Screaming Cowbirds. Even after observing Screaming Cowbirds repeatedly enter the nest of Bay-winged Cowbirds, he assumed that the ten eggs in the nest were those of Bay-winged Cowbirds. He attributed the extra eggs to at least one other Bay-winged Cowbird laying in the nest, although he never observed more than two Bay-winged Cowbirds at the nest. Later in that season, he discovered juvenile Screaming Cowbirds being escorted by a small flock of Bay-winged Cowbirds; at this point, he realized the parasitic mode of Screaming Cowbirds.

Feeding

Screaming Cowbirds are omnivorous ground foragers, but they may consume more seeds than other items. Their food habits are similar to Bay-winged Cowbirds, but they forage near stables more frequently. However, their association with cattle is not as noticeable as Shiny Cowbirds' (Ridgely and Tudor 1989). Friedmann (1929b) found stomach contents consisting of only seeds and corn in some Screaming Cowbird individuals he examined; in others he found corn, seeds, rice, flies, locusts, coleopteran larvae, moths, and caterpillars.

Habitat

The habitat of Screaming Cowbirds is similar to Bay-winged Cowbirds: open woodlands, grasslands, farms, pastures, suburban landscapes, and stables. As Bay-winged Cowbirds extend their range across deforested areas of Brazil, Screaming Cowbirds follow (Fraga 1986; Willis and Oniki 1985).

Mating System

Screaming Cowbirds appear to be mostly monogamous. Early ornithologists (Hudson 1920; Friedmann 1929b) drew this conclusion by observing the consistency with which birds appeared in pairs; monogamy was later confirmed with use of color bands by Mason (1987). An occasional unmated satellite male may successfully gain temporary access to females, but he may not be further associated with them (Mason 1987). Monogamy in a brood parasite is of great interest as one of the most common explanations for the advantages of monogamy is that reproductive success is higher with both parents helping to raise the young. The advantages of monogamy in

Screaming Cowbirds are not clear and warrant further research but may include mate guarding, facilitating detection of and access to host nests, and possible territorial defense.

Several ornithologists have remarked on the strong pair formation in Screaming Cowbirds (Friedmann 1929b; Willis and Oniki 1985; Fraga 1986), and Mason (1987) suggested that pair formation is stronger in Screaming Cowbirds than in numerous nonparasitic species. Males almost always follow and guard females within 2 m, perching above them during nest inspections and often performing the song spread display. Pairs are stable through the breeding season (Mason 1987), but pairing behavior is observed only during the breeding season (Fraga 1986). Screaming Cowbirds visit potential host nests in pairs, and individuals of established pairs even enter traps together (Mason 1987).

Courtship

Courtship displays are primarily arboreal and are not given as frequently as by other parasitic cowbirds. Infrequency of courtship displays may be due, in part, to strong pair bonds and to monomorphic plumage. Friedmann (1929b) suggested that males may be less stimulated to perform courtship displays if females are similarly colored than if they are markedly different. During courtship displays, the male bows forward with his wings held out in a horizontal position, erects his body and head feathers, and quivers all over. He then emits a squeaky *dzeee* note followed by a two note *pe-tzee*. He may or may not jump up and down while emitting the ending notes (Friedmann 1929b). Friedmann's (1929b) impression was that their courtship season curiously ends about one month before the breeding season begins. Perhaps the pair bond is cohesive enough to diminish the need for courtship displays, and perhaps the close and constant following of the male behind the female serves the same function as a courtship display and helps maintain the pair bond.

Territories

Reports of territorial behavior among Screaming Cowbirds vary considerably. Friedmann (1929b) insisted that Screaming Cowbirds establish and maintain territories. However, his observations are not particularly convincing, especially in light of some more recent work on banded individuals. Trapping data and observations of color-banded individuals suggest that Screaming Cowbirds have a strong site attachment (Mason 1987), and perhaps this is what Friedmann interpreted as territorial behavior. Fraga (1986) also found no evidence of territories and, instead, regarded the spaces occupied as home ranges. Additionally, he found overlapping home ranges.

Small flocks and gregarious nest searching also suggest that territories are not established and defended by pairs. Fraga (1986) rarely observed fighting among Screaming Cowbirds. In experiments with mounted Screaming Cowbirds and playbacks of their vocalizations, he observed neither aggression nor agonistic behavior toward mounted specimens by conspecifics near nests of Bay-winged Cowbirds or in roosting areas. Although Screaming Cowbirds may not be territorial in the strict sense of defending an established area against all conspecifics, they do compete for access to Bay-winged Cowbird nests. On numerous occasions, Hoy and Ottow (1964) observed pairs of Screaming Cowbirds battling after a pair tried to enter a nest.

For parasitic species, our traditional concepts of territorial behavior may need to be altered. Usually, we look for signs of advertising, or singing and expulsion of conspecifics other than the mate or potential mate. For brood parasites, these behaviors may not be as important in defining territoriality as some other behaviors. While the presence of intruders in the territory of a nesting species may be disadvantageous, for brood parasites, extra individuals or "visitors" to another's home range may sometimes be advantageous. Visitors may reveal the location of available nests that were previously undiscovered by the "residents" (individuals regularly occupying the site or home range). The visitors will benefit if they are undetected by the residents. If they are undetected and the resident never finds the nest, the visitor's activities would not alter the success of the resident. If a resident does discover a nest through the visitor's activities or if a visitor uses a nest already known to the resident, the resident always has the option of puncturing the visitor's egg, rendering it inviable. In addition to other behaviors, egg-puncturing behavior may help to define territoriality in brood parasites. This also makes territoriality more difficult to investigate in brood parasites than in nesting species, and perhaps this is why results of territory investigations vary so widely. Fraga (1986) reported an observed egg puncture rate of 9.5% among Screaming Cowbirds, and this may reflect territorial behavior even though other behaviors, such as expulsion and fighting, were weak or absent.

Song and Calls

Hudson (1920) assigned the common name to Screaming Cowbirds for their noisy characteristics. Some who have worked with Screaming Cowbirds seem to find the name appropriate; others (Ridgely and Tudor 1989) do not. Call notes of two distinct types are emitted by both sexes. A harsh scream, or "chuck" note, which may serve numerous functions, is emitted upon alighting, taking flight, during flight, when perched in trees, and even when fighting. A second call, the "plunk" call, sounds like a metal stake being lightly tapped by another piece of metal. It is used rarely (Friedmann 1929b), and the function is not well understood. Hudson (1920) interpreted it as a warning call; ironically, he described this as the only mellow expression in the Screaming Cowbird's vocal repertoire. Friedmann (1929b) questioned this interpretation and rarely heard the call; Fraga (1986) suggested that Screaming Cowbirds do not emit alarm calls.

Hudson (1874) described Screaming Cowbirds as lacking a song but, instead, as singing unmelodious notes. Perhaps Hudson was describing call notes. The song is given by males in song spread courtship displays, but Hudson (1920) suggested a duet by male and female. Friedmann (1929b) described the first note as explosive, ascending, low, short, and loud. It is closely followed by a two syllable *pe-tzeee*, the first syllable being very short and the second syllable being longer and drawn out; the interval between the syllables is barely perceptible.

Social Organization

Screaming Cowbirds are usually observed in pairs or small flocks of up to ten individuals (Hoy and Ottow 1964). Fraga (1986) reported a maximum flock number of 38 individuals, and this was during a rainy winter day. Hudson (1920) reported that flocks rarely exceed six individuals.

Movements

Screaming Cowbirds are quite sedentary, and color-banded individuals may be observed in the same area throughout the year (Fraga 1986). Minimal home ranges for Screaming Cowbirds are between 15 and 25 ha (Fraga 1986).

Sex Ratio

The sex ratio, at least in some areas of the Screaming Cowbirds' range, does not differ from unity (Mason 1987).

Breeding Season

The breeding season of Screaming Cowbirds is the same as for Bay-winged Cowbirds—approximately October through March. However, their egg laying is not perfectly synchronized with Bay-winged Cowbirds as they commonly lay their eggs in advance of their hosts.

Nonbreeding Season

During the nonbreeding season, Screaming Cowbirds are often seen in mixed flocks with Bay-winged Cowbirds. In 70% of Fraga's (1986) sightings of Screaming Cowbirds, they were in flocks associated with Bay-winged Cowbirds. He attributed the association to a predator avoidance strategy by Screaming Cowbirds. Apparently, Screaming Cowbirds do not give alarm calls in response to avian predators, whereas Bay-winged Cowbirds do. Fraga (1986) witnessed 34 attacks on mixed flocks of Screaming Cowbirds and Bay-winged Cowbirds by Aplomado Falcons (*Falco femoralis*), and alarm calls emitted by Bay-winged Cowbirds were effective in immobilizing the flock. Because Aplomado Falcons primarily seize birds in flight, immobilization is probably an effective predator avoidance strategy, and Screaming Cowbirds may benefit by their association with Bay-winged Cowbird flocks. During the nonbreeding season, they may also commonly be seen foraging in rice fields with Shiny Cowbirds.

Parasitism

Screaming Cowbirds are curious about nests and pay frequent visits to nests in all stages. Nest inspections by Screaming Cowbirds occur throughout the day (Mason 1987), and curiosity about nests may start long before their breeding season. Fraga (1992) reported an average of 2.5 visits to Bay-winged Cowbird nests per hour (1992) and up to nine visits an hour. During 30 hours of observation, Mason (1987) observed one visit per 43 minutes. Screaming Cowbirds continue visits during the nestling stage to a lesser degree (one visit per 164 minutes) but probably do not harm nestlings (Fraga 1986), as sometimes occurs by Brown-headed Cowbirds. Screaming Cowbirds parasitize Bay-winged Cowbird nests at natural sites and nest boxes with similar frequency (Fraga 1986).

Host Egg Puncturing

Ample evidence suggests that Screaming Cowbirds regularly puncture eggs. Screaming Cowbirds were undoubtedly responsible for a considerable portion of the 39.1% of Bay-winged Cowbird eggs that either were punctured or disappeared in Fraga's (1986) study. In addition to observations of natural nests, to investigate egg-puncturing behavior by Screaming Cowbirds, Fraga (1986) placed two nest boxes containing five eggs of two unspecified passerines plus five eggs of Screaming Cowbirds near Bay-winged Cowbird nests. In both boxes, he observed punctured eggs immediately after visits by Screaming Cowbirds. Similarly, in nest checks of natural nests, immediately following visits by Screaming Cowbirds, he observed punctured host eggs. The difference in clutch sizes between parasitized and unparasitized nests provides enough additional evidence to conclude that Screaming Cowbirds puncture host eggs.

Fraga (1986) reported that in addition to puncturing and/or removing Bay-winged Cowbird eggs, 9.5% of Screaming Cowbird eggs were punctured, and he attributed at least some of this egg puncturing to conspecifics. While egg puncturing may reduce host sibling competition and provide food resources for Screaming Cowbirds, chances for nest desertion are increased.

Eggs and Egg Laying

Screaming Cowbird eggs vary in background colors of white, cream, grayish white, pinkish white, light brown, greenish gray, and bluish green, and in markings of reddish brown, brown, and greenish brown. Eggs with a bluish or greenish tint are common in northwest Argentina (Hoy and Ottow 1964). Variation within individuals is not well understood, but Friedmann (1929b) reported it to be low. Mean egg measurements are 23.2 mm × 17.8 mm, $n = 255$, weighing 4.0 g, $n = 117$ (Fraga 1983). Screaming Cowbird eggs may be easily distinguished from Bay-winged Cowbird eggs in a given nest, but if eggs of both species were mixed in a series, determining identity would be extremely difficult because of the wide variety in both species (Friedmann 1929b). Compared to eggs of Bay-winged Cowbirds, Screaming Cowbird eggs are rounder, and the predominant colors are pinks and reds (Fraga 1983). Predominant colors of Bay-winged Cowbird eggs are grays and whites (Fraga 1983). Compared with Bay-winged Cowbird eggs, Screaming Cowbird egg maculation tends to be more uniform with less contrast, but this is not a reliable characteristic. Fraga (1983) suggested that the presence of scrawls,

Table 4.4

Distribution of Screaming Cowbird Eggs Found in Nests of Bay-winged
Cowbirds from Buenos Aires Province

Number of eggs	1	2	3	4	5–6	7–12
Number of nests	13	15	13	12	7	9

Source: Fraga 1986.

or thick dark lines, is the most useful characteristic in distinguishing between Screaming and Bay-winged Cowbird eggs. Scrawls are more abundant
in Screaming Cowbird eggs, particularly in eggs of cool-colored backgrounds
that may be more easily confused with Bay-winged Cowbird eggs.

Screaming Cowbirds lay their eggs at one-day intervals, and the total
number of eggs laid during the season is not known. Hudson (1920) suggested that it was only five eggs. Friedmann (1929b) agreed with the estimate, based on his belief that Screaming Cowbirds establish territories and
would probably not lay eggs outside their territories. In a histological study,
Davis (1942) determined that if more than five eggs are laid, there is a rest
period between clutches of five.

Screaming Cowbird eggs are often laid prematurely (before host eggs are
laid). When this occurs, Screaming Cowbirds eggs are typically ejected or
the nest is deserted; hence, the eggs are wasted. In addition to laying in
nests prematurely, Screaming Cowbirds also waste their eggs by laying in
abandoned nests (Mason 1980; Fraga 1986). Perhaps the wide range of time
between completion (or guarding) of nests and commencement of egg laying by Bay-winged Cowbirds makes it difficult for Screaming Cowbirds to
synchronize their activities with their hosts. While premature laying is common, the degree of this egg wastage varies; Mason (1980) reported that 87%
of Screaming Cowbird eggs were laid prematurely, while Fraga (1986) reported that only 14.3% were laid prematurely.

Screaming Cowbirds frequently engage in multiple parasitism (table 4.4)
and dump nesting (P. Mason, pers. comm.). Hoy and Ottow (1964) reported that up to 12 females may have parasitized one dump nest, and Fraga
(1972) reported a Bay-winged Cowbird nest with all four host eggs punctured and seven Screaming Cowbird eggs. Hoy and Ottow (1964) observed
that in northwestern Argentina, where Screaming Cowbirds are very abundant, parasitized nests of Bay-winged Cowbirds are likely to contain from 6
to 20 Screaming Cowbird eggs.

Screaming Cowbirds lay their eggs in the morning. Their eggs take approximately 12 days to incubate by Bay-winged Cowbirds and under the same conditions will hatch out only eight hours before the first Bay-winged Cowbird egg hatches (Fraga 1986). Hatching failure of Screaming Cowbird eggs is high. Out of 59 incubated Screaming Cowbird eggs, 24 (40.7%) failed to hatch; this compares to 8 of 86 (9.3%) Bay-winged Cowbird eggs that failed to hatch (Fraga 1986). Reasons for the high failure rate are not clear. Thirteen of the 24 eggs were laid after the Bay-winged Cowbird laying period, and removing these from the sample, plus three in which timing was unknown, results in 8 of 43 (18.6%). Therefore, inappropriate timing only partly explains the high failure rate.

Success in the Nest

Screaming Cowbirds are not as successful as their hosts relative to the number of eggs laid and number of eggs incubated. Only 31 of 257 (12.1%) Screaming Cowbird eggs hatched in Fraga's (1986) study, and this figure is perhaps inflated because many Screaming Cowbird eggs may have been ejected before he was able to detect them. Success for Screaming Cowbirds is much higher during the nestling phase (61.3%, Fraga 1986). Whereas the egg ratio (Bay-winged Cowbirds:Screaming Cowbirds) of incubated parasitized clutches was 1.5:1, the brood ratio was 2.1:1.

Young

Development. Screaming Cowbirds hatch out naked with their eyes closed. At two to three days, they begin to open their eyes, and by seven to eight days, pin feathers emerge. They hatch out at an average of 3.4 ± 0.7 g ($n = 15$) and grow slightly more rapidly than Bay-winged Cowbirds (Fraga 1986). They are approximately 43 mm in length (Friedmann 1929b), and their predicted asymptote weight is 45.3 g, compared to 36.7 g for Bay-winged Cowbirds (Fraga 1986). The maximum weight attained by an individual Screaming Cowbird by the twelfth day is 52.5 g (Fraga 1986). Screaming Cowbirds fledge from the nest when they are 12–16 days old (14 days average, $n = 15$, Fraga 1986). Fledglings less than 13 days old are able to run and climb but not fly.

Behavior. As with Bay-winged Cowbirds, Screaming Cowbirds utter a peep call at a frequency range of 4–8 kHz (Fraga 1986). They begin to emit stereotypic calls at about eight to nine days in the range of 2–9 kHz, with two to nine similar elements (Fraga 1986); these contact calls are similar to those emitted by Bay-winged Cowbirds. Begging vocalizations are similar to Bay-winged Cowbirds and continue through the postfledging period until independence. Unlike Bay-winged Cowbirds, Screaming Cowbirds beg indiscriminately and will beg for food from humans handling them (Friedmann 1929b). Ejecting host nestlings from the nest is probably extremely rare, but Fraga (1986) relayed the details of one incident in which he believed two Screaming Cowbird nestlings ejected a Bay-winged Cowbird host sibling from the nest on two separate occasions.

Fledglings. Screaming Cowbirds depend exclusively on their host parents for food for at least for three weeks postfledging and may depend on them for up to five weeks. Fraga (1986) calculated that only 32.8% of Screaming Cowbird juveniles survived beyond 35 days after fledging; however, he believed that this figure underestimated actual survival because it was based on resighting of individuals, and some individuals may have dispersed before day 35. In Fraga's (1986) study, adult Bay-winged Cowbirds outnumbered Screaming Cowbirds by a ratio of 3.7:1, whereas the ratio of fledglings was 3.3:1, suggesting that postfledging mortality may be generally higher among Screaming Cowbirds.

After Screaming Cowbirds begin to molt into their black plumage, they associate with conspecifics and join either small flocks of adults or small flocks composed primarily of juveniles (Fraga 1986). These molting juveniles often emit recognizable versions of adult calls in a squeaky voice (Fraga 1986).

Plumage and Molt

Freshly hatched nestlings have mousey gray down. The juvenile plumage, acquired by a complete postnatal molt, is indistinguishable from that of Bay-winged Cowbirds. Young begin their postjuvenile molt (into the black adult plumage) within approximately 39 days (range: 34–44 days) after hatching (Fraga 1986). Fraga reported a 74-day-old individual to be half black with rufous wings and another individual 133 days old to be completely black with rufous wings. Molting occurs once a year, after the breeding season.

Specialist Parasitism by the Screaming Cowbird on the Bay-winged Cowbird

Few factors that may have favored evolution of host specificity have been heretofore identified, and mechanisms for maintenance of host specificity are unknown. Advantages of depending exclusively on one host are questionable, and such a state, whether primitive or derived, may not necessarily represent an optimal condition. Specialist parasitism may, in fact, represent a suboptimal or a compromised condition in which the parasite's existence precariously depends on only one other species. Nevertheless, from an evolutionary perspective, it is useful to look for any selective pressures that might have favored a specialist strategy and to search for specific adaptations.

In a teleonomic study, Fraga (1984) proposed that Bay-winged Cowbirds may be more beneficial to Screaming Cowbirds than other potential hosts because Bay-winged Cowbirds remove botflies (*Philornis* spp.) from both their own and Screaming Cowbird nestlings. Botfly larvae burrow into the skin and feed on the blood and tissues of nestlings and can be an important source of nestling mortality. Chances of mortality increase with the number of botfly larvae, with six to seven often being lethal (Fraga 1984; Mason 1985). Apparently, few birds do, or are able to, remove larval botflies, but the intense propensity of Bay-winged Cowbirds to allopreen in general may facilitate the behavior of removing ectoparasites. Although removal of botflies is unquestionably advantageous to Screaming Cowbirds, ectoparasitism may not have been a major selective agent for evolution of host selection.

Mechanisms by which Screaming Cowbirds maintain host preference have not to my knowledge been investigated. Determination of whether individual Screaming Cowbirds parasitize Bay-winged Cowbirds by a genetic predisposition, because they learn through an imprinting process, or a combination of genetic influence and learning would be of tremendous interest and would provide us with some clues regarding how exclusively dependent Screaming Cowbirds are on Bay-winged Cowbirds. The maintenance of host specificity could be investigated through experimental cross-fostering of eggs into nests of other potential hosts with subsequent observation of the young that fledge from these nests. A similar question is raised regularly with Brown-headed Cowbirds. The problem, albeit a stiff challenge, is not nearly as difficult with Screaming Cowbirds for several reasons: (1) they are sedentary and site faithful; (2) they are philopatric, so chances of resighting them are excellent if they survive; (3) they are perhaps easier to observe than Brown-headed Cowbirds; and (4) because they are specialist parasites, concentrated efforts can be made in searching for their eggs in the nests of Bay-winged Cowbirds and experimental host(s).

Competition with the Shiny Cowbird

Shiny Cowbirds parasitize Bay-winged Cowbirds with a much lower frequency than they do other hosts. Friedmann (1929b) proposed two major reasons for this: (1) Bay-winged Cowbirds primarily use the nests of Leñateros, which rarely serve as host to Shiny Cowbirds; and (2) most Shiny Cowbirds are toward the end of their egg laying period when Bay-winged Cowbirds commence breeding. Therefore, competition with Shiny Cowbirds is probably minimal.

Degree of Similarity or Mimesis in Eggs and Young

Adult Screaming Cowbirds do not mimic Bay-winged Cowbirds in plumage, vocalization, or behavior. However, eggs and young are remarkably similar, and these similarities have been attributed to both mimicry (Hudson 1920; Lack 1968) and common descent (Friedmann 1929b). Egg morphology of Bay-winged and Screaming Cowbirds varies and overlaps considerably, making distinction between the two difficult. However, they can be fairly reliably distinguished by a trained observer. In a sample of 137 hatched eggs for which identity could be confirmed, Fraga (1983) made only one error in identifying the eggs to species. In general, Screaming Cowbird eggs are heavier, more spherical, and thicker than those of Bay-winged Cowbirds. Bay-winged Cowbird eggs are more distinctly pointed at one end, whereas both ends of Screaming Cowbird eggs are more rounded. Hoy and Ottow (1964) suggested that using the relative shell thickness (q = length of egg in mm × width of egg in mm/weight of eggshell in mg) is a reliable method of distinguishing between the two species; they reported q to be 1.62 (1.49–1.79) for Bay-winged Cowbirds and 1.28 (1.15–1.50) for Screaming Cowbirds, with little overlap. However, the relative thickness of the shell is not a particularly useful parameter in the field. Fraga (1983) and P. Mason (pers. comm.) believe that the most useful characteristic in distinguishing Screaming Cowbird eggs from Bay-winged Cowbird eggs in the field is the nature of markings. Fraga (1983) suggested that scrawls are more abundant in Screaming Cowbird eggs.

Early investigators reported Screaming Cowbird nestlings and fledglings to be indistinguishable from Bay-winged Cowbirds, with regards to plumage, behavior, and vocalizations. Fraga (1979) described how host and parasitic young can be distinguished during their first five days. Bay-winged Cowbird skin is orange, and the bill is pinkish with a conspicuous dark tip. Screaming Cowbird skin is pink; the bill is pinkish but lacks the dark subterminal tip. Skin differences are no longer noticeable after four to five days; Friedmann (1929b) noticed that the pink skin of Screaming Cowbirds may change as early as at two days. As older nestlings, Screaming Cowbird bills appear paler, but this may not be consistent. As fledglings, Screaming Cowbird bills blacken, while Bay-winged Cowbird bills become browner, with light tips eventually becoming black by two to three months. It is unknown whether these differences occur across the entire ranges of the two species. By these identification methods, Fraga (1986) failed to identify the species correctly in only 2 of 137 cases. He also found no significant differences in vocalizations of young nestlings between Bay-winged Cowbirds and Screaming Cowbirds. After seven days, while the stereotypic calls remain similar, begging vocalizations can vary considerably (Fraga 1986), but begging calls vary within the species, and heterospecifics in a particular nest may or may not have similar begging vocalizations.

Under favorable environmental conditions and equivalent hatching, Screaming Cowbird nestlings are heavier than Bay-winged Cowbirds. However, many factors, such as hatching order, number of helpers at the nest, food resources, and brood size, may alter weight; therefore, size difference is not a reliable characteristic for distinguishing between host and parasite nestlings. Older fledgling Screaming Cowbirds may be distinguished by their larger size; they eventually become larger than their foster Bay-winged Cowbird parents (Hudson 1920; Fraga 1979). Additionally, as older fledglings, Bay-winged Cowbirds may begin to solicit allopreening, whereas this behavior is not observed in Screaming Cowbirds (Fraga 1979).

Fraga (1986) believed that the resemblance of plumage, begging vocalizations, and other behaviors (whether it is mimicry or common descent) to be important to the survival of Screaming Cowbirds. Observations of nests naturally parasitized by Shiny Cowbirds suggested that they do not enjoy high success in nests of Bay-winged Cowbirds; only one Shiny Cowbird succeeded in fledging, and once it left the nest, it was generally unattended by its host parents for extended periods of time. Fraga attributed this, in part, to the fact that it remained within 25 m of the nest, whereas its two host siblings moved approximately 80 m away.

In a series of nestling cross-fostering experiments to compare responses of adult Bay-winged Cowbirds to Screaming Cowbird nestlings and Shiny Cowbird nestlings, who look, vocalize, and behave dissimilarly, Fraga (1986) reported some intriguing results. He conducted a small series of two different experiments. In one experiment type, he simply transplanted Shiny Cowbirds (nestlings and eggs) into nests of Bay-winged Cowbirds. Although Shiny Cowbirds developed normally in all nests, they were not observed beyond two days postfledging; because their Bay-winged Cowbird nestmates were observed, Fraga presumed the Shiny Cowbird fledglings dead. By comparison, 7 of 19 (36.8%) Screaming Cowbirds, and only 11 of 66 (16.7%) Bay-winged Cowbirds, were not observed beyond two days postfledging.

In another experiment, he placed Shiny Cowbird, Screaming Cowbird, and Bay-winged Cowbird nestlings, taken from other nests, outside of, but close to, Bay-winged Cowbird nests. During one trial, Shiny Cowbirds were fed by Bay-winged Cowbird parents at a higher rate than either Screaming Cowbirds or Bay-winged Cowbirds, perhaps due to their persistent begging vocalizations. In another trial lasting two hours, however, a Shiny Cowbird placed outside the nest of a Bay-winged Cowbird was ignored, even though a Screaming Cowbird placed outside the nest was fed eight times, and three Bay-winged Cowbird nestlings and one Screaming Cowbird nestling inside the nest were fed 13 times. Another trial produced similar results. It is tempting to suggest that the similarity of Screaming Cowbird young to Bay-winged Cowbirds may infer an advantage on Screaming Cowbirds, particularly after the nesting period; however, further experimental research is needed in this area, and perhaps Fraga's (1986) important work will provide an impetus for future research.

Is It an Exclusive Relationship?

While numerous investigators have reported finding Screaming Cowbird eggs only in nests of Bay-winged Cowbirds (Hudson 1974; Fraga 1986), others have provided evidence of Screaming Cowbird eggs in the nests of other hosts. Hoy and Ottow (1964) reported that Red-fronted Thornbirds (*Phacellodomus rufifrons*) are more frequently parasitized than Bay-winged Cowbirds by Screaming Cowbirds in northwestern Argentina; they attributed the high rate of parasitism to high visibility of Red-fronted Thornbird nests compared to the more concealed nests of Bay-winged Cowbirds. Hudson (1874) observed Brown-and-Yellow Marshbirds feeding what he originally thought was a Bay-winged Cowbird, but in retrospect he believed it to be a Screaming Cowbird. Numerous other hosts in addition to Bay-winged

Table 4.5

Hosts of Screaming Cowbirds in Addition to Bay-winged Cowbirds

Species/Source

Rufous Hornero (Ovenbird) *(Furnarius rufus)*: Girard in Pereyra 1938
Red-fronted Thornbird *(Phacellodomus rufifrons)*: Hoy and Ottow 1964
Antshrike sp. *(Thamnophilus major)*: Girard in Pereyra 1938
Great Kiskadee *(Pitangus sulphuratus)*: Girard in Pereyra 1938
Chalk-browed Mockingbird *(Mimus saturninus)*: Girard in Pereyra 1938
Brown-and-Yellow Marshbird *(Pseudoleistes virescens)*: Grant 1911, 1912; Mermoz and
    Reboreda 1996
Chopi Blackbird *(Gnorimopsar chopi)*: H. Sick in Ridgely and Tudor 1989; Fraga 1996
Yellow-winged Blackbird *(Agelaius thilius)*: Grant 1912
Finch sp. *(Sicalis pelzelni)*: Girard in Pereyra 1938

Cowbirds have been reported (table 4.5). It is not known how reliable these reports are, but even with occasional use of other hosts by certain individuals, Screaming Cowbirds, in general, can unequivocally be considered a specialist parasite. Interestingly, Ridgely and Tudor (1989) reported that Screaming Cowbirds have extended their range through Paraná, Brazil, where apparently Bay-winged Cowbirds do not exist. Here, they parasitize Chopi Blackbirds *(Gnorimopsar chopi*, Ridgely and Tudor 1989).

Suggestions for Further Research

I suggest that research on the following questions would contribute significantly to our overall knowledge of Bay-winged and Screaming Cowbirds.

Bay-winged Cowbirds

1. Helpers at the nest—Do scarce food resources and narrow habitat requirements contribute to the maintenance of helping behavior? Does productivity of young increase with increasing group size?
2. What behaviors establish and maintain pair bonds in Bay-winged Cowbirds? Allopreening is one possible behavior that can be investigated.
3. What, if any, is the level of CNP among Bay-winged Cowbirds?
4. What are the functions of allopreening in wild populations?
5. What factors affect territory size?
6. In general, Bay-winged Cowbirds need to be studied throughout their

range; for example, very little is known about the disjunct *fringillarius* population in Brazil, and it would be useful to obtain knowledge about that population before and if the *badius* race extends to the *fringillarius* range.

Screaming Cowbirds

1. What are the advantages of pair stability and monogamy?
2. How many Screaming Cowbird eggs are laid by an individual per season?
3. What defines territoriality in Screaming Cowbirds?

Relationship between Bay-winged Cowbirds and Their Parasites

1. What are some advantages of host specificity, and what maintains the host selection strategy?
2. Does the presence of an adult parasite at the nest have any effect on parasitic egg acceptance by Bay-winged Cowbirds; that is, how accurate are artificial nest parasitism experiments in determining rejection behavior?
3. Unlike Screaming Cowbirds, are Shiny Cowbirds largely abandoned by their Bay-winged Cowbird foster parents after they fledge the nest?

Summary

Bay-winged and Screaming Cowbirds are both ground foragers and can be found in numerous habitats. They are presently expanding their ranges through areas of landscape modification in Brazil. They are both nonmigratory and gregarious, and form small flocks during the nonbreeding season. Both types of cowbird are monogamous, and while pair bonds are strong within a breeding season, pair bonds rarely last between seasons. Screaming Cowbirds have a simple courtship while Bay-winged Cowbirds appear to have no courtship at all. Bay-winged Cowbirds maintain territories during the breeding season, but they are cooperative breeders, recruiting helpers at the nest after incubation to assist in caring for and protecting the young. Territoriality of Screaming Cowbirds is not well understood. Nest-building behavior in Bay-winged Cowbirds is reduced; some individuals build a rather elaborate cup nest, some build a more simple nest, while others take over nests of other species. Screaming Cowbirds are parasitic almost exclusively on Bay-winged Cowbirds. Parasitism by Screaming Cowbirds is often intense, and multiple parasitism is common. Bay-winged Cowbirds often reject the eggs of Screaming Cowbirds, particularly when the nests contain many parasitic eggs or when the nest is parasitized before commencement of their own egg laying.

Chapter 5

Giant Cowbirds

Relatively little is known about Giant Cowbirds, possibly because their hosts are difficult to study. Although their oropendola and cacique hosts (Icterinae) nest in dense colonies, which can facilitate data collection, the nests are typically very high and placed at the ends of thin branches. Hanging basket nests present an additional challenge because simple mirror and pole observation methods are not always effective. Some hosts also regularly nest in association with formidable stinging insects, which can be a daunting experience even for the most determined researcher. Smith (1968) removed nests from branches to examine them and placed them back with the use of ladders, long telescoping poles, sticky tape, and contact cement; later (Smith 1979; Fleischer and Smith 1992) he used a United States Air Force "high lift" truck with a hydraulically articulated arm that could reach to 22 meters.

Description

Giant Cowbirds are sexually dimorphic in size and plumage. As their name implies, they are much larger than other cowbirds; males are 33–38 cm and average 120 g; females are 28–33 cm and average 74 g (Smith 1979). Size may vary geographically; Haverschmidt (1966) documented three females from Suriname that weighed an average of 129 g. Size dimorphism averages 24% (Orians 1985). Giant Cowbirds have long rounded tails, black feet, and black bills. Iris color varies from yellow to orange-red (Ridgely and Tudor 1989) to ruby red (Skutch 1954). Males are entirely black with a violet sheen and a conspicuous neck ruff, which contributes to a small-looking head. Females are dark brown to black; the edges of their body feathers may be darker and slightly glossier, giving them a scaled appearance, but they lack the overall glossy appearance of males. Both sexes have an unusual bill, unlike any other cowbird. The maxilla is expanded into a flat frontal shield or casque, rounded at the base, and it is remarkably similar to the bill of Yellow-rumped

Caciques *(Cacicus cela)*. The nasal opening is on the side, rather than on the top as with other cowbirds. Giant Cowbirds' flight is noisy as wind passes through the flight feathers, producing a rushing or whirring sound. They have a characteristic flight of several flaps to a soar.

Other common English names of Giant Cowbirds include Rice Grackle and Rice Oriole. In Spanish, common names include Tordo Gigante (Birkenstein and Tomlinson 1981; Navas et al. 1991), Tordo Mantequero (Birkenstein and Tomlinson 1981), Arrocero (Birkenstein and Tomlinson 1981), Boyero Negro Grande (Olrog 1963), Vaquero Gigante (Howell and Webb 1995), Pájaro Vaquero Grande (Birkenstein and Tomlinson 1981), and Oropéndola Negra (Smith 1982).

Distribution and Subspecies

Friedmann (1929b) mentioned three subspecies: *S. o. oryzivora*, *S. o. violea*, and *S. o. mexicana*. Blake (1968) recognized only two subspecies: *S. o. oryzivora*, inhabiting eastern Panama, Trinidad, and South America through eastern Peru to the south, north and east Bolivia, eastern Paraguay, northeast Argentina, and western and south Brazil (Ridgely and Tudor 1989), and *S. o. impacifa*, occupying the Caribbean coast of southern Mexico and Central America to western Panama (fig. 5.1). *S. o. violea* and *S. o. mexicana* are synonyms for *S. o. impacifa*. Giant Cowbirds are locally common, and their range is probably restricted to the range of cacique and oropendola hosts.

Natural History

Feeding

Giant Cowbirds sometimes forage alone, but more usually in small flocks of 2–12 individuals (Robinson 1988). Occasionally, larger flocks are observed; for example, Robinson (1988) observed a flock of 80 foraging on a beach in Peru. Giant Cowbirds have generalized foraging habits; however, they differ from other cowbirds in foraging tactics and diet. They are omnivorous, but while other cowbirds consume mostly arthropods and grains, Giant Cowbirds consume arthropods, fruit, and nectar. Robinson (1988) described three distinct foraging tactics: (1) terrestrial foraging—gleaning prey from water surface, sand, and leaves, and probing into piles of wood; (2) foliage searching—searching branches, leaves, and stripping bark; and (3) foraging on the backs of mammals, such as Capybaras *(Hydrochoerus hydrochoerus)*, picking off biting horseflies. Giant Cowbirds also turn over stones along rivers (Skutch 1954).

Habitat

Giant Cowbirds inhabit forest edges and adjacent open and semiopen areas. They are common along rivers and lake shores, cane fields, and other cultivated fields where they forage.

Movements

Giant Cowbirds are not known to be migratory, but Smith (1979) suggested that they may be partially migratory, as they are rarely observed outside of the breeding season in Panama; Orians (1985) lists them as having some nomadic movements during the nonbreeding season.

Mating System and Courtship Displays

The mating system of Giant Cowbirds is promiscuous. Courtship displays typically occur in large flocks on the ground, usually in open areas (Smith 1979). Males stiffly strut among the females with goose steps, chests puffed out, and heads drawn back. Planting himself directly in front of a female and standing tall, a male then erects his neck ruff and body feathers and draws his head down, often resting his bill against his breast feathers. He may also engage in some bobbing and leg flexing. Skutch (1954:132) was apparently greatly amused by the courtship displays of Giant Cowbirds, as he described males as "taken up with their own importance . . . [having] an air of ludicrous pomposity . . . swollen with insolence and pride."

Vocalizations

Little is known about Giant Cowbird vocalizations, perhaps because they are extremely quiet compared to other cowbirds. At times they utter various clucks, chatters, and nasal whistles.

Preening Solicitation

Giant Cowbirds solicit preening in a manner similar to other cowbirds. Chapman (1928) noted that Giant Cowbirds solicited preening from their

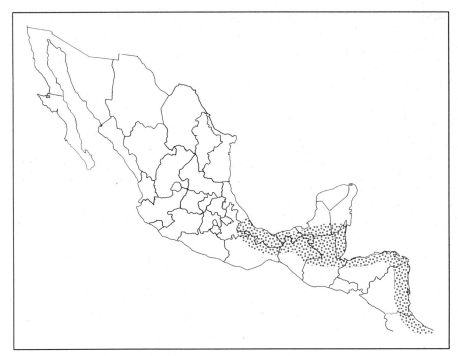

Fig. 5.1. Range of Giant Cowbirds.

Chestnut-headed Oropendola *(Psarocolius wagleri)* hosts by bowing their heads with nape feathers erected. Harrison (1963) observed captive Giant Cowbirds solicit preening from African Blackhead Plovers *(Sarciophorus tectus)*. Payne (1969) visited a captive Giant Cowbird that performed the display toward humans. Payne was pleased to oblige the cowbird by poking his fingers through the cage and scratching the bird's neck. He tried poking his finger through in several other places of the cage, and wherever he poked his finger through, the bird sidled up to it and bowed its head for scratching.

Eggs

Giant Cowbird eggs are coarsely grained and dull or very slightly glossy. They measure 34.0 mm × 25.5 mm (*n* = 34 eggs, Haverschmidt 1966), but Kreuger (in Schönwetter 1983) collected much more elliptical eggs of 37.9 × 21.7 mm, 5.9 g and 38.1 × 24.4 mm, 6.9 g. Giant Cowbird eggs are polymorphic, but Fleischer and Smith (1992) suggested that individual

females likely lay a consistent type. The background color varies: white, greenish white, pale blue, and pale bluish green. Eggs may be immaculate or marked with various spots and scratches of brown, dark brown, or black (Haverschmidt 1966). Fleischer and Smith (1992) suggested that eggshells of Giant Cowbirds are much thicker and rougher than those of their hosts. Data from Rahn et al. (1988) indicate that eggshells of Giant Cowbirds are not as thick relative to egg size as eggshells of other cowbird species, but they are slightly thicker than eggshells of their hosts. Eggs are typically incubated in 10–13 days, which is roughly five to seven days earlier than their hosts (Smith 1968).

Young

Upon hatching, the skin of the Giant Cowbird is white and covered with dark gray down. The rictal flanges are yellow, and the beak is curiously white, like that of several of their hosts (Chestnut-headed Oropendola and Montezuma Oropendola, *Gymnostinops montezuma*, Penard and Penard 1910). Their eyes open earlier than those of their host siblings—about 48 hours after hatching (Smith 1968). By two weeks, their plumage is black but their bills are still white. By four weeks, the white bill begins to darken at the sides (Crandall 1914), and by four months, the bill is entirely black or black with only faint white traces. Adult sheen is attained during their second prebasic molt at one year (Howell and Webb 1995); therefore, they are still in their juvenile plumage through the first breeding season (Orians 1985). As with other cowbirds, Giant Cowbird nestlings eagerly beg for food.

Parasitism

Goeldi (1894, cited in Friedmann 1929b) was the first ornithologist to fully recognize and document that Giant Cowbirds are obligate brood parasites. Others, however, had previously noted Giant Cowbird eggs in nests of other species without fully realizing the extent of the parasite-host relationships (Friedmann 1929b).

Hosts

Giant Cowbirds regularly parasitize only oropendolas and caciques (Icterinae, Smith 1968; table 5.1). Lehmann (in Friedmann 1963) reported a couple

Table 5.1

Reported Hosts of Giant Cowbirds

Species/Source

Crested Oropendola [Yellowtail] *(Psarocolius [Ostinops] decumanus [cristatus])*: Goeldi 1897; Friedmann 1929b; Young 1929; Schäfer 1957; Smith 1968

Green Oropendola *(Psarocolius [Xanthornus] viridis)*: Friedmann 1963

Russet-backed Oropendola *(Psarocolius angustifrons)*: Robinson 1988

Chestnut-headed Oropendola (Wagler's Oropendola) *(Psarocolius [Zarhynchus] wagleri)*: Crandall 1914; Friedmann 1929b; Chapman 1930; Smith 1968, 1983; Fleischer and Smith 1992

Montezuma Oropendola *(Psarocolius [Gymnostinops] montezuma)*: Crandall 1914; Skutch 1954; Smith 1968

Yellow-rumped Cacique *(Cacicus cela [persicus])*: Penard and Penard 1910; Friedmann 1929b; Young 1929; Smith 1968; Fleischer and Smith 1992

Red-rumped Cacique *(Cacicus haemorrhous)*: Friedmann 1963

of Green Jays *(Cyanocorax yncas)* feeding a fledgling Giant Cowbird for several consecutive days, but he did not observe parasitism in the nest.

Oropendolas and caciques nest in colonies, which vary in size from a few to a hundred or more nests. Three major explanations for colonial nesting among birds are commonly cited: (1) improved foraging conditions, (2) protection from predators, and (3) limited nesting sites. Colonial nesting may also reduce risks of brood parasitism because the presence of more adults provides a greater possibility of detection and defense against intruding parasites. Some of the same breeding strategies that provide protection against predators can be exploited by Giant Cowbirds. For example, synchronous breeding may provide clues to cowbirds regarding the appropriate timing of parasitism. Webster (1994) found a correlation between the number of female hosts lining their nests and the number of visits per hour by Giant Cowbirds.

While colonial nesting may improve foraging efficiency, when all or most of the colony leaves to forage, nests are vulnerable to brood parasitism. Robinson (1988) found that Russet-backed Oropendolas *(Psarocolius angustifrons)* foraged together, leaving the colony unattended and prone to parasitism; Yellow-rumped Caciques, on the other hand, rarely left the colony unattended and were never parasitized. In a colony of Montezuma Oropendolas, most cowbird visits occurred during the late morning when most potential hosts were away from their nests (Webster 1994). Feekes (1981) observed that when females were out foraging together, males often darted out of the forest edge to chase Giant Cowbirds away from the colony.

Aggressive attacks by oropendolas and caciques are often effective in driving away brood parasites, but some Giant Cowbirds can be stoic and remain determined to enter nests no matter how aggressively they are repelled. Chapman (1930) suggested that Giant Cowbirds become more persistent as their need to lay an egg increases. Webster (1994) reported that cowbirds were sometimes tolerated in a tree by a colony of Montezuma Oropendolas and were chased only when they approached nests. By contrast, Chapman (1930) observed that when Giant Cowbirds entered a nest tree, they were attacked by the entire colony of Chestnut-headed Oropendolas. In a mixed colony of Yellow-rumped Caciques and Russet-backed Oropendolas, Robinson (1988) observed that caciques chased Giant Cowbirds from unattended oropendola nests, suggesting a benefit to oropendolas by their nesting association with caciques. Giant Cowbirds may visit a colony numerous times before they can successfully enter a nest. For example, Robinson (1988) documented that of 88 visits by Giant Cowbirds to a Russet-backed Oropendola colony in Peru, only six females were able to enter nests. Similarly, in Costa Rica, only seven Giant Cowbirds were able to successfully enter Montezuma Oropendola nests in 83 observed attempts (Webster 1994).

Male Participation

Smith (1979) rarely observed male Giant Cowbirds at host colonies in Panama, whereas in Peru, Robinson (1988) observed males facilitating parasitism by distracting hosts. Giant Cowbirds visit host colonies either singly or in small groups (Chapman 1930; Robinson 1988; Webster 1994). When they visit in small groups, males may distract hosts by approaching within a meter, giving displays similar to courtship, and encouraging chases; meanwhile, females may enter nests (Robinson 1988). Females also opportunistically enter nests while other females are chased away by the nest owner.

Adult Interactions with Fledglings

Little is known about interactions between adult and fledgling Giant Cowbirds. Interestingly, Schäfer (in Haverschmidt 1966) observed, without a doubt, a fledgling Giant Cowbird being fed by an adult Giant Cowbird.

Multiple Parasitism

Multiple parasitism is fairly common, at least in some areas; Smith reported that in Panama, 42.1% (n = 1,503 nests) of host nests contained more than one Giant Cowbird egg. Electrophoretic and color analyses revealed that more than one female laying in an individual nest is common (Fleischer and Smith 1992). Dump nesting, on the other hand, is not common; most parasitized nests contain only one to two parasitic eggs. Friedmann (1963) doubted Kuschel's (1896) report of six eggs found in one nest, but Kuschel did not actually mention this in the note Friedmann cited; however, Skutch (1954) mentioned one nest with six Giant Cowbird eggs and noted that it was unusual.

Effects of Parasitism

In some areas, Giant Cowbirds appear to have minimal impact on host populations (Crandall 1914), but impacts for individual hosts may be very high. Robinson (1988) found that while Russet-backed Oropendola females fed only one fledgling (whether host or parasite), only 3 of 24 (12.5%) fledglings were Giant Cowbirds. The low impact that some investigators have reported may be primarily due to the relatively large size of Giant Cowbirds' hosts. Even the size differences between Giant Cowbirds and their smaller cacique hosts are not as pronounced as in some other host-parasite relationships. Giant Cowbird nestlings are able to successfully compete with larger oropendola siblings because they hatch out earlier and develop more rapidly.

Parasitism on any one host species may vary geographically, depending on differing nesting strategies and perhaps alternative host availability. In Panama, Fleischer and Smith (1992) documented a higher rate of Giant Cowbird parasitism on Yellow-rumped Caciques (54.2%, n = 24) than Chestnut-headed Oropendolas (25.8%, n = 31), and Smith (1968) reported that 111 Giant Cowbirds were successfully raised among 173 Yellow-rumped Cacique and Montezuma Oropendola nests. In contrast, Yellow-rumped Caciques in Peru are rarely parasitized, whereas Russet-backed Oropendolas are parasitized; Robinson (1988) attributed this to Yellow-rumped Cacique nest holes that are too small for Giant Cowbirds to enter. Yellow-rumped Caciques are also much more aggressive toward Giant Cowbirds than are Russet-backed Oropendolas. These geographic differences in host nesting strategies may reflect the curious relationship of a tenuous mutualism between parasite and host in Panama.

Advantages of Being Parasitized by the Giant Cowbird

Smith's (1968, 1979, 1980) well-known work in Panama on the advantages of being parasitized has been cited numerous times as an example of complex ecological interactions among host, brood parasite, stinging and biting bees or wasps, and ectoparasitic botflies (*Philornis* spp.). He found that, under certain circumstances, oropendolas and caciques had a greater chance of reproductive success if their nests were parasitized by Giant Cowbirds.

Background of Caciques and Oropendolas

Caciques and oropendolas often nest in association with formidable insects, such as biting bees (*Trigona* spp.) and stinging wasps (*Protopolybia, Steleopolybia,* and *Syboeca* spp., Haverschmidt 1966; Smith 1968, 1979), which is fairly common in the tropics and can be an effective defense against predators (Wunderle and Pollock 1985). Smith (1979, 1980) stressed that in this complex bird/insect nesting association it is the birds, not the insects, that choose the association. Experimental removal of insect nests resulted in abandonment by bird colonies if investment was low (prior to egg laying); however, once egg laying had commenced, removal of insect nests had no effect (Smith 1980). In addition to discouraging vertebrate predators from entering a colony, wasps and bees may keep ectoparasitic botflies out of a colony. In Smith's study, colonies associated with wasps or bees were rarely parasitized by botflies, whereas those without the association were heavily parasitized. Experiments revealed that Hymenoptera detect the presence of botflies by wing noise and odor, responding to them as if they were parasitoids (Smith 1983).

Botfly Parasitism

According to Smith (1968, 1983), the primary source of nestling mortality among oropendola and cacique colonies in Panama was botflies. Adult botflies lay their eggs on nestlings, and the larvae burrow into the skin, nourishing themselves from the tissues and fluids of the birds. Eventually, they crawl out of the skin; an infestation of seven to ten late instar larvae appears to be lethal.

Giant Cowbird nestlings preen their host nestmates, removing botflies at all developmental stages. Stomach analysis of Giant Cowbird nestlings revealed the remains of botfly eggs, larvae, and adults (Smith 1968). Stomach analysis of oropendola and cacique nestlings indicated no reciprocal removal of botflies from Giant Cowbird nestlings, but Smith noted that Giant Cowbirds were rarely parasitized by botflies. He suggested that because Giant Cowbird nestlings were active, snapping at anything that moved close to them, this probably accounted for their general lack of botfly parasitism in his study. Removal of botflies by Giant Cowbird nestlings is undoubtedly advantageous to their hosts. Among colonies unprotected by wasps or stinging bees, Smith (1968) found a striking dichotomy of botfly parasitism: host chicks raised with Giant Cowbird nestlings were parasitized by botflies at a rate of only 8.4% (n = 676), whereas chicks raised in the absence of Giant Cowbird nestlings suffered a much heavier rate of botfly parasitism (90.1%, n = 424).

Success of Host with and without Parasite

In Smith's study, parasitism appeared to be advantageous only in colonies unprotected by aggressive insects. In colonies lacking protection of wasps or bees, the probability of successfully fledging at least one host was roughly three times greater in nests parasitized by Giant Cowbirds than nests without Giant Cowbird nestlings (Smith 1968). This was not the case in colonies protected by wasps and bees; in these colonies, nests parasitized by Giant Cowbirds fledged fewer hosts than nests without Giant Cowbird nestlings.

While the presence of Giant Cowbird chicks may appear to be advantageous in some situations, they may depress success of their hosts by affecting host clutch size. Smith (1968) reported a clutch size of 1.8 eggs (n = 1,768 nests) in nonparasitized nests and 1.3 eggs (n = 1,502 nests) in parasitized nests. He suggested that if parasitism occurred within 24 hours of the first host egg, the host rarely laid a second egg and absorbed any remaining follicles. Although direct evidence is lacking (Smith 1968), Giant Cowbirds may, in some situations, outcompete their host siblings for food.

Success of Host and Parasite with and without Association of Stinging Insects

Although oropendolas and caciques in Panama appeared to prefer an association with wasps or bees, these colonies were slightly less successful (15%) than colonies unprotected by wasps or bees (22%, Smith 1968). In his study, Smith identified three possible disadvantages of nesting in a colony with wasps or bees: (1) nests tended to be crowded around insect nests, and the weight of all the bird nests sometimes causes branches to snap off; (2) the insects often deserted their nests before the birds fledge, leaving them unprotected from botflies; and (3) the insects typically bred late, shortening the birds' season by as much as two months compared to colonies lacking associations with insects. Nesting without protection of insects, however, has disadvantages that may be just as weighty. The advantage of being parasitized by Giant Cowbirds is dubious, particularly because some parasitized hosts raise only a parasite, and clutch size is depressed. Also, nesting in association with formidable insects may not be a matter of selection; appropriate trees lacking insect nests appeared to be scarce (Smith 1968). Recently planted palms are relatively predator resistant, but individual leaves have a short life span, lasting only as long as the nesting period of oropendolas and caciques. When leaves die and become desiccated, they fall, along with any attached nests.

Dichotomy of Behavior

Oropendolas and caciques in colonies associated with insects were more aggressive toward Giant Cowbirds than were those in colonies lacking the protection. They had nothing to gain and everything to lose from being parasitized, and they generally ejected all nonmimetic eggs from their nests. In contrast, colonies unprotected by insect nesting associations generally accepted nonmimetic eggs and were fairly passive when cowbirds attempted to gain access to host nests. Giant Cowbirds also displayed a dichotomy of behavior, depending on the host colony behavior and insect nesting association.

Egg Ejection by Caciques and Oropendolas. Smith (1968) found that response toward experimental eggs was a colonywide reaction; that is, all individuals of a colony either accepted or rejected particular egg models. When egg ejection occurred, it was by way of puncture ejection, which left a characteristic spiked hole in the eggshell. Spiked cowbird eggs were far more numerous beneath colonies associated with stinging and biting insects than under colonies without the insect association (Smith 1979).

In order to determine the basis for the colonywide acceptance/rejection behavior, Smith (1979) performed a set of experiments and reported only on females. He switched eight 17-day-old host chicks between colonies; foster mothers readily accepted transplanted chicks. Generally, these birds returned to the colony from which they fledged two years later, which is the

Table 5.2

Responses (to Experimentally Added Rose-Colored Eggs) of Breeding Chestnut-headed Oropendolas and Yellow-rumped Caciques That Had Been Cross-Fostered as Nestlings

Original Colony Foster colony	Response	Years Following Switch							
		1	2	3	4	5	6	7	8
Discriminator									
264 to same and other discriminator colonies									
	Discriminator	0	35	41	31	54	70	91	88
	Nondiscriminator	15	1	9	0	3	3	9	9
536 to nondiscriminator colonies									
	Discriminator	0	11	13	9	17	18	12	10
	Nondiscriminator	19	74	90	63	102	112	111	107
Nondiscriminator									
223 to same and other nondiscriminator colonies									
	Discriminator	0	6	6	2	12	11	9	14
	Nondiscriminator	4	34	39	30	46	77	76	89
453 to discriminator colonies									
	Discriminator	0	37	54	61	81	104	100	106
	Nondiscriminator	33	9	15	15	17	23	14	13

Source: Smith 1979.

typical age of first breeding. However, some birds bred as yearlings; all yearlings were nondiscriminators, regardless of their real or foster mother's behavior. In the nests of adult females that had been cross-fostered as nestlings, Smith placed rose-dyed eggs to test acceptance response; if the experimental eggs were still in the nest after two days, he scored them as accepted. He also observed how females reacted to intruding cowbirds. With only a few exceptions, cross-fostered individuals assumed the behavior of the colony, regardless of their real mother (table 5.2). Apparently the rejection response is learned. Experimental addition of wasp nests to colonies previously lacking insect association did not alter the responses (Smith 1980).

Skutch (1954) made an interesting observation that fledgling cowbirds were not in groups of host Montezuma Oropendolas at the end of the nesting season. He did, however, observe a single female caring for her young foster cowbird, and he mused over whether the foster mother led a lonely life because the colony would not tolerate the young cowbird. This observation has not been reported by others, and N. Smith (pers. comm.) observed that Giant Cowbirds do stay with host flocks. It would be of great interest to specifically look for this behavior in the occasional accepter individual in a colony of rejecters to determine whether the dichotomy of acceptance/rejection behavior carries over through the fledgling stage.

Cowbird Egg Mimesis and Behavior. Smith (1968) described five Giant Cowbird egg types in Panama; four types resembled each of four hosts, and a fifth type was a generalized nonmimetic egg. Smith (1968:690) noted that "when associated with the colony in which they were laid, it was clear that *Scaphidura* had evolved a rather elaborate egg mimicry." Females that laid nonmimetic eggs were generally aggressive toward their hosts, moved about host colonies unabashed in small groups, and often laid more than one egg per nest (usually two or three), whereas females that laid mimetic eggs patiently waited for hosts to leave the nest and usually laid only one egg per nest. It is unclear from Smith's papers whether he followed individual Giant Cowbirds to associate behavior with egg morphology, which would be a difficult undertaking, but I gather from his (above) statement that the association was generalized within colonies. In other words, in discriminator colonies (protected by insects), Giant Cowbird eggs appeared to be mimetic, hosts were aggressive, and Giant Cowbirds sneaked about the colony. In nondiscriminator colonies (lacking insects), Giant Cowbird eggs were nonmimetic, hosts were passive toward Giant Cowbirds, and Giant Cowbirds were bold as they moved about the colony.

Mimicry Revisited

Smith (1968) suggested that because Giant Cowbirds evolved rather elaborate egg mimicry, there must have been a strong selective advantage for parasitizing colonies associated with insects. He further suggested that predators and the unpredictability of palm tree leaves were unequivocally factors in evolution of egg mimicry. However, the current data do not clearly support this. There is no difference in the probability of a Giant Cowbird fledging between the discriminator colonies (0.76 fledglings per nest, $n = 433$ nests) and nondiscriminator colonies (0.73 fledglings per nest, $n = 682$ nests).

During Smith's (1968) egg model experiments, he noted that discriminator hosts ejected dissimilar eggs within five minutes of discovery. If hosts discriminate against all dissimilar eggs, then only similar eggs would remain, those which would appear to mimic host eggs. When the ejection response is so rapid, ejection of dissimilar eggs would likely occur before an investigator could detect dissimilar eggs. Smith further noted that spiked Giant Cowbird eggs were far more numerous under colony trees associated with insects, but he did not describe the morphology of these ejected shells. The shell morphology of these ejected eggs would be an important piece of evidence in determining whether true mimesis exists. If discriminators generally accept mimetic Giant Cowbird eggs and eject dissimilar Giant Cowbird eggs, one would expect a large number of dissimilar ejected eggs below the colony. Furthermore, Fleischer and Smith (1992) later determined that multivariate analysis of egg measurements and markings revealed significant, nonoverlapping differences between Giant Cowbirds and two major hosts, which does not support Giant Cowbird egg mimicry. Further work and more complete reporting is necessary to determine if mimicry is operational and, if it exists, how it is maintained.

Other Studies

Others have looked for similar interactions in other regions but have not found the kind of mutualism that Smith described. In Costa Rica, Fraga (1989) found no nesting association between Montezuma Oropendola colonies and insects, and in Peru, Robinson (1988) found no association between oropendola and cacique colonies and wasps (but see Robinson 1985). In Costa Rica, Webster (1994) found that Montezuma Oropendolas did not associate with social stinging insects, but Chestnut-headed Oropendolas and Yellow-rumped Caciques nested in association with Hymenoptera. Botfly parasitism was apparently heavy among Montezuma Oropendolas. In both studies, host success was lower than in Smith's study in Panama; females rarely fledged more than one young in Robinson's and Webster's studies. Therefore, raising a parasite would confer no advantage whatsoever. Even though botfly infestation was heavy and no colonies were associated with wasps or other stinging insects in Peru, Robinson (1988) never observed hosts passively allowing Giant Cowbirds to enter nests.

Summary

Giant Cowbirds differ from other cowbirds in numerous ways: their sexual size dimorphism is more pronounced, they consume more fruit, they are quiet, their eggshells are not as thick, and their bill morphology is strikingly different from other cowbirds and more similar to caciques. They range from southern Mexico to northeastern Argentina and specialize in parasitizing oropendolas and caciques. Giant Cowbirds are nomadic and have a promiscuous mating system. Multiple parasitism is fairly common, with two to three eggs per nest. The impact of Giant Cowbird parasitism on host populations is generally rather minimal, due to low rates of parasitism, but can devastate nesting success of individuals.

In Panama, Giant Cowbird nestlings remove botflies from their host siblings, conferring a dubious advantage to being parasitized. Nesting near stinging or biting insects also reduces risks of botfly parasitism for hosts. In colonies associated with insects, oropendolas and caciques eject dissimilar Giant Cowbird eggs and are aggressive toward intruding parasites; in these colonies, cowbirds sneak about the colony. In colonies lacking insect nesting associations, Giant Cowbirds boldly approach nests; the hosts are relatively passive toward them and accept parasitic eggs dissimilar to their own.

Bronzed Cowbirds

The parasitic habits of Bronzed Cowbirds were first published in 1861 by Owen (cited in Friedmann 1964), but their mode of reproduction was known long before to natives of Guatemala (Friedmann 1964). Bronzed and Giant Cowbirds can probably be considered the least known cowbirds, and for many years, we had many mistaken impressions of their breeding biology and social behavior. Carter (1984, 1986) recently contributed significantly to our understanding, but much remains to be learned of this brood parasite.

Subspecies

The history of taxonomy and nomenclature of Bronzed Cowbirds is fraught with confusion. Historically, Bronzed Cowbirds were separated into two species: Arment's Cowbird (*M. armenti*) and Bronzed Cowbird (*M. aeneus*). Arment's Cowbird, or Colombian Red-eyed Cowbird, was once thought to be extremely rare and was known only by a few specimens. Currently, it is considered a subspecies of *M. aeneus* (Blake 1968). Adding to the confusion, the name *M. a. aeneus* (*Psarocolius aeneus*, Wagler) was inadvertently used for the northwestern subspecies (now known as *M. a. loyei*, see Parkes and Blake 1965). Blake (1968) recognized four subspecies: *M. a. loyei* (type locality [*M. aeneus milleri*], Tucson, Arizona), *M. a. assimilis* (type locality [*Collothrus a. assimilis*], Acapulco, Guerrero, Mexico), *M. a. aeneus* (type locality, Mexico City),[1] and *M. a. armenti* (type locality, Cartagena, Colombia).

Throughout our short history of knowledge of Bronzed Cowbirds, numerous changes have been proposed based on one or a few unique characteristics. In 1887, Ridgway proposed that what we currently recognize as two subspecies, *M. a. aeneus* and *M. a. loyei*, be separated from *Molothrus* based on the presence of elongated neck feathers and four outer primaries with emarginated inner webs; he revived the genus name, *Callothrus*. Friedmann (1927, 1929b) added to this list of characteristics the hairlike quality of the neck and breast plumage. Later, however, these characteristics were found to be shared to some degree by other cowbirds.[2] The outer primaries are slightly sinuated in Screaming Cowbirds, at least in some of the Shiny Cowbird subspecies (Parkes and Blake 1965), and Bay-winged Cowbirds (fig. 6.1), and a slight development of a neck ruff appears to be evident in at least some Shiny Cowbirds (Friedmann 1929b). Parkes and Blake (1965) suggested that Bronzed Cowbirds have no unique morphological or behavioral characteristics that justify a monotypic genus assignment, and they proposed to return it to *Molothrus*. They also proposed that *armenti* should be considered a subspecies of *M. aeneus*.

Description

Bronzed Cowbirds have been commonly known as Red-eyed Cowbirds, Red-eye, Red-eyed Bronzed Cowbirds, and Glossy Cowbirds (also a common name for Shiny Cowbirds). Among the common Spanish names are Tordo, Tordito, Bandito de Cola, Analgo, Tordo Gallito (Friedmann 1929b), Vaquero Ojirojo (Howell and Webb 1995), Tordo Ojirojo (Birkenstein and Tomlinson 1981), Tordo Mantequero, and Pájaro Vaquero, the last two of which are also names used for Giant Cowbirds (Birkenstein and Tomlinson 1981).

Bronzed Cowbirds are sexually dimorphic in size and plumage. Among the *M. a. aeneus* race, adult males average 69 g, and adult females average 57 g. Among males of all subspecies the chin, throat, crown, and nape are dark brown with a consistent sheen of gold or bronze; wings and tail are black with various iridescent shades of purple, blue, and green. The bill is deep but long compared with other molothrine cowbirds. The feet and bill are very dark brownish black. In *M. a. loyei*, the upper back, chest, and belly are also dark brown-bronze, and the rump, upper-tail coverts, and under-tail coverts are black with various iridescent shades of purple, blue, and green. In *M. a. aeneus* and *M. a. assimilis*, the dark brown-bronze extends through the rump and most of the upper-tail and under-tail coverts. The iris is red in summer and brownish orange to orange in winter. The neck feathers are fairly elongated, forming a conspicuous ruff that gives the bird a top-heavy appearance.

The plumage of females varies tremendously among subspecies, but they all lack the intense sheen and conspicuous neck ruff of males. The feet and bill are brownish black, and the iris is brownish orange to orange, similar to males in winter but lighter. In *M. a. loyei* females, the chin is a light mousey

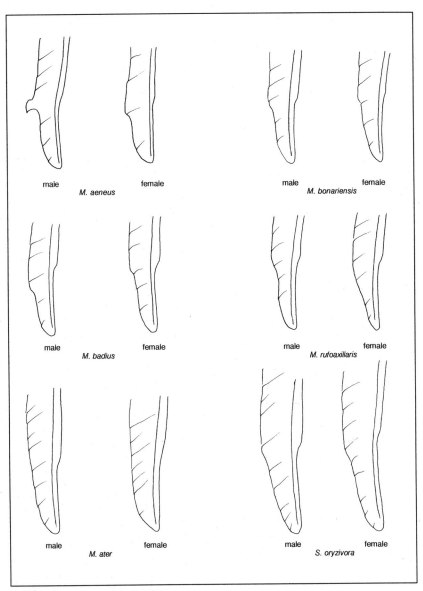

male	female	male	female
M. aeneus		*M. bonariensis*	
male	female	male	female
M. badius		*M. rufoaxillaris*	
male	female	male	female
M. ater		*S. oryzivora*	

Fig. 6.1. Eighth primary of each cowbird species.

brown with or without faint streaking of darker brown. The throat and breast are the same shade but slightly darker than the chin, and the streaking is slight but more obvious. The belly and under-tail coverts are more heavily streaked and may be a slightly different shade of mousey brown (reddish compared with the yellowish of the chin, throat and breast); this difference in shade may form a rather sharp delineation. The crown, nape and upper back are dark grayish brown; on each feather, the central portion is brownish black, resulting in a scaled appearance. The rump, tail and upper-tail coverts, and wings and upper-wing coverts are medium brownish gray (darker than belly and lighter than upper back). The central portion of these feathers are darker brown. The under-wing coverts are mousey brown like the background color of the belly. *M. a. assimilis* females are medium brown all over, but the dorsal side is slightly darker with a tinge of purple-green iridescence. *M. a. aeneus* is similar to *M. a. assimilis*, but dark brown to sooty black.

Distribution and Abundance

Bronzed Cowbirds range from central Arizona, southwestern New Mexico, and southern Texas through most of Mexico and Central America, to central Panama and the Caribbean coast of Colombia (fig. 6.2). Their breeding and wintering ranges have expanded north and east over the last century. The first Bronzed Cowbird in Florida was recorded in 1962 (Hoffman and Woolfenden 1986), and the first specimen was collected in Florida in 1968 (Matteson 1970). Numerous sightings were reported throughout the southern states in the 1970s and early 1980s (Robbins and Easterla 1981), and reports continued to arrive during the 1990s in Louisiana (Purrington 1993; Stedman 1993), Mississippi (Jackson 1993), Alabama (Jackson 1992), and Florida (Langridge 1992, 1993; Ogden 1992; West and Wamer 1993a, 1993b). The most significant changes have occurred in California, Arizona, and Texas. Bronzed Cowbirds may have first appeared in Arizona in 1909 (Visher 1909) and in the southwestern portion of New Mexico in 1947 (Phillips et al. 1964). Although they are not abundant in Arizona, Bronzed Cowbirds have become increasingly more common (Johnson and Roer 1968). Friedmann (1925) indicated that they were fairly common in the lower Rio Grande Valley of Texas but that their range extended only into southern Texas and extreme southeastern Arizona (Friedmann 1929b). Reports of individuals during the breeding season have also steadily increased in the New Orleans, Louisiana, area (Stewart 1976; N. Newfield in Robbins and Easterla 1981;

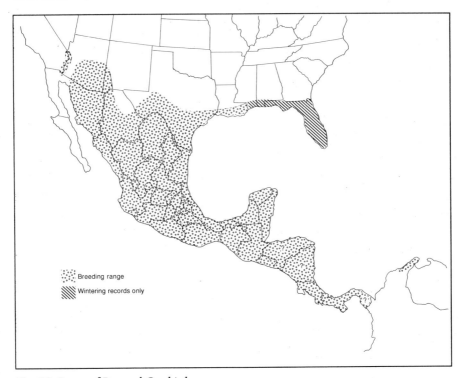

Fig. 6.2. Range of Bronzed Cowbirds.

Purrington 1993). Their range is not well known in the Río Colorado Delta of Mexico; Howell and Webb (1995) suggested that they be looked for in this area.

The range of the Bronzed Cowbird is less well understood in their southern extremes (de Schauensee 1964). They now inhabit central Panama (Howell and Webb 1995), at least to the Panama Canal area where they are seen regularly (N. Smith, pers. comm.). Presently, they appear to occupy the Caribbean coast of Colombia, but Friedmann (1957) mentioned that Carriker and Dugand both searched that area and failed to find any individuals. De Schauensee (1964) mentioned that Bronzed Cowbirds were recorded only from Leticia on the north bank of the Amazon. Ridgely and Tudor (1989), however, emphasized that they do not occur in Leticia and that such reports were erroneous due to a mix-up in a live bird shipment, which was undoubtedly the same shipment from Leticia mentioned by Friedmann (1957).

Although the range of Bronzed Cowbirds has extended slightly north and east, their overall abundance in the contiguous United States has not increased significantly. According to the Breeding Bird Survey, the trend from 1966 through 1994 was an average increase of 2.1% per year. The abundance of the *armenti* race in Colombia has either increased or was historically poorly understood. Early in our recorded history of Bronzed Cowbirds, the *armenti* race was reported as one of the rarest birds in the world (Friedmann 1929b). Ridgely and Tudor (1989) listed it as locally uncommon to fairly common in Colombia.

M. a. *armenti* is an isolate population in Colombia. M. a. *aeneus* ranges from extreme southern central Texas and the lower tier of the United States through eastern and southern Mexico through Central America to central Panama. M. a. *loyei* occupies central Arizona, California, and southwestern New Mexico through central Sonora, west Chihuahua, south through Sinaloa and Nayarit, and west through Durango. M. a. *assimilis* (Lesser Bronzed Cowbird) ranges from Jalisco through Colima, western Michoacán, western Guerrero, and Puebla, to the Pacific slope of Oaxaca and Chiapas, where it intergrades with M. a. *aeneus*.

Natural History

Feeding

Bronzed Cowbirds often forage alone, but they may also feed in small flocks, sometimes together with other blackbirds. They feed on beaches and in cattle pastures where they may pick insects from the backs of large domestic animals. As with Giant Cowbirds, they turn stones, searching underneath for edible matter. Bronzed Cowbirds are generally omnivorous, consuming a variety of seeds and arthropods (Carter 1984). Carter found snails in the stomachs of most females he examined; these undoubtedly supplied the large amount of calcium needed to produce eggs.

Habitat

Bronzed Cowbirds can be found in a variety of habitats: open and semiopen fields, agricultural fields, forest patches, and suburban landscapes. They are widely distributed in lowlands and altitudes up to 2,743 m (Bent 1958). They apparently occupy high altitudes primarily during the breeding season and descend to warmer climates for the winter. Generally, they are not found in heavily forested areas. As with other cowbirds, they invade formerly forested areas that have been extensively cleared.

Mating System and Sex Ratios

The mating system and variation within the mating system of Bronzed Cowbirds is little understood. In the *M. a. loyei* population, Dickey and van Rossem (in Bent 1958) reported that harem flocks, composed of one old strutting male and four or five females, were stable throughout the breeding season, strongly suggesting polygyny. Others have also found flocks with males far outnumbering females (Merrill 1877; Friedmann 1929b; Bent 1958). Clotfelter (1995) reported a male-biased sex ratio of 1.6:1 in southern Texas, and he reported a sex ratio of 1.4:1 from Carter's (1984) study, with an operational sex ratio (all adult females and after-second-year males) near unity. The data Clotfelter reported from Carter's study, however, needs to be clarified. The sex ratio of 1.4:1 was for hatched-year birds; the operational sex ratio was 1.2:1.[3] Considering all adults, the sex ratio from Carter's (1984) study was far more male biased at 4:1 (144:36), suggesting high female mortality. Carter (1984) identified the mating system in south Texas as promiscuous. N. Smith (pers. comm.), however, almost always observed flocks in Panama composed of a 1:1 sex ratio; he has also observed pair spacing and suspected monogamy.

Yearling males appear to be subordinate to older males and arrive on their breeding grounds in Texas at least one week after older males and females (Carter 1984). Whereas at least some yearling females breed, yearling males probably do not breed; their testes during the breeding season averaged less than half the size (0.02 g, *n* = 194) of older males' testes (0.04 g, *n* = 60, Carter 1984). Only 3.2% (*n* = 62) of adult females were not in reproductive condition during the breeding season, suggesting that yearling females probably breed (Carter 1984).[4] Males courted females indiscriminately both within their defended territories and in foraging areas; however, they did not defend females outside of their defended territories (Carter 1984). Carter (1984) suggested that promiscuity is the most likely mating system in Bronzed Cowbirds for several reasons: (1) males and females are both emancipated from parental duties, (2) food resources are abundant and not defensible, and (3) parasite densities are high, and host densities are low.

Territories

Males clearly establish singing trees, at least in some areas (Friedmann 1929b), but it is less clear that they actually establish exclusively defended territories. In southern Texas, females and yearling males do not appear to be territorial, whereas older males have more or less defined territories or dominions but defend them only on a part-time basis (Carter 1984). Distant food supplies undoubtedly affect whether a territory is defended on a full-time or part-time basis. When males are not defending their dominions, they wander about or move to distant foraging grounds. Why males defend an area as large as 5 ha within a promiscuous mating system is of considerable interest.[5] Carter (1984) suggested that the social organization of Bronzed Cowbirds is leklike and that maintenance of territories is for attracting and mating with females. In lek and dominion systems, spacing is variable; the main difference between them is the larger size of the dominion compared to individual spacing on leks. Another difference is that females may also lay their eggs within a dominion. Carter suggested that the distinction between leks and dominions is somewhat arbitrary, and Bronzed Cowbirds appear to be intermediate on a continuum between the two.

Conspecific egg-piercing behavior may negate the need for female territorial behavior in the traditional sense (chapter 4), as it reduces conspecific competition. It is perhaps significant that cowbirds exhibiting the most egg-piercing behavior (Shiny, Bronzed, and Screaming) appear to be less territorial than Brown-headed Cowbirds who engage less frequently in egg piercing. Egg piercing could be considered territorial behavior if it occurs within a defined area, but this has not to my knowledge been investigated.

Home Range

Little is known about Bronzed Cowbird home ranges and how variation, if it occurs, is related to variation in the mating system. Females have extensive home ranges, averaging 176 ha ($n = 6$), that overlap considerably with other females (Carter 1984). However, these home ranges are based upon observations during the day; nocturnal readings revealed that females roost in cattail marshes with other blackbirds at least 2.4 km from diurnal home ranges. Males wander over large areas and travel to distant foraging grounds, but Carter (1984) estimated defended areas of males to be approximately 5.3 ha. Both males and females leave their breeding areas periodically during the day to travel to distant foraging sites.

Courtship

Courtship displays vary, but during most displays, males inflate themselves by erecting feathers on their head, neck, and breast, with their neck ruff forming a halo appearance; they bring their spread tails forward and under, spread their wings, and stiffly prance around females, rocking back and forth with rapid and audible wing quivering. They may press their bills to their chests, bow forward, bob up and down, flex their legs, and emit guttural bubbling notes and/or squeaky whistles that are variously described as thin and high or low. Sometimes they hover directly above females in an aerial display (Clotfelter 1995). Although the mating system may be polygynous, males display to only one female at a time. Often, females feed on the ground while being courted and show no intrasexual competition for males. Females usually do not follow males after a courtship display, but males are as equally likely as not to follow females after a courtship display (Clotfelter 1995).

Migration and Flocking

Bronzed Cowbirds appear to be partially migratory, at least within part of their range. The most northern population moves an unknown distance south (Lowther 1995). Some altitudinal migration also takes place (Bent 1958). In Texas, Bronzed Cowbirds arrive in February to occupy traditional roosts and disperse to breeding grounds in mid-April; in September they migrate south (Lowther 1995). During migration, they form flocks with other blackbirds, and juveniles start flocking with adults by late August (Carter 1984).

Songs and Calls

Bronzed Cowbirds have a high-pitched tinny call. Howell and Webb (1995) noted that they sound somewhat like European Starlings. Friedmann (1929b) described the song as similar to that of Brown-headed Cowbirds, but with the beginning guttural notes deeper and individual notes shorter; he wrote it as *ugh gub bub tse pss tseee*. While the song is confined to males, "chuck" notes are emitted by both males and females. Friedmann (1929b) associated this chuck note with foraging.

Plumage and Molt

Juvenile plumage, which is a sooty gray with paler, distinctly streaked under parts, is acquired by a complete postnatal molt; females are slightly lighter than males. The first winter plumage is acquired by a complete, or nearly complete, postjuvenile molt in the fall. In this plumage, young males and females both look similar to adult females. The first nuptial plumage is acquired mostly by wear, but males also molt some new bronze feathers, primarily on the neck. This partial nuptial molt may also involve the head and breast. Adult male plumage is not acquired until the first postnuptial molt, which takes place in September and October. Therefore, second-year individuals can be distinguished from older birds, and they generally have a plumage intermediate between juveniles and after-second-year birds. Adult Bronzed Cowbirds have one complete molt annually, the postnuptial molt, which occurs from July into the fall. In the spring, during a prenuptial molt, they replace only some feathers, primarily on the back, breast, and head.

Hosts

Our early history indicated that orioles were the most frequently selected hosts of Bronzed Cowbirds. Friedmann (1929b) reported that approximately 75% of all Bronzed Cowbird hosts were orioles. In 1963, Friedmann still considered Bronzed Cowbirds to be rather selective, and many of the hosts on the growing list of 52 hosts were considered accidental. Friedmann and Kiff (1985) listed 77 species to be parasitized by Bronzed Cowbirds and then considered Bronzed Cowbirds to be generalists. Clotfelter and Brush (1995) added two new species to the growing list: Yellow-billed Cuckoos and Western Kingbirds (*Tyrannus verticalis*). Stewart et al. (1988) found a nest of a Slaty Vireo (*Vireo brevipennis*) containing a dead Bronzed Cowbird nestling and a vireo egg. Although Friedmann (1929b) and Friedmann (1963) listed Yellow-throated Brush-Finches (*Atlapetes gutturalis*) and Skutch (1960), Rowley (1962), and Friedmann (1963) listed Yellow-green Vireos (*Vireo flavoviridis*) as hosts, these species were not mentioned in Friedmann and Kiff (1985). This brings the list up to 82 known host species and 32 species known to be successful fledging Bronzed Cowbirds (table 6.1); only 11.1% are orioles.

Table 6.1

Known Victims of Bronzed Cowbirds

Order Cuculiformes, Family Cuculidae
  Yellow-billed Cuckoo, *Coccyzus americanus* (Clotfelter and Brush 1995)
Order Columbiformes, Family Columbidae
  Mourning Dove, *Zenaida macroura*
  White-winged Dove, *Zenaida asiatica*
  Common Ground-Dove, *Columbina passerina*
Order Passeriformes, Family Tyrannidae
  *Tropical Kingbird, *Tyrannus melancholicus*
  *†Couch's Kingbird, *Tyrannus couchii*
  *†Western Kingbird, *Tyrannus verticalis* (Clotfelter and Brush 1995)
  †Scissor-tailed Flycatcher, *Tyrannus forficatus*
  Sulphur-bellied Flycatcher, *Myiodynastes luteiventris*
  Social Flycatcher, *Myiozetetes similis*
  †Great Kiskadee, *Pitangus sulphuratus*
  Rose-throated Becard, *Pachyramphus aglaiae*
Family Corvidae
  *Green Jay, *Cyanocorax yncas*
Family Troglodytidae
  Rufous-naped Wren, *Campylorhynchus rufinucha*
  *Bewick's Wren, *Thryomanes bewickii*
  Happy Wren, *Thryothorus felix*
  Banded Wren, *Thryothorus pleurostictus*
  *Carolina Wren, *Thryothorus ludovicianus*
  Sinaloa Wren, *Thryothorus sinaloa*
  *Plain Wren, *Thryothorus modestus*
Family Muscicapidae, Subfamily Polioptilini
  White-lored Gnatcatcher, *Polioptila albiloris*
Family Muscicapidae, Subfamily Turdinae
  Black-billed Nightingale-Thrush, *Catharus gracilirostris*
  *Orange-billed Nightingale-Thrush, *Catharus aurantiirostris*
  Russet Nightingale-Thrush, *Catharus occidentalis*
  Clay-colored Robin, *Turdus grayi*
  Rufous-backed Robin, *Turdus rufopalliatus*
Family Mimidae
  *†Northern Mockingbird, *Mimus polyglottos*
  Tropical Mockingbird, *Mimus gilvus*
  *Long-billed Thrasher, *Toxostoma longirostre*
  †Curve-billed Thrasher, *Toxostoma curvirostre*
  LeConte's Thrasher, *Toxostoma lecontei*
Family Vireonidae
  Slaty Vireo, *Vireo brevipennis* (Stewart et al. 1988)
  Plumbeous Vireo, *Vireo plumbeus*
  *Red-eyed Vireo, *Vireo olivaceus*
  Yellow-green Vireo, *Vireo flavoviridis*
Family Emberizidae, Subfamily Parulinae
  *Tropical Parula, *Parula pitiayumi*
  *Golden-cheeked Warbler, *Dendroica chrysoparia*

Table 6.1 continued

Painted Redstart, *Myioborus pictus*
Rufous-capped Warbler, *Basileuterus rufifrons*
Yellow-breasted Chat, *Icteria virens*
Family Emberizidae, Subfamily Thraupinae
  *Red-crowned Ant-Tanager, *Habia rubica*
  Red-throated Ant-Tanager, *Habia fuscicauda*
  *Flame-colored Tanager, *Piranga bidentata*
  Hepatic Tanager, *Piranga flava*
  *Summer Tanager, *Piranga rubra*
  *Red-headed Tanager, *Piranga erythrocephala*
  *Scarlet-rumped Tanager, *Ramphocelus passerinii*
Family Emberizidae, Subfamily Cardinalinae
  Black-headed Grosbeak, *Pheucticus melanocephalus*
  *Northern Cardinal, *Cardinalis cardinalis*
  Pyrrhuloxia, *Cardinalis sinuatus*
  Blue Grosbeak, *Guiraca caerulea*
  Painted Bunting, *Passerina ciris*
Family Emberizidae, Subfamily Emberizinae
  *Song Sparrow, *Melospiza melodia*
  *Rufous-collared Sparrow, *Zonotrichia capensis*
  Black-chested Sparrow, *Aimophila humeralis*
  Stripe-headed Sparrow, *Aimophila ruficauda*
  Rusty Sparrow, *Aimophila rufescens*
  *Spotted Towhee, *Pipilo maculatus*
  *Canyon Towhee, *Pipilo fuscus*
  White-throated Towhee, *Pipilo albicollis*
  Rusty-crowned Ground Sparrow, *Melozone kieneri*
  *White-faced Ground-Sparrow, *Melozone biarcuatum*
  *White-eared Ground-Sparrow, *Melozone leucotis*
  *Olive Sparrow, *Arremonops rufivirgatus*
  Black-striped Sparrow, *Arremonops conirostris*
  White-naped Brush-Finch, *Atlapetes albinucha*
  *Yellow-throated Brush-Finch, *Atlapetes gutturalis*
  White-collared Seedeater, *Sporophila torqueola*
  Blue Seedeater, *Amaurospiza concolor*
Family Emberizidae, Subfamily Icterinae
  Chestnut-headed Oropendola, *Psarocolius wagleri*
  Yellow-winged Cacique, *Cacicus melanicterus*
  Yellow-billed Cacique, *Amblycercus holosericeus*
  Spot-breasted Oriole, *Icterus pectoralis*
  Altamira Oriole, *Icterus gularis*
  Streak-backed Oriole, *Icterus pustulatus*
  *Hooded Oriole, *Icterus cucullatus*
 *†Bullock's Oriole, *Icterus bullockii*
  *Orchard Oriole, *Icterus spurius*
  Black-vented Oriole, *Icterus wagleri*
  *Audubon's Oriole, *Icterus graduacauda*

Table 6.1 continued

Scott's Oriole, *Icterus parisorum*
*Red-winged Blackbird, *Agelaius phoeniceus*

Hosts known to have raised Bronzed Cowbirds are identified with an asterisk, and victims known to eject
Bronzed Cowbird eggs are identified with a dagger.
Source: Most species were listed by Friedmann and Kiff 1985.

Host Selection

Not all hosts are equal in their ability to foster Bronzed Cowbirds, and
some hosts are probably inappropriate as they are not known to raise the
parasites. Carter (1986) suggested that Bronzed Cowbirds shy away from
cavity nests and domed nests, but Great Kiskadees (*Pitangus sulphuratus*)
and Banded Wrens (*Thryothorus pleurostictus*) are parasitized by Bronzed
Cowbirds (Rowley 1962; Friedmann 1963; Friedmann and Kiff 1985).

Parasitism by Bronzed Cowbirds may be intense in some areas and during
some years. Carter (1986) found Bronzed Cowbird parasitism to be 100%
among nests of Green Jays, Long-billed Thrashers (*Toxostoma longirostre*),
Northern Mockingbirds, Red-winged Blackbirds, Northern Cardinals, and
Olive Sparrows (*Arremonops rufivirgatus*). Such high rates of parasitism may
be detrimental to parasites as well as their hosts because as conspecific competition increases, reproductive success decreases. Fischer (1980) noted that
although Bronzed and Brown-headed Cowbirds were abundant on his study
site in Texas, Curve-billed Thrashers (*Toxostoma curvirostre, n* = 76 nests)
and Long-billed Thrashers (*n* = 14 nests) were not parasitized by either
species.

Many regular Bronzed Cowbird hosts are large relative to the major hosts
of Brown-headed Cowbirds and (to a lesser degree) Shiny Cowbirds. Selection of large hosts has been suggested as a possible character displacement
reducing competition in areas of sympatry with Brown-headed Cowbirds.
Large hosts tend to have large nests that are easy for Bronzed Cowbirds to
find and easy to gain access to. Predators may also find these nests more
easily; however, large birds may be more successful in defending their nests
against predators. Likewise, larger birds may be more aggressive toward intruding parasites, but aggression may also provide cues to cowbirds regarding location of available nests and quality of hosts' age and experience. If
aggression provides cues to parasites, aggression by hosts may increase the
risks of multiple parasitism during conditions when parasite abundance is
high and host populations are low. Aggression often includes vocalizations

and commotion, which can attract other parasites. Bronzed Cowbirds can reduce competition from larger host nestlings by piercing host eggs, and this behavior may have evolved in response to increased competition from larger host nestlings. It may not be coincidental that Bronzed Cowbirds prefer larger hosts and pierce host eggs more frequently than Brown-headed Cowbirds do.

Nest Searching

Females search for potential host nests alone or in small groups of up to six individuals, but Carter (1986) never observed them traveling in groups during the dawn hours when they lay their eggs.

Breeding Season

Friedmann (1929b) reported that Bronzed Cowbirds do not begin breeding in Texas until mid-May and in Arizona until late May, a month later than Brown-headed Cowbirds in the same area. Carter (1984, 1986) reported the breeding season to begin in Texas the first week of May, which is earlier than Friedmann (1929b) reported, but still later than the beginning breeding of some hosts. Lowther (1995) reported that egg laying begins in mid-April.

Eggs and Egg Laying

Bronzed Cowbird eggs are immaculate pale blue, bluish green, greenish blue, or nearly white. Occasionally, they are maculated with scattered brown speckles (Harrison 1978). They vary in shape from ovate to fairly round, but are generally more ovate and slender than other molothrine eggs. The eggshell is thick, rather glossy, and finely granulated. Measurements vary only slightly across subspecies (Schönwetter 1983; Lowther 1988; Rahn et al. 1988).

Eggs of individuals are similar in size, shape, and color (Friedmann 1929b). Carter (1986) did not quantify variation in color and size, but his field impression was that there was little variation. Incubation of Bronzed Cowbird eggs ranges from 11 to 13 days; Carter (1984) reported an average incubation of 11 days. Bronzed Cowbirds lay their eggs at dawn (Carter 1984). The number of eggs that Bronzed Cowbirds lay over the breeding season is not known (Niles 1970), but Carter (1984) suggested that females do not lay continuously throughout the three month laying season.

Table 6.2

Number and Percent of Host Eggs and Cowbird Eggs Pierced in Nests
Parasitized by Cowbirds in Texas

	Number of Parasitized Nests	Number of Nests with Pierced Host Eggs	Number of Nests with Pierced Cowbird Eggs
Northern Mockingbird	16	9 (56.3%)	4 (25.0%)
Long-billed Thrasher	20	1 (5.0%)	2 (10.0%)
Green Jay	5	1 (20.0%)	0 (0.0%)
Red-winged Blackbird	2	0 (0.0%)	0 (0.0%)
Northern Cardinal	3	2 (66.7%)	3 (100%)
Olive Sparrow	6	0 (0.0%)	0 (0.0%)
Total	52	13 (25.0%)	9 (17.3%)

Source: Carter 1984, 1986.

Reduced Host Clutch Size

Friedmann (1929b) adamantly claimed that Bronzed Cowbirds do not remove host eggs, but data from other studies strongly suggest that Bronzed Cowbirds do remove host eggs (Rowley 1962; Carter 1986). Clutch size was significantly reduced in parasitized nests for all six host species that Carter (1986) investigated, varying from 33% of clutches reduced in Red-winged Blackbirds to 91% of clutches reduced in Northern Cardinals.

Egg Puncturing

Pierced eggs in parasitized nests are common, at least in southern Texas (Carter 1986). Carter assumed that Bronzed Cowbirds were responsible for all pierced host eggs; however, his data are particularly interesting and worth further consideration. He reported that a total of 25% of parasitized nests had at least one host egg pierced (table 6.2). The distribution of pierced host eggs among species, however, is not what would be expected by chance ($P < 0.005$, $\chi^2 = 18.11$). Most egg piercing occurred in Northern Mockingbird and Northern Cardinal nests. The sample size of Northern Cardinal nests is too small for speculation (albeit worth consideration in future studies), but what is most interesting is the discrepancy between rates of pierced host eggs in Northern Mockingbird and Long-billed Thrasher nests. Northern Mockingbirds appear to exhibit the unusual behavior of being intermediate between accepters and rejecters (Cruz et al. 1989; Post et al. 1990; P. Mason, pers. comm.). They often allow another egg to remain in their nests for several days before ejection. It is not known whether Northern Mockingbirds eject eggs by puncturing them or grasp ejecting them, but if they puncture eject (or at least attempt to), they may damage their own eggs in the process.

The nonrandom distribution of pierced host eggs among parasitized nests, together with the knowledge that Northern Mockingbirds engage in some amount of egg ejection, may suggest that Northern Mockingbirds pierce more of their own eggs than do Bronzed Cowbirds. Presuming the possibility exists of Northern Mockingbirds damaging their own eggs, removing them from Carter's sample would reduce egg piercing to only 11.1% of host nests. An alternative explanation is that individual Bronzed Cowbirds have a propensity to engage in host egg piercing and also preferentially parasitize certain species, in this case, Northern Mockingbirds. Yet another possibility is that Long-billed Thrashers remove damaged eggs from their nests more rapidly than Northern Mockingbirds do, and egg piercing may, therefore, be largely undetected.

Bronzed Cowbirds probably puncture conspecific eggs, as well as host eggs. Conspecifics in a nest represent competition at least equal to if not greater than competition from host nestlings. Carter (1986) reported that 17.3% of parasitized nests contained at least one punctured cowbird egg (table 6.2), and 31.7% ($n = 41$) of the total cowbird eggs laid in these nests were punctured. By contrast, he found 66.7% ($n = 33$) of host eggs in 13 nests punctured; this percentage remains the same even when removing Northern Mockingbirds from the sample (six of nine punctured host eggs) because of the large number of pierced host eggs within a nest and small clutches. Carter (1986) assumed that Bronzed Cowbirds could distinguish their own eggs from host eggs but could not distinguish their own from other Bronzed Cowbird eggs, so he suggested that pierced cowbird eggs should represent egg laying by multiple females. In other words, a Bronzed Cowbird exhibiting optimal behavior should pierce eggs of conspecifics if she has not previously laid in a given nest.

Our lack of understanding of host reactions to Bronzed Cowbird eggs in this population makes any conclusions of Bronzed Cowbird responses toward conspecific eggs speculative at best because it is not entirely clear whether hosts also contribute to egg puncturing, damaging their own eggs in the process. Experimental additions of Bronzed Cowbird eggs could help elucidate whether particular hosts attempt to puncture Bronzed Cowbird eggs and whether they damage their own eggs during ejection attempts. Such experiments would be especially revealing among nests that are not naturally parasitized (many hosts in Carter's study were parasitized at rates of, or close to, 100%) because Bronzed Cowbirds might attempt to puncture experimental eggs.

Bronzed Cowbird Nestlings

Newly hatched Bronzed Cowbird nestlings have orange-pink skin and yellowish bills and feet. The rictal flanges are creamy white, and the inside of the mouth is reddish. Their down is mousey gray. Like other cowbirds, their development is rapid; feathers in sheaths emerge in all tracts by four days. By five days, their eyes open, and feathers emerge from their sheaths. Begging responses are vigorous, with wing quivering and body trembling. Friedmann (1929b) described nestlings as restless, picking at nest linings and taking considerable interest in their surroundings. Due to their large size and rapid development, Bronzed Cowbirds often outcompete their host nestmates and become sole occupants of a nest. Growth rate constants of Bronzed Cowbird nestlings are similar to those of host nestlings in Texas, but these hosts were generally larger (Carter 1984). Not all Bronzed Cowbirds that hatch are successful. If a Bronzed Cowbird hatches before the host eggs or within 36 hours of the host eggs, it has a far better chance of fledging than if it hatches later. In nests of large hosts, if a Bronzed Cowbird hatches 48 hours or more beyond the host hatching, it will not survive (Carter 1986).

During the first day of hatching, Bronzed Cowbird nestlings are usually silent (Carter 1986). By two days, they begin to vocalize with soft peeps, while some hosts are still silent (Carter 1986). The average begging rate for a four- to six-day-old Bronzed Cowbird in Texas was 50 calls per minute (Carter 1984). Thurber and Villeda (1980) noted differences in begging behavior of nestling Bronzed Cowbirds and suggested that contrasting behavior may arise in the nests of different hosts. They also astutely observed that when they touched a Bronzed Cowbird nestling, it adopted a threatened posture and erected its light supraorbital feathers that resembled little horns.

Bronzed Cowbirds do not eject host nest mates or host eggs. Larger hosts are often able to raise their own young with a Bronzed Cowbird in the nest, but smaller hosts often fail to raise their own. Bronzed Cowbirds remain in the nest for about 10–12 days. Carter (1986) reported two distinct nestling size groups corresponding to gender differences; both males and females were approximately 70% of adult weight by the time of fledging.

Bronzed Cowbird Fledglings

After fledging, Bronzed Cowbirds remain in the vicinity of the nest for a few days. They are cared for by their foster parents for at least two weeks after fledging. After fledging, food provided by parents can vary depending on the host species, mostly by the addition of plant material by some hosts. Up to 54% of a Bronzed Cowbird's diet may consist of vegetable matter (Carter 1984, 1986). Once Bronzed Cowbirds reach independence, up to 76% of their diet may consist of a variety of seeds (Carter 1986).

Multiple Parasitism and Dump Nests

Single parasitism seems to be the general rule in some areas of the Bronzed Cowbird range, but numerous reports of multiple parasitism and dump nests exist. In Morelos, Mexico, and in Texas, Rowley (1962) and Carter (1986), respectively, found multiple parasitism to be the rule.

Concepts of dump nests are not altogether consistent. Friedmann et al. (1977:65) presented a fairly good concept of dump nests as being nests "in which excessive numbers of eggs of multiple hens are deposited. Since these nests are either old ones, already deserted by their builders, or since the unacceptably large numbers of parasitic eggs cause the hosts to desert, these eggs are wasted." Lowther (1995:8) defined dump nesting as, "egg not laid in nest of 'suitable' host . . . occurs when eggs are laid on ground or in abandoned nests, in nests of wholly inappropriate hosts (e.g., doves), or perhaps in those nests ultimately receiving large number of cowbird eggs."

The majority of dump nests undoubtedly result from egg laying by more than one female, but occasionally a nest that fits all other criteria of a dump nest may result from a single female. Not all nests with excessive numbers of parasitic eggs are necessarily old, already deserted, or destined to be deserted. Therefore, although the majority of eggs in dump nests are unequivocally "wasted," it is not a foregone conclusion. Clotfelter and Brush (1995) provided two excellent examples of dump nests that defied at least some of the criteria in the definitions put forth by Friedmann et al. (1977) and Lowther (1995). Both nests were actively attended and incubated by Green Jay hosts. One host incubated the clutch (11 Bronzed Cowbird eggs and two host eggs) long beyond the normal incubation period; the other host incubated the clutch (nine Bronzed Cowbird eggs and two host eggs) to the point where three Bronzed Cowbird eggs hatched. The latter nest was preyed upon just after the eggs hatched, so it is unknown how many more would have hatched. Even though nests such as these are quite unusual, a better working definition of dump nests is needed in order to include exceptional nests and eliminate ambiguities.

I would consider any nest with four or more parasitic eggs to be a dump nest. Working within the concept of Friedmann et al. (1977), nests with four or more parasitic eggs are likely to be deserted, and they are also likely to be the result of more than one female. Additionally, four parasitic eggs are likely to outnumber host eggs. A working definition may, therefore, simply be a nest with at least four parasitic eggs, usually laid by more than one female.

Carter (1986) found dump nests to be common. He reported that 16 of 86 (18.6%) parasitized nests contained five or more Bronzed Cowbird eggs, and in 12 nests, more than one cowbird egg appeared in a single day. Bronzed Cowbirds may parasitize the same nests more than once when parasitism has reached a saturation point; most of the preferred hosts in Carter's study that were multiply parasitized were parasitized at a rate of 100%.

Carter (1986) noted that dump nesting is more common among Bronzed and Shiny Cowbirds than Brown-headed Cowbirds and suggested differential predation within their ranges may play a role in this apparently wasteful strategy. If a female finds a nest that she intended to parasitize was depredated when she returns to lay her egg, she must lay her egg somewhere else. Laying in a nest with another cowbird egg may be more appropriate than laying the egg on the ground or in a nest that has just been depredated. If predation is high, then many females must do this many times, and dump nesting may be more common. Carter also pointed out that predation tends to be higher in the tropics (Skutch 1949; but see Oniki 1979), within the range of Bronzed and Shiny Cowbirds, than in temperate zones, where most Brown-headed Cowbirds breed.

Alternatively, dump nesting may be influenced by differences in social organization (A. Dufty, pers. comm. in Carter 1986). Dump nesting may be less common among Brown-headed Cowbirds because they are more territorial and may exclude other females. Interestingly, many reports of Brown-headed Cowbird dump nesting are in areas of recent range expansion or increased abundance. In areas where parasites and hosts have coexisted for a long time, dump nests may be more unusual. For Bronzed Cowbirds, most dump nests have been recorded in south Texas. Although Bronzed Cowbirds have long been breeding in extreme southern Texas where Carter conducted his research, they are at the edge of their breeding range. Conditions may be marginal beyond, and the populations at borders of the range may experience numbers beyond the saturation level (enough hosts to maintain the parasite population) from time to time when the populations from the interior of the range experience changes and push outward. It would be worth investigating whether the social behavior of these individuals breeding at the edge of their range differs from other populations and whether dump nests are more common at other boundaries of their range. The unusual number of dump nests in Texas may also reflect a bias in reporting, as information is spotty throughout the range of Bronzed Cowbirds.

Competition with Giant Cowbirds and Brown-headed Cowbirds

The range of Bronzed Cowbirds overlaps with both Giant Cowbirds and Brown-headed Cowbirds. The only host that Bronzed and Giant Cowbirds share in common is the Chestnut-headed Oropendola (Friedmann and Kiff 1985). Rowley (1962) and Friedmann et al. (1977) suggested that finding both Bronzed and Brown-headed Cowbird eggs in the same nest is common. Among hosts in which Bronzed and Brown-headed Cowbird eggs have been found in the same nest are Northern Mockingbird (Carter 1986), Yellow-breasted Chat (*Icteria virens*, Friedmann et al. 1977), Northern Cardinal (Carter 1986), Painted Bunting (*Passerina ciris*, B. Peer through P. Lowther, pers. comm.), Rusty-crowned Ground-Sparrow (*Melozone kieneri*, Friedmann 1963), Olive Sparrow (*Arremonops rufivirgatus*, Carter 1986), Hooded Oriole (*Icterus cucullatus*, Gilman in Bent 1958; Friedmann 1963), Orchard Oriole (*Icterus spurius*, Merrill 1877; Friedmann et al. 1977), and Red-winged Blackbird (P. Lowther, pers. comm.). Why Bronzed Cowbirds do not parasitize the nests of some species within their range is not understood. Brown-headed Cowbirds successfully parasitize numerous hosts not victimized by Bronzed Cowbirds (table 6.3). Interestingly, these are all smaller hosts and may be less preferred by Bronzed Cowbirds. Eggs of Bronzed Cowbirds are larger than those of Brown-headed Cowbirds, but it is unknown whether species that are not parasitized by Bronzed Cowbirds respond differently to Bronzed and Brown-headed Cowbird eggs.

Summary

Four subspecies of Bronzed Cowbirds range from central Arizona, southwestern New Mexico and southern Texas, through most of Mexico and Central America, to central Panama and the Caribbean coast of Colombia. Their range has extended north and east over the last few decades, but accord-

Table 6.3

Hosts of Brown-headed Cowbirds Not Parasitized by Bronzed Cowbirds but within the Range of Bronzed Cowbirds

Olive-sided Flycatcher	Yellow Warbler
Western Wood Pewee	Yellow-rumped Warbler
Western Flycatcher	Black-throated Gray Warbler
Say's Phoebe	Grace's Warbler
Vermillion Flycatcher	Common Yellowthroat
Horned Lark	Red-faced Warbler
Barn Swallow	Abert's Towhee
Blue-gray Gnatcatcher	Rufous-winged Sparrow
Black-tailed Gnatcatcher	Chipping Sparrow
Phainopepla	Lark Sparrow
White-eyed Vireo	Five-striped Sparrow
Bell's Vireo	Eastern Meadowlark
Black-capped Vireo	Western Meadowlark
Hutton's Vireo	House Finch
Warbling Vireo	

ing to the Breeding Bird Survey, the trend for the United States from 1966 through 1994 is a nonsignificant average increase of 2.1% per year. Bronzed Cowbirds are generally omnivorous and are often found in association with cattle and other large grazing mammals. They are found in a variety of habitats, including pastures, open and semiopen fields, agricultural fields, forest patches, and suburban landscapes. A variety of mating systems has been reported, including promiscuous, polygynous, and suggestions of monogamy. Sex ratios are generally male biased. Yearling males are usually subordinate to older males and probably do not breed, whereas second-year females probably breed. Females have large home ranges and do not appear to defend territories, but the prevalence of egg piercing may represent territorial behavior. Males in some areas of their range defend small territories, at least on a part-time basis, and Carter (1984) suggested a leklike social organization. Bronzed Cowbirds appear to be at least partially migratory, but their movements are not well understood. Bronzed Cowbirds parasitize at least 82 species.

Brown-headed Cowbirds

The dubious reputation of and subsequent hatred for Brown-headed Cowbirds often arises from uninformed sources. Typical of popular news articles about Brown-headed Cowbirds, Cone (1993:1a) wrote numerous erroneous statements: "Killer cowbirds are preying on the delicate warblers like crows stalking Tippi Hedren in 'The Birds.' . . . Cowbirds feverishly hunt down songbirds. . . . The fast-hatching cowbird chicks push songbird eggs overboard."

This reputation appears to be common, and it is somewhat understandable. Brown-headed Cowbirds are often held accountable for the recent decline of North American songbird populations and are frequently accused of being lazy, wretched, and immoral. Abhorrence and disrespect for Brown-headed Cowbirds, however, dates much further back than our recent knowledge of declining songbird populations. Pearson and Burroughs (1917:246) wrote in an account of Bronzed Cowbirds, "The Red-eyed Cowbird is a handsomer bird than that feathered wretch, the [Brown-headed] Cowbird . . . but this fine apparel has no effect upon the bird's habits, which apparently are as reprehensible as those of her northern relative." Later, Pearson et al. (1959:354) wrote, "Cowbirds are social outcasts, for as a rule, other birds do not associate with them. The males are without song and the sounds they make in spring are described by Chapman as 'guttural bubblings produced with apparently nauseous effort.'"

Obviously, the abhorrence of Brown-headed Cowbirds was so deeply rooted that little effort was exerted to understand them better or even to conduct a small amount of literature review. With regards to the comments of Pearson et al. (1959), Brown-headed Cowbirds have long been known to be gregarious in mixed species flocks, and Friedmann (1929b) described their song long before Pearson et al. made their odious statements. But still the hatred for them continues: after a Brown-headed Cowbird recently showed up in Scotland (McKay 1994), Brewer (1995:157) urged that any

Table 7.1

Common Names of Brown-headed Cowbirds

English	
Blackbird	Sheep Bird
Brown-headed Blackbird	Sheep Blackbird
Brown-headed Oriole	Shifty-bird
Brown-headed Starling	Shiny Eye
Buffalo Bird	
Buffalo Pecker	**Spanish**
Chucking Cowbird	Enmantecado
Chuckold	Gaycamaya Roja (Birkenstein and Tomlinson
Clodhopper	1981)
Coffee-headed Blackbird	Tongonito
Common Cowbird	Tordito
Cow Blackbird	Tordo
Cow Bunting	Tordo Negro (Birkenstein and Tomlinson 1981)
Cow Oriole	Vaquero Cabecicafé (Howell and Webb 1995)
Cow Troupial	
Cowbird	*M. a. ater*
Cowpen	Large Cowbird
Cowpen Bird	Northern Cowbird
Cowpen Bunting	
Cowpen Finch	*M. a. artemisiae*
Cuckold	Nevada Cowbird
Cuckoo Bunt	Northwestern Cowbird
Lazy Bird	Sagebrush Cowbird
Little Blackbird	
Parasite	*M. a. obscurus*
Prairie Blackbird	California Cowbird
Salt Bird	Dwarf Cowbird

other Brown-headed Cowbirds that appear in Europe "be instantly shot, without debate or delay."

Others have more appreciation for Brown-headed Cowbirds; they view them as an integral part of our ecosystem and as respectable survivors with an ingenious mode of reproduction. Brown-headed Cowbirds have been called numerous names, including arch-villains (Holmes 1993) and pests of the highest order (Neff 1926), but more reflective of reality, they have also been called scapegoats (Smith 1994). Brown-headed Cowbirds have been assigned numerous common names, descriptive of times and habits (table 7.1). The parasitical nature of Brown-headed Cowbirds was first reported in print in 1809 by Wilson (cited in Friedmann 1964), although the habit may have been well known to residents within the area of the bird's habitat.

Subspecies

Three subspecies of Brown-headed Cowbirds are currently recognized: *M. a. ater* (Boddaert), *M. a. artemisiae* (Grinnell), and *M. a. obscurus* (Gmelin). *M. a. ater* breeds in the east, *M. a. artemisiae* breeds in the northwest, and *M. a. obscurus* occupies the southwest and along the Pacific coast through Washington and Vancouver. Assignment of subspecies for adults is based on bill morphology and a series of measurements (Oberholser 1974), with wing length perhaps being the most important and reliable because it is a large, unambiguous measurement. The bill of *ater* is distinctly more conical than *obscurus* or *artemisiae*. Following is a key to the subspecies for adults:

1. Body metallic black, head brown. . . . males
   1a. Upper mandible deeply curved; wing chord 104.9–113.8 mm—*ater*
   1b. Upper mandible straight; wing chord > 110.0 mm—*artemisiae*
   1c. Upper mandible straight; wing chord < 104.6 mm—*obscurus*
2. Body and head brown, breast may have faint streaks. . . . females
   2a. Upper mandible curved; wing chord 96.5–101.1 mm—*ater*
   2b. Upper mandible straight; wing chord 97.1–104.9 mm—*artemisiae*
   2c. Upper mandible straight; wing chord < 90.9 mm—*obscurus*

Juvenile males can usually be distinguished from juvenile females by the contrast of primaries compared with darker underwing coverts, and in areas of nonoverlapping subspecies, wing length may be useful to distinguish between males and females (USFWS 1986). Nestlings of the *obscurus* race may be distinguished from both *ater* and *artemisiae* by bright yellow rictal flanges (Rothstein 1978a). *Ater* and *artemisiae* both have white rictal flanges, and only a few dubious intermediate forms have been reported (Rothstein 1978a). Apparently this characteristic involves a single locus with two alleles. Because no intermediates are known, even from areas of free interbreeding, one allele is most likely dominant (Fleischer and Rothstein 1988; Ortega and Cruz 1992b).

The significance of nestling rictal flange coloration is not well understood. Rothstein (1978a) suggested that perhaps the most important hosts in the southwest have yellow rictal flanges and that character displacement with Bronzed Cowbirds, who have white rictal flanges, may be operational because *obscurus* is (or was at the time) the only race to overlap with another cowbird and the only race to display a unique rictal flange color.

Aging

For bird banders, population ecologists, and ethologists, distinction between second-year (SY) and after-second-year (ASY) birds is of vital importance. Selander and Giller (1960) suggested that 97% of SY males retain some juvenile feathers in their underwing coverts. USFWS (1986), therefore, suggested that if the underwing coverts are uniformly black and no juvenile body feathers are present, the individual should be aged as ASY. While retention of juvenile feathers in the underwing coverts appears to be reliable for determining that a male is SY, uniformly black underwing coverts do not reliably indicate ASY status (Ortega et al. 1996). In addition to the color of underwing coverts, comparison between primaries and secondaries may be used to assess age (Ortega et al. 1996). Following is a key for aging males.

1. Underwing coverts with brownish gray feathers*—SY
2. Underwing coverts totally dark (excluding bottom row)*
    2a. Primaries and secondaries uniformly dark—ASY
    2b. Primaries brown or browner than secondaries**—SY

To my knowledge, we have no reliable methods for distinguishing SY females from ASY females in the field. Selander and Giller (1960) described ASY females as being darker, less striped ventrally, and less mottled dorsally, with considerable overlap.

Distribution and Abundance

The Brown-headed Cowbird is a short grass/edge species. That is, they are apparently limited to areas where at least some short grass is available for foraging and where perches provide them with sites from which to search for host nests. The amount of short grass necessary may not be great and may be provided by very small forest gaps, lawns, berms, and small fields or meadows scattered throughout forests. Brown-headed Cowbirds are typically found in association with large herds of grazing mammals that flush up insects.

---

* Do not include bottom row as these are brownish gray in all age classes.
** The amount of brown varies from the tips (10–20 mm) to browner throughout.

Historic Distribution

Prior to the arrival of Europeans, no cattle existed in North America. At that time, Brown-headed Cowbirds presumably followed large migratory herds of Bison and other large native grazing mammals. Bison not only flushed up insects, but also trampled down tall grasses which might have been otherwise unsuitable for cowbirds. In the early days of European settlement in North America, Brown-headed Cowbirds primarily occupied grasslands of the midcontinent.

Although scattered reports of Brown-headed Cowbirds in the east exist, the birds appeared to be entirely missing from large unbroken forest tracts (Mayfield 1965b). Many avian census accounts of the eighteenth century from eastern states do not mention Brown-headed Cowbirds, and the bird did not appear in the tenth edition (1758) of Linnaeus' "Systema Naturae" (in Mayfield 1965b). In western North America, Brown-headed Cowbirds occupied areas east of the Cascades in Washington and Oregon, and east of the Sierra Nevada in the Great Basin (Rothstein et al. 1980). If cowbirds existed west of the Sierra Nevada prior to this century, they were probably in small numbers and inconspicuous.

Conditions Favoring Range Expansion

Habitat alterations occurred with human settlement and the introduction of livestock. Large tracts of forest were cleared for cultivation, and tallgrass prairies were converted to range lands. Trees were planted in areas previously devoid of them, which provided cowbirds with perches. Many parts of the east became suitable habitat for cowbirds, and cowbirds switched their association with Bison to an association with cattle, sheep, and horses.

Eastern North America. By 1790, the mideastern states were well populated and had been under cultivation for many generations. Brown-headed Cowbirds were fairly common in Pennsylvania and New York by the late 1700s, whereas they appeared to be absent from these two states in the mid-eighteenth century.

In the southeastern states, the range of Brown-headed Cowbirds was spotty even by the mid-twentieth century. The first breeding record in the Carolinas was in 1933 (Pearson et al. 1959). From 1933 through the mid-1950s, reports were still few and scattered, but by the mid-1960s, Brown-headed Cowbird parasitism was common throughout the Carolinas (Potter and Whitehurst 1981). In the 1930s, Brown-headed Cowbirds irregularly wintered throughout Florida (Sprunt 1954) and Alabama (Webb and Wetherbee 1960); however, no definite breeding records existed at this time. By 1957, the breeding range included the entire continental United States excluding Florida. As the 1957 American Ornithologists' Union (A.O.U.) *Check-list of North American Birds* was in press, Monroe (1957) reported the first unquestionable breeding record for Florida. Webb and Wetherbee (1960) reported that Brown-headed Cowbirds were breeding in northern Florida by 1958, and throughout Alabama by 1960.

Northern States. Brown-headed Cowbirds did not become established in Ohio until the mid-nineteenth century, and according to Mayfield's (1965b) interpretation of Fothergill's unpublished notes, they may have been scantily present in southern Ontario in the early nineteenth century. During the twentieth century, Brown-headed Cowbirds unquestionably expanded into Ontario (Snyder 1957). However, Snyder suggested that they may have first arrived around 1889. According to Mills (1957), the first breeding record of Brown-headed Cowbirds on Nova Scotia did not occur until 1933, but Friedmann (1929b) included southern Nova Scotia in his range map.

Western North America. Cowbirds rapidly expanded west between 1900 and 1930. Laymon (1987) suggested that a major influx of cowbirds occurs within ten years after an area is invaded. Laymon (1987) listed numerous references of first breeding records at various locations in California from 1862 to 1960. Law (1910) was probably the first to report Brown-headed Cowbirds breeding in the Los Angeles area. Cowbirds were first reported in the San Bernardino Valley of California in 1918 (Hanna 1918), and they invaded the Yosemite Valley by 1934 (Rothstein et al. 1980). Grinnell and Miller (1944) suggested that cowbirds were absent from the entire Sierra Nevada range in 1940. However, there are scattered records of parasitism in Mono County, east of Yosemite National Park, in the 1920s and 1930s (Dixon 1934; Rowley 1939; Friedmann 1963; Friedmann et al. 1977).

Present Distribution and Abundance

Presently, Brown-headed Cowbirds inhabit the entire United States, northern Mexico, and much of Canada (fig. 7.1). An occasional individual is observed outside North America (McKay 1994), but this is rare. Data from the Breeding Bird Survey (BBS) are useful in evaluating population trends. No

Fig. 7.1. Range of Brown-headed Cowbirds. The double hatched lines indicate nonbreeding range.

avian census technique is completely error free, and cowbirds present a few additional problems that may increase error (B. Peterjohn, pers. comm.). Cowbirds often congregate in flocks, even during the breeding season, and sporadic observation of flocks may inflate or deflate population estimates. Males also have a high pitched call and song that may be undetected during a survey.

In 1970, a few years after the BBS was initiated, results suggested that the center of highest abundance was the midcontinent extending from North Dakota through Oklahoma (Van Velzen 1972). This center of high abundance appears to have remained relatively stable. Overall, cowbird populations appeared to increase from 1966 through 1976 and decrease from 1977 through 1984 (Peterjohn and Sauer 1993). From 1966 through 1992, the Brown-headed Cowbird population has declined 0.9% overall, with a slight increase in the United States and a slight decrease in Canada.

East. Brown-headed Cowbirds are abundant throughout most of the east and are found even in large forest tracts in Illinois. They are not as abundant in Florida at this point in time, but their populations may increase rapidly there as they have only recently invaded Florida.

West. Brown-headed Cowbirds appear to be well established throughout the Sierra Nevada, but Verner and Ritter (1983) found them absent from unlogged forests in the western Sierra Nevada. In the southeastern Sierra Nevada, Rothstein et al.(1980) found cowbirds along 25 of 29 (86.2%) survey routes and in 39% of their counting sites. In the west-central Sierra Nevada, they detected cowbirds at all six survey routes and in 17.7% of their counting sites.[1] The difference in cowbird abundance between the west-central and southeastern areas may reflect the sites selected; the southeastern sites were selected because they represented good cowbird habitat. Fewer cowbirds occupied areas far from human influence and areas close to extensive stretches of unlogged forests (Rothstein et al. 1980). They suggested that scattered pack stations, meadows, towns, and campgrounds may provide suitable foraging for cowbirds.

Gene Flow of *Obscurus* into *Artemisiae* Populations

Historically, during the breeding season, *ater* occupied the eastern United States, *obscurus* was found in the extreme southwest, and *artemisiae* occupied the northwest. The northward expansion in the west has been by the *obscurus* race. The *obscurus* cowbirds did not stop where they met *artemisiae* but, instead, penetrated the *artemisiae* population and bred freely with them. In these zones of contact, cowbirds display characteristics intermediate between *obscurus* and *artemisiae*. The rate of gene flow of *obscurus* into the *artemisiae* population has been both unhindered and very rapid (Laymon 1987; Fleischer and Rothstein 1988; Ortega and Cruz 1992b). Presently, it is not known how broad the zones of contact are, but data from parts of California, Colorado, and the Pacific Northwest suggest that they may be several hundred kilometers in some areas. Using wing length, rictal flange color, and vocal dialects, Fleischer and Rothstein (1988) documented extensive gene flow of *obscurus* into the *artemisiae* population eastward across the Sierra Nevada. A few years later, Fleischer et al. (1991) reported that the mitochondrial DNA of cowbirds collected at eight areas in California and Nevada showed distinct patterns that confirmed extensive gene flow between the subspecies in the Sierra Nevada.

Friedmann's (1929b) map indicated the absence of *obscurus* in Colorado, and Rothstein's (1978a) map clearly indicates lack of *obscurus* in Colorado and the Pacific Northwest (fig. 7.2). Behle (1985) reported *obscurus* in Utah only in the extreme southwest. I found evidence of *obscurus* genes in north-central Colorado by 1985 (Ortega and Cruz 1992b), and from 1992 through 1996, J. Ortega and I found the population in southwest Colorado to be approximately 85% intergrades between *obscurus* and *artemisiae*. Of 20 Brown-headed Cowbird nestlings hatched in southwest Colorado, 40% had yellow rictal flanges.

The gene flow of *obscurus* into the *artemisiae* population has apparently been rapid. Historic specimens from southwestern Colorado exist in small numbers, but if the 1957 A.O.U *Check-list* range was accurate, in 35 years, our study site changed from no reports of *obscurus* to a nearly completely intergraded population. Where documentation has been more thorough, Laymon (1987) found that *obscurus* expanded north and west in California between 1900 and 1930 at a rapid rate. Fleischer and Rothstein (1988) calculated the rate of *obscurus* gene flow in California, as evidenced by yellow rictal flange color, to be an average of 5.5 km per generation over 20 years.

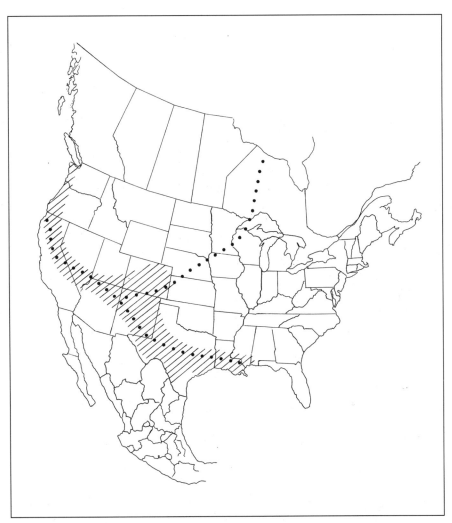

Fig. 7.2. Range of Brown-headed Cowbird subspecies, with *M. a. ater* occupying the east, *M. a. artemisiae* occupying the northwest and Great Plains, and *M. a. obscurus* occupying the southwest and west coast. The solid dots indicate the ranges according to Rothstein (1978a), based upon the 1957 edition of the A.O.U. *Check-list*. The hatched lines indicate actual and probable zones of interbreeding.

Natural History and Ecology

Habitat

Brown-headed Cowbirds are generally associated with disturbed habitats that offer high concentrations of food. They are found in a wide variety of habitat types and commute readily between feeding and breeding sites. They are common in forest edge, riparian zones, thickets, prairies, fields, cattail marshes, pastures, orchards, and suburban landscapes. Mayfield (1965b) suggested that Brown-headed Cowbirds will enter forests only momentarily for nest searches and in social contexts. This seems to imply that they will not penetrate into the interior of large forest tracts, but the more recent work by Scott Robinson in Illinois demonstrates that Brown-headed Cowbird parasitism is heavy even in the center of large forest tracts (chapter 9).

Feeding

Brown-headed Cowbirds forage mainly on the ground, often in association with large grazing ungulates that stir up insects. They feed primarily on arthropods and seeds, and females consume large amounts of mollusc shells. Stomach analysis in Michigan revealed a diet of grasshoppers, leafhoppers, beetles, caterpillars, and small seeds, with 59% (n = 26 birds) of stomach contents consisting of insects before 16 July (Payne 1965). In winter, Brown-headed Cowbirds feed heavily on grains (Payne 1973b). In a more widespread (20 states) stomach analysis study throughout the year, Beal (in Savage 1897) found that the Brown-headed Cowbird diet overall consisted of 28% animal matter and 72% vegetable matter.

Sex Ratio

The sex ratio of Brown-headed Cowbirds probably is close to unity at hatching and fledging (Hill 1976b; Weatherhead 1989). Among adults, the sex ratio is almost always male biased; the 1.3:1 and 1.4:1 that Dufty (1982a) and Dufour and Weatherhead (1991), respectively, reported are low relative to other studies (table 7.2).[2] The imbalance may be partially attributed to the more secretive nature of females. Sex ratio data are often from trapping records, and females may be more trap shy than males. Elliott (1980) reported a sex ratio of 6.3:1 among trapped birds, but he found a sex ratio of 1.1:1 from census data. Similarly, Dufour and Weatherhead (1991) caught a higher proportion of females in mist nets than in decoy traps. By contrast,

Table 7.2

Locations and Sex Ratios Reported for Brown-headed Cowbirds

Location	M:F	Method	n	Source
Breeding season				
Arizona: Phoenix	2.9:1	trapped	71	Neff 1943
California: S Sierras	6.0:1	pack stations		Rothstein et al. 1980
California: southern	2.3:1	trapped	562	Beezley and Rieger 1987
California: Fresno Co.	3.3:1	trapped	125	Rothstein et al. 1987
California: east central	1.8:1	trapped	716	Yokel 1989a
California: east central	2.1:1	census	235	Yokel 1989a
Kansas	1.1:1	census	506	Elliott 1980
Kansas	6.3:1	trapped	44	Elliott 1980
Michigan: south	1.2:1	trapped	89,000	DeCapita 1993
New Hampshire	2.2:1	trapped	377	Kennard 1978
New York: Broome Co.	1.3:1	trapped	88	Dufty 1982a
New York: Broome Co.	1.4:1	trapped	65	Dufty 1982a
New York: Broome Co.	2.1:1	trapped	66	Dufty 1982a
Ontario: London	2.0:1	trapped	143	Darley 1971
Ontario: eastern	1.4:1	trapped	1,862	Dufour and Weatherhead 1991
Ontario: eastern	1.3:1	trapped	242	Teather and Robertson 1986
Breeding and nonbreeding season				
California: San Diego Co.	2.3:1	trapped	562	Beezley and Rieger 1987
Ohio	3.4:1	trapped	28,726	Burtt and Giltz 1976
Nonbreeding season				
California: Colusa Co.	3:1	trapped	10,000	Crase et al. 1972
North Carolina: Wake Co.	1.5:1	trapped	197	King and West 1988
Texas: Brazos Co. and Houston Co.	3:1–8:1	trapped	75,000	Arnold and Johnson 1983

Yokel (1989a) reported that sex ratios were similar between trapped and censused birds. The population that Burtt and Giltz (1976) trapped was male biased, but females reentered the trap more often than males. Placement of traps may also attract more or fewer females. Traps placed in foraging sites, such as dairies or stockyards, may attract a disproportionate number of males. Alternatively, male-biased ratios may be real, indicating higher mortality among females. Why there is geographic discrepancy in sex ratios is enigmatic. Sex ratios may be less male biased in some areas as a consequence of reduced stress on females defending smaller areas. Alternatively, females may emigrate from areas of high stress and low host densities to more profitable breeding grounds, leaving a male-biased sex ratio in other areas.

Home Range

Home range sizes of Brown-headed Cowbirds vary considerably and may be correlated with density of host nests and abundance of cowbirds. Rothstein et al. (1984) found breeding home ranges in the eastern Sierra Nevada (68 ha)[3] to be among the largest of passerines, exceeded only by Common Ravens (*Corvus corax*), while Dufty (1982a) reported a mean home range of 20.4 ha in New York. Female home range averaged 9.89 ± 2.69 ha in Teather and Robertson's (1985) study in Ontario, and Darley (1968) reported ranges of only 4.5 ha for females and 6.6 ha for males in Ontario, which is the smallest ever reported. Dufty (1982a) suggested that Darley may not have been able to visually track these birds throughout their entire range, but the smaller ranges may have reflected a more dense host population. Host nests are clearly the most important resource within the breeding home range, and density of hosts undoubtedly dictates the home range size of brood parasites.

Rothstein et al. (1984) calculated that the expected home range for a nonparasitic bird the size of Brown-headed Cowbirds is only 1–3 ha. They suggested an analogy between home ranges of Brown-headed Cowbirds and predators, such as accipiters who search for many of the same passerines. Indeed, they found the home range of Brown-headed Cowbirds similar to accipiters in Craighead and Craighead's (1956) study.

Diurnal Movements and Activities

Virtually all studies have demonstrated similar diurnal activities among Brown-headed Cowbirds. Much of the morning is occupied with egg laying, searching for nests, courtship, mating, and establishment and maintenance of male dominance hierarchies; afternoons are typically spent foraging. Diurnal movements, however, differ and may depend on landscape patterns. Throughout much of their range, breeding and foraging sites are close because of abundant edge habitat and widespread agricultural and pastoral activities. In these areas, Brown-headed Cowbirds may move little during the day.

In other areas, appropriate breeding sites may be distant from preferred foraging sites. In these situations, Brown-headed Cowbirds breed and forage in disjunct areas, switching from the solitary activity of egg laying and nest searching to gregarious foraging flocks. Rothstein et al. (1984) found that five radio-tagged females and four of eight radio-tagged males in the eastern Sierra Nevada consistently spent their mornings at breeding sites and commuted 2.1–6.7 km to prime feeding sites, consisting of corrals and bird feeders, where they spent the remainder of the day; just before dark, females would return to their breeding sites, where they presumably roosted for the night.

From 84 radio-tagged females in Illinois and Missouri, Thompson (1994) found that, generally, females spent mornings in forests and shrub-sapling habitats, and use of croplands, feedlots, and short grass habitats increased throughout the day. Breeding locations were tightly clustered and some-times distinct from feeding locations, which were widely dispersed. Females moved an average of 1.2 km (0.03–7.34 km) between breeding and feed-ing locations. Generally, they moved greater distances between roosting and breeding or feeding locations than between breeding and feeding locations, with some cowbirds moving up to 10 km (Thompson 1994). Thompson found that 57% ($n = 1,351$ observations) of foraging observations occurred in association with livestock.

Territories

The concept of territories has a history dating back at least as far as Aris-totle (Nice 1953). Definitions vary from brief statements to long discourses, but what is almost always explicitly stated is that the area or territory, with its resources, is defended. For nonparasitic species, defended resources are often mates, food, and nest sites. The adult sex ratio of all Brown-headed Cowbird studies is male biased; therefore, males are usually not a limited resource for females. Brown-headed Cowbirds congregate at feeding sites, and these food resources are not known to be defended. Breeding sites, on the other hand, include host nests, which are an important resource. If exclusive use of nests in an area is advantageous, then female territorial be-havior might be expected. Nonexclusive use of host nests could result in a high incidence of multiple parasitism, which reduces reproductive success. Darley (1983) and Dufty (1982b) proposed that females may be territorial specifically to avoid multiple parasitism.

Female Territoriality. Friedmann (1929b) believed females to be territorial, but his opinions were based on impressions of unmarked individuals. In New York, Dufty (1982a) found that nonfeeding ranges of mated individuals were essentially nonoverlapping with other mated individuals, and females aggressively defended their ranges from other females. In experimental play-backs of female chatter and male song, Dufty (1982b) found that females re-

sponded aggressively to female chatter but not to male song, further suggesting female territoriality. In Ontario, Darley (1983) reported that, although
female ranges overlapped, most aggressive encounters occurred at mutual
boundaries. Similarly, Teather and Robertson (1986) observed overlapping
ranges in Ontario; females were aggressive toward each other, but aggression was more intense at the centers of their ranges. Yokel (1989a) found
females to be neither territorial nor aggressive toward each other in California. Laskey (1950) observed many fights among Brown-headed Cowbirds,
but she found no evidence that the birds were defending territories.

There appears to be a continuum of female behavior ranging from highly
territorial with exclusive home ranges to overlapping ranges with little to
no aggression. Variation may depend on the defensibility of host nests. The
ability to defend a territory may be influenced by many factors, including
(1) how large the breeding home range is, which may be dictated by density
of host nests; (2) how much time can be spent defending the area; (3) how
much help males provide in defending an area; and (4) cowbird density.

Home ranges are smaller and host density is higher in the northeast where
territoriality has been observed than in the Kansas prairies and California
where Elliott (1980) and Yokel (1989a), respectively, found no evidence of
female territorial behavior. Teather and Robertson (1985) made several predictions based upon female spacing patterns in relation to host nest density:

1. Where host density is low, these resources may not be defensible because they are spread out over a very large area. Elliott (1980) rarely
   observed female aggression in Kansas where host nests are spread out
   over large areas.
2. Where host density is high, defense is feasible because nests are more
   concentrated. In the northeast, Dufty (1981) found both more diverse
   and abundant hosts in concentrated areas.
3. Assuming there is a cost to aggression, females should increase defense
   in nest-rich areas and reduce defense in nest-poor areas. Teather and
   Robertson's (1985) observations of higher female aggression while feeding in woodland swamps than at feeding stations and other communal
   foraging sites seemed to support this prediction; however, they did not
   report on the time of day aggression most frequently occurred in these
   two contexts.

In addition to feeding gregariously in stockyards, pack stations, and pastures, females must find mollusc shells, which provide a source of calcium.
These are often found in riparian areas that may also be within their breed-

ing areas. It is not known what time of day females search for mollusc shells in various locations, but these requirements may also influence territorial behavior to some degree. If the breeding area is located within a riparian zone and if the females search for shells in the mornings, when territorial defense is most likely to occur, they may be preoccupied with foraging and spend less time in territorial defense. Yokel's (1989a), Darley's (1983), and Teather and Robertson's studies were all conducted in either riparian zones or swampy areas. Alternatively, if calcium resources are within their breeding ranges, females may spend more time in these areas during the afternoons.

Those who reported the highest degree of territorial behavior also reported the strongest male-female pairs (Dufty 1981, 1982a; Friedmann 1929b). The male role in female territorial defense is not clear, nor is it entirely clear why males would defend territories against other females. However, when intruding females lay eggs fertilized by other males, it would benefit a male to defend his mate from other females who may lower her (and his) reproductive success through multiple parasitism. Dufty (1981) found that females and males both responded to playbacks of female chatter, but he felt that the role of males was passive, that is, they were merely following their mated females. Even if the male role in female-female aggressive encounters is passive, the presence of a male may be a more effective deterrent than a female by herself.

Cowbird density may also influence female territorial behavior. Yokel (1989a) found a positive correlation between cowbird density and aggression in two sites that differed in cowbird density. However, he commented that territoriality does not always correlate with cowbird density because female territoriality was reported from areas with the lowest density (Dufty 1982b) and the greatest density (Darley 1983). Smith and Arcese (1994) reported that Mandarte Island, Vancouver, was consistently occupied by only a few females over a 16-year period; they suggested that numbers may have been limited by aggressive female territoriality.

Dufty (1982a) predicted that three patterns would emerge as a consequence of differences in social/spatial patterns: (1) in defensible areas (e.g., the northeast), morphology of cowbird eggs should be similar within territories with overlap at territorial boundaries; (2) dissimilar eggs should be found in local areas of low host density; and (3) multiple parasitism should be lower in defensible areas (northeast) than in nondefensible (prairie) habitats. The incidence of multiple parasitism does seem to be more prevalent in prairie states than in the northeastern states, but several investigators have recently found a high incidence of multiple parasitism in host-rich areas of the midwest (Robinson 1992; Donovan et al. 1993; Robinson et al. 1993; Trine 1993). Alternatively, multiple parasitism may be observed along mutual boundaries in areas where females defend a range or territory; in host-rich areas where ranges tend to be smaller, the incidence of multiple parasitism might be greater, especially when only the center of the range is defended, because the edge to area ratio is greater than in large areas with low host density. Searching for trends between host density and territoriality may also be confounded if females remove eggs of other cowbirds from host nests, which can be considered territorial behavior and may be difficult to detect. Clearly, more research needs to be conducted in other areas before we may understand what ecological conditions favor territoriality in females, but patterns to look for include: (1) breeding home range size, (2) density of host nests, (3) how much time can be spent defending a territory, (4) habitat, (5) foraging strategies, (6) how much help males provide in defense, (7) cowbird density, and (8) incidence of multiple parasitism or conspecific egg puncturing.

Male Territoriality. Resources that males could theoretically defend are food, mates, and host nests. The latter would probably occur only with monogamous pairs and has not been definitively observed. As with females, males gather in gregarious flocks to forage, and defense of these resources is not known. Males could display territorial behavior in defense of females, but this is only likely to occur where females have an established defended territory. In the study that clearly demonstrated home ranges defended by females at the boundaries, Dufty (1982a) did not observe territorial behavior by males, even though they had essentially nonoverlapping home ranges. The nonoverlapping home ranges may have resulted from monogamous males following their mates through a female-defended territory. Darley (1982) reported that male ranges overlapped considerably, and they did not defend their ranges; most male-male aggressive encounters occurred well within the boundaries, not at the boundaries. In prairie habitat in Kansas, Elliott (1980) found three males had essentially nonoverlapping home ranges, but he observed no territorial defense behavior. In the eastern Sierra Nevada, Rothstein et al. (1984) found little evidence of territories in breeding habitat. Payne (1965) observed flocking throughout the summer and found no evidence of defended territories in Michigan. Male Brown-headed Cowbirds often have established singing trees (Friedmann 1929b; Payne 1973b; Elliott 1980; Darley 1982), but singing trees are used simultaneously by many males (Elliott 1980; pers. obs.).

Males may find it more economical to guard females rather than to patrol and defend large areas. Territoriality by males is also not as feasible in parasitic species as nonparasitic species because activities are not focused around a single nest. Although Brown-headed Cowbird males are generally not territorial, they do establish dominance hierarchies, in which dominant males presumably are the most attractive to females and consequently gain access to the limited female population.

Dominance Hierarchies

Male-biased sex ratios result in competition for limited females. Females exhibit a high degree of mate choice, and their choice may be based upon assessment of rank within male dominance hierarchies. Male Brown-headed Cowbirds form dominance hierarchies, but it is not a site-based dominance. Often, dominant males have greater access to females both in studies of wild populations (Laskey 1950; Darley 1982) and captive populations (Darley 1978; Rothstein et al. 1986). In studies of dominance hierarchies, rank is usually determined by the number of displacements, song spreads, and flight whistles.

With male-biased sex ratios, one could predict that younger males may have delayed breeding and would generally be subordinate to older males. Numerous studies of other species have demonstrated that older males dominate younger males (Searcy 1979; Parsons and Baptista 1980; Post 1992; but see Robinson 1986). During the breeding season, SY males appear to have testes large enough to be considered in breeding condition (Scott and Middleton 1968; Payne 1973b; Dufty and Wingfield 1986a; O'Loghlen and Rothstein 1993), but they may not have equal opportunities to mate. In Dufty and Wingfield's (1986a) study, the mean testes weight of SY males was similar to those of ASY males, which suggests breeding activity, but they also found that luteinizing hormones and androgens, particularly testosterone, were generally higher for longer in ASY males than in SY males. Payne (1973b) and Scott and Middleton (1968) found that SY males had significantly smaller testes than ASY males, but there was overlap in size. While some studies have suggested equal or near equal mating opportunities between SY and ASY males in wild populations (Ankney and Scott 1982; Teather and Robertson 1986) and captive populations (Darley 1978; Rothstein et al. 1986), others have shown higher mating success potential among ASY males (Payne 1973b; Yokel 1989b). In general, ASY males are dominant over SY males (Yokel 1989b). ASY males were also dominant to SY males in a restricted food situation (Weatherhead and Teather 1987) and with restricted female access (Teather and Weatherhead 1995).

ASY males are, on average, larger than SY males, and Yokel (1989b) found males that won contests tended to be both larger and older. Even among only ASY males, larger individuals won more contests (Yokel 1989b). Teather and Weatherhead (1995), by contrast, found size to be unimportant in gaining dominance status. Dominant males arrived on breeding grounds earlier and engaged in more contests in Yokel's (1989b) study. Mated males are generally dominant over unmated males (Darley 1982); this is particularly noticeable in the presence of the female the dominant male is mated to. Males were also more likely to win contests if the mate was present (Yokel 1989b).

Male-male aggression during peak establishment of dominance hierarchies is positively associated with testosterone. Male companionship has an important effect on testosterone levels, and Dufty and Wingfield (1990) unexpectedly found that seeing another male is probably more important than hearing another male. In captivity, grouped males had higher testosterone levels than isolated males, and males with devocalized companions had higher testosterone levels than isolated tutored males who had male songs broadcast to them two hours daily. In an earlier study, Dufty and Wingfield (1986b) found males isolated visually but not vocally had a more pronounced response than what was found in isolated tutored males in the 1990 study; they attributed this to males in the earlier study being exposed to male vocalizations all day long and to an unnatural increase in photoperiod.[4] The important effects males have on each other is also evident from higher testosterone levels of males in high density populations compared with those in low density populations (Dufty 1993).

Singing is apparently important in establishment of Brown-headed Cowbird dominance hierarchies, but once ranks are established, they may be maintained through nonvocal means (Dufty 1986). Whether a male gains mating success because he is dominant or whether dominance is gained after achieving mated status is not clearly understood, especially in the wild. Teather and Weatherhead (1995) tested this in a captive population by removing the dominant male and allowing a subordinate male exclusive access to a female. When the dominant male was returned after two days, the originally subordinate male remained dominant in only 5 of 19 (26.3%) trials. The presence of females could have influenced dominance switching if exposure to females resulted in elevated levels of testosterone. When the dominant males were removed, they may have experienced a decrease in testosterone levels, whereas subordinate males may have experienced increased levels of testosterone; for some, it may have been sufficient to maintain dominance when the originally dominant male was returned. Rothstein (1972b) found that in a captive situation, singing was primarily by dominant males, but if the dominant male was removed, subordinates began to sing or increased their singing.

Perch song, flight whistles, and size have been suggested as possible cues for females to assess the quality of males (Yokel and Rothstein 1991). Given that Dufty and Wingfield (1990) found testosterone levels of males were affected more by seeing another male than hearing another male and that ASY males tend to dominate SY males, it would also be instructive to investigate potential visual cues, such as under-wing coverts that are exposed during song spread, that both males and females can use to assess age.

Courtship Display

Courtship displays are varied in actions and intensity but usually include ruffling of head, neck, chest, and back feathers, bowing, and song. Often the wings are lifted into what is known as the song spread; the tail may also be spread and pulled up or down. The display may or may not end with a bill wipe. This display is used to court females as well as to display toward other males. These displays can be terrestrial or arboreal. The displays last only for about 4–5 seconds, but may be repeated many times.

Mating System

A great deal of variation has been reported for Brown-headed Cowbird mating systems. Ankney and Scott (1982) claimed that the mating system was poorly understood, and varying reports resulted from observer bias. It is unclear why they believed that variations were due to observer bias, particularly when they pointed out that sex ratio also varies. One might expect that mating systems vary with sex ratios and different sets of ecological conditions. Methods of assessing the mating system have also varied from casual observations to more sophisticated radiotelemetry and observations of copulations in marked populations.

Methods of Assessing Mating Systems. In addition to using different behaviors to evaluate the mating system of Brown-headed Cowbirds, observations have been made at different times of day, which makes comparison diffi-

Table 7.3

Methods and Behaviors Used to Determine the Mating System of Brown-headed Cowbirds

Study	Time of Observation	Marked Individuals	Mate Guarding	Following	Consortships (Courtship)	Copulations	Observed Together
Wild populations							
Ankney and Scott 1982	a.m. feedlot						✓
Darley 1982	unspecified	✓	✓		✓		✓
Dufty 1982a	a.m. & p.m.	✓		✓			✓
Elliott 1980	unspecified	✓				✓	
Laskey 1950	a.m. & p.m.	✓	✓			✓	
Teather and Robertson 1986	a.m. & p.m.	✓		✓	✓		✓
Yokel 1986, 1989a	a.m.	✓		✓	✓	✓	✓
Captive studies							
Darley 1978	a.m.	✓	✓				
Eastzer et al. 1985	a.m.	✓	✓		✓	✓	
Rothstein et al. 1986	a.m. & p.m.	✓		✓	✓		
West et al. 1981		✓		✓	✓	✓	

cult (table 7.3). Although most copulations occur during the morning, data may be biased if observations are not also made in the afternoon. Dominant males tend to relax their positions in afternoons while they are foraging, which could allow afternoon copulations by subordinate males. Time of observations within the breeding season is also important because there may be a higher degree of disorder earlier in the season, and some individuals may switch mates during the course of the breeding season (Yokel 1986).

Mate guarding has often been used synonymously with following, but distinctions should be made because the two behaviors may be used in different contexts. During mate guarding, an individual quickly positions himself/herself between the one being guarded and the intruder and attempts to remain between the two. Yokel (1989a) suggested that following may not necessarily imply mate guarding but, instead, may be an opportunity for males to demonstrate their quality to females. He pointed out that in Darley's (1982) and Teather and Robertson's (1986) studies, following also occurred between nonmated males and females. Mate guarding has been considered as evidence of monogamy in numerous studies (Laskey 1950; Darley 1978, 1982). However, mate guarding may also hint that the mating system tends toward polygyny, or at least that excess males may opportunistically obtain copulations. Teather and Robertson (1986) found that when males and females were seen together, males followed females on 83% of departures, whereas females followed males on only 13.8% of departures. Similarly, Dufty (1982a) found males followed females on 86% of departures, whereas females followed males on only 16% of departures.

Yokel (1986) viewed consortship as an interaction involving courtship behavior. A distinction must be made between this and an association during foraging or other maintenance behavior. However, consortships do not necessarily indicate that a pair is mated or that the system is monogamous, especially in the beginning of the breeding season. Many investigators have used the number of times individuals are seen together to assume mating relationships, but these observations might be best used only as additional support to more concrete evidence, such as copulations, following, and guarding. Ideally, copulations of marked birds should also be witnessed, but these have been used in only a few studies.

Variation in Mating Systems. Most studies suggest a monogamous or mostly monogamous mating system (table 7.4). Reports of mating systems within captive populations also vary, but this may be a consequence of different study methods and conditions. For example, the sex ratio in captive popu-

Table 7.4
Location, Mating System, Sex Ratio, and Habitat of Brown-headed Cowbirds

Location of Study	Mating System	Sex Ratio	Female Territorial	Average Home Range (ha)	Habitat	Source
Wild populations						
California: SE	M	2.2:1	no		Riparian/desert scrub	Yokel 1989a
California: SE	M	2.0:1	no		Riparian/sage, open conifer	Yokel 1989a
California: SE	M	6-3:1–2.5:1 [a]	no		Riparian	Yokel 1986
New York	M	1.6:1	yes	20.4	Deciduous forest; old fields	Dufty 1982a
Ontario	MM & P				Feedlots	Ankney and Scott 1982
Ontario	MM & P	1.5:1	yes	0.4–25.0	University campus	Darley 1982
Ontario	MM & P	1.3:1	no	9.89		Teather and Robertson 1986
New York	M & PA					Friedmann 1929b
Tennessee	M				Suburban landscape	Laskey 1950
California	P				Open oak woodland	Payne 1973b
Kansas	Pr	1.1:1	no	"large"	Prairie	Elliott 1980
California: Sierras	No pairing	3:1–6:1		68	Open coniferous forest	Rothstein et al. 1980, 1984
Captive populations						
	M	1.7:1				Darley 1978
	M	1:1				Rothstein et al. 1986
	P					West et al. 1981
	Pr	2.1:1				Eastzer et al. 1985

[a] Sex ratio was 1.1:1 among censused birds and 6.3:1 among trapped birds.
M = monogamous; P = polygynous; Pr = promiscuous; MM & P = mostly monogamous with some polygyny or bigamy; M & PA = monogamous and polyandrous.

lations (and wild populations) may have a dramatic effect on the mating system. Eastzer et al. (1985) observed 29 males and 14 females in a captive situation and reported promiscuous relationships. Rothstein et al. (1986) reported monogamy in a captive population of equal sex ratio. They suggested that this situation should have favored polygyny; however, in a population of equal sex ratios, all individuals theoretically have the opportunity to mate, and this could favor monogamy. Rothstein et al. (1986) suggested that captive studies can produce rigorous results if used in a comparative manner between populations. As an example, they pointed out that birds taken from different populations will display the same tendencies in captivity because the introduced artifacts are likely to be similar. However, if one introduces sex ratios that are different from those found in the local wild population, this may alter behavior and produce results that are neither applicable to nature nor comparable to other captive studies. The best way to avoid this problem is to duplicate the sex ratio of the population the birds came from.

Assortative Mating. It is not known if mates are selected based upon subspecies in the wild. However, under captive situations, assortative mating appears to occur based upon social experience (Freeberg 1996). The high degree of apparent free breeding between and among subspecies in the zones of contact suggests that assortative mating does not occur as frequently in the wild as it does in captive populations.

Factors That May Affect the Mating System. The mating system of Brown-headed Cowbirds may be influenced by sex ratio, habitat, home range, territoriality, host density, and cowbird density. Ankney and Scott (1982) suggested that males try to pair, but some are unable to do so because of the skewed sex ratio; unpaired males are successful at sneaking copulations, and casual observations of marked or unmarked birds could mislead researchers.

Variation in the mating system of Brown-headed Cowbirds has often been attributed to differences in habitat, host density, cowbird density, and home range (table 7.4). Elliott (1980) noted that the northeastern United States generally has a lower density of cowbirds and a higher density and diversity of host species as compared with midwestern prairie habitat, where he observed a promiscuous pattern. The home ranges of three males in midwestern prairie habitat were essentially nonoverlapping; a female's home range extended through the home ranges of two males, implying that mate guarding or following did not occur, or that these behaviors were ineffective. He suggested that in a prairie habitat, where host density is l(w, cowbirds may have to search larger areas for host nests; therefore, defense of territories is not feasible, and large female ranges weaken the ability of males to guard females. Contrasting with the northeast, in prairie habitat no distinctions could be determined between breeding and feeding areas. Also, cattle grazed on the prairie, creating a moving resource. Rothstein et al. (1984) found no evidence of pair bonds in the eastern Sierra Nevada, where there are large expanses and long commuting distances between breeding and foraging sites.

What Do Females Have to Gain from Males? Females control the resource (host nests) that is probably the primary selective pressure that defines the mating system. Female Brown-headed Cowbirds may look for a couple of qualities in males: (1) desirable traits he can pass on to her offspring, and (2) how he can increase her reproductive success. Although females in southeastern California were courted by numerous males (0–14), they copulated with only one male, clearly indicating female mate choice (Yokel and Rothstein 1991). However, Teather and Robertson (1986) suggested that because males follow females, males maintain the pair bond. Yokel and Rothstein (1991) believed that dominance is used as a cue to assess genetic quality, and females may not choose mates until they have witnessed numerous contests. But if all females look for is genetic quality, why do all females not mate with the best male? Yokel and Rothstein (1991) suggested that if females spend too much time looking for the one best male, it would divert them from consuming activities such as nest searching and foraging. Additionally, the territoriality of females may prevent them from coming into contact with the one best male.

Females may have little to gain by being faithful to a male. If he successfully guards her, this is a good trait to pass on to her sons; however, the traits of a successful sneaker might be equally good. Although females may benefit from males driving away other pestering males while she forages, guarding males may incur a cost while she searches for nests. For monogamy to evolve, the cost of accepting a male's constant presence may need to be less than the cost of repelling him. However, females do commonly drive males away while they search for nests. Courtship by males appeared to Yokel and Rothstein (1991) as a form of harassment to females, but they found no evidence that males protected their mated females from harassment by other males. Males do not provide food or nesting resources to females, and there is no evidence that they provide protection from predators; therefore, Yokel and Rothstein (1991) concluded that females may gain nothing more than genes from males.

Differences in habitat, cowbird population size, host densities, sex ratio, and feasibility of mate guarding may partially explain differences in mating systems of Brown-headed Cowbirds. Lower competition (fewer cowbirds) and more abundant resources (host nests) in the northeast may make the defense of smaller territories profitable. Higher densities of cowbirds might decrease organization within dominance hierarchies, thereby making promiscuity the likely mating system (Yokel 1989a). Yokel (1986) pointed out that monogamy may appear to be an unlikely mating system for cowbirds because there is no readily apparent adaptive value for females, and even though they have ample opportunities to mate with other males, they usually do not.

Wittenberger and Tilson (1980) compiled five hypotheses to explain the evolution of monogamy in birds:

1. "Monogamy should evolve when male parental care is both nonshareable and indispensable to female reproductive success."
2. "Monogamy should evolve in territorial species if pairing with an unavailable unmated male is always better than pairing with an already mated male."
3. "Monogamy should evolve when males are less successful with two mates than with one."
4. "Monogamy should evolve even though the polygyny threshold is exceeded if aggression by mated females prevents males from acquiring additional mates."
5. "Monogamy should evolve in nonterritorial species when the majority of males can reproduce most successfully by defending exclusive access to a single female."

The first three hypotheses are not applicable to Brown-headed Cowbirds, which provide no parental care and are not always territorial. The fourth hypothesis is not supported because females generally do not guard or follow their mates. The fact that Yokel (1989a) demonstrated female choice sheds doubt on the fifth hypothesis.

Another factor that may influence the mating system of Brown-headed Cowbirds, but to my knowledge has not been considered, is the potential problem of inbreeding and recognition of close kin. Brown-headed Cowbirds are site faithful, and both males and females return to their natal breeding grounds, at least in some areas (Ortega and Ortega, unpubl. data). Recognition of close kin may be difficult because individuals are raised in the absence of parents and siblings. Theoretically, an individual may have several to many siblings and parents returning to the same location to breed year after year. It is not known if individuals avoid inbreeding, and if they do, what mechanisms are employed in the recognition or avoidance process. If recognition is by some obscure means (comparing traits of others with their own), a female may select a mate with very different traits from her own and remain mated with him rather than risk mating with closely related males. Alternatively, if females are not able to recognize close kin, promiscuity might spread the risks of mating with close kin. Several investigators have noticed a high percentage of inviable Brown-headed Cowbird eggs compared with host eggs, and perhaps these eggs are the product of closely related parents. It is not known whether both males and females return to their natal breeding grounds throughout their range and whether this may be associated with differences in mating systems.

It is clear that the mating system of Brown-headed Cowbirds is flexible, probably as a consequence of different ecological pressures, and by pigeonholing the mating system, we may be overlooking valuable information (Mock 1983), such as which individuals females avoid mating with. For example, Yokel (1989a) pointed out that females had ample opportunities to mate with other males yet usually did not, but who were these other males? More work needs to be conducted with consistent methods and expanded questions throughout the cowbirds' range before we can understand variation in the mating system and why it appears to be predominantly monogamous.

Migration

Brown-headed Cowbirds are more migratory than other cowbirds, generally migrating between wintering and breeding grounds. Friedmann (1929b) depicted the advance of spring migration for *ater* beginning 1 March from northern Mexico and the southern United States, reaching their most northern destination by 10 April. Lowther (1993) calculated this movement as about 40 km per day. Friedmann (1929b) suggested that temporal migration patterns of *artemisiae* are contradictory and inappropriate to chart. Migration of *obscurus* is also not well understood as they winter throughout much of their breeding range.

Brown-headed Cowbirds tend to arrive on their breeding grounds later than related species and depart earlier. Cowbirds may depart from their breeding grounds abruptly and before hosts have noticeably declined in late July or early August (Rothstein et al. 1980; pers. obs.). A general trend exists for males to arrive earlier than females, vagrants to arrive before residents, and older males to arrive before younger males (Friedmann 1929b; Darley 1982, 1983). Less is apparently known regarding sex and age classes leaving breeding grounds. In southwestern Colorado, ASY males depart first, followed by SY males; females and hatch-year birds remain for a week or two beyond when the majority of males leave; only an occasional adult male lags behind. Rothstein et al. (1980) reported a similar trend, with adult males leaving first.

Populations that winter in the same areas disperse widely to breeding grounds; likewise, individuals that share breeding grounds disperse widely to wintering grounds. Brown-headed Cowbirds also winter within much of their breeding range, and some individuals move very little between wintering and breeding grounds. Individuals banded during winter in Texas were recovered during subsequent breeding seasons from coast to coast and as far north as Ontario (table 7.5). Additionally, individuals do not always winter in the same location every year, and some move to other areas within a winter. Individuals banded during the breeding season and captured in Texas during the winter show a pattern of widespread distribution from breeding grounds to common wintering grounds (Coon and Arnold 1977).

In general, during late summer and fall many small flocks of Brown-headed Cowbirds form and move southward. These flocks join other flocks as migration progresses southward, and the larger units continue to move south as weather and feeding conditions dictate. Coon and Arnold (1977) found that wintering roosts of Brown-headed Cowbirds increased after every major cold front.

Rothstein et al. (1980) suggested that cowbirds leave their breeding grounds earlier at higher altitudes. Birds banded in the Sacramento Valley, California, during fall and winter, were recovered during spring and summer in Oregon, Washington, and British Columbia, and throughout the year in California, suggesting an altitudinal migration (Crase et al. 1972). Crase et al. (1972) also suggested that among Brown-headed Cowbirds that winter in California, *artemisiae* individuals migrate north to breed, whereas *obscurus* individuals remain in California. However, the museum specimens I have inspected from the Pacific Northwest suggest free interbreeding between *artemisiae* and *obscurus* on the west slope of the Cascades; most specimens are closer to *obscurus* in size. On Mandarte Island, off Vancouver, British Columbia, Smith and Arcese (1994) reported only yellow-flanged nestlings, indicative of *obscurus*.

Table 7.5

Number and Location of Brown-headed Cowbirds Recaptured on Breeding Grounds That Were Banded in Texas during Winter 1932–1968 and in Brazos County, Texas, 1969–1974

Recovery Location	Following Breeding Season 15 April–15 Aug.	Subsequent Breeding Seasons 15 April–15 Aug.	Same Winter or Spring	Subsequent Winters, Fall or Spring
Alabama	1			
Arkansas	13	9	3	
California	1			
Illinois	3	5	1	
Indiana	1	1		
Iowa		1		
Kansas	12	3		
Kentucky	2	3		
Louisiana			2	8
Manitoba			1	
Michigan	4	4		
Minnesota	1			
Mississippi		1		
Missouri	2	3		1
New York	1			
Ohio	1	2		
Oklahoma	4	2	6	6
Ontario		3		
Pennsylvania	2			
Tennessee	5	1		1
Texas	3		13	
Virginia				1
Wisconsin	1	2		1

Source: Coon and Arnold 1977.

Site Fidelity and Homing

Brown-headed Cowbirds appear to be more site faithful to their breeding grounds than to their wintering grounds. In Dolbeer's (1982) study, the average distances between breeding season banding locations and breeding season recovery locations of Brown-headed Cowbirds were 44 km ($n = 169$) for females and 95 km ($n = 68$) for males, whereas the average distances between winter banding location and winter recovery location were 244 km ($n = 39$) for females and 288 km ($n = 85$) for males. Arnold and Johnson (1983) banded over 75,000 cowbirds and reported fidelity to wintering areas. Many others have observed high rates of return to breeding sites among years (Laskey 1944; Kennard 1978; Darley 1982; Dufour and Weatherhead 1991; Smith and Arcese 1994). In southwest Colorado, J. Ortega and I have found a return rate of 16.8% ($n = 804$) for males and 4.2% ($n = 260$) for females. Brown-headed Cowbirds have a tendency to enter decoy traps over and over again, and trapping can reveal whether individuals remain in a particular area during certain seasons.[5] Burtt and Giltz (1970) reported that 11.5%–12.7% of the individuals they banded in Ohio reentered their trap.

Homing is also strong among Brown-headed Cowbirds (Gillespie 1930; Fox 1940; Neff 1943). For example, a Brown-headed Cowbird released 172 km north of the original trapping site returned two days later; the same bird taken 296 km away returned after a week (Fox 1940). Of 87 Brown-headed Cowbirds released 6–13 km from a banding station, 60 (69%) returned, and of 60 birds released 21–177 km from the banding station, 19 (31.7%) returned (Neff 1943).

Site fidelity may have several advantages and disadvantages. Familiarity with an area and other members of the population may facilitate breeding success and increase chances of obtaining higher status within dominance hierarchies. For populations that display territorial behavior, familiarity with an area can also facilitate establishment of territories in subsequent years. Dufty (1982a) found the territories of some returning birds to be similar to prior breeding seasons. Darley (1982) found a high rate of return by males to the same study area, but only 5 of 12 (41.7%) that returned to the same area in the next year mated successfully, indicating more to success than being familiar with an area. For females, familiarity may also enhance nest searches because many hosts also return to the same territories to breed. Disadvantages of both males and females returning to their natal breeding sites may include recognition and avoidance of mating with close kin.

Wintering Populations

Brown-headed Cowbirds gather in large aggregations with various other icterines and European Starlings during the winter months. Estimated roosting flocks average approximately nine million in Louisiana; one flock was estimated at an astounding 38 million (Ortego 1993). In Oklahoma, Goddard (1971) reported that Brown-headed Cowbirds constituted 38.7% of blackbird roosting flocks in the winter of 1964–1965, but in the following winter, they constituted only 12.3% of the roosting population. Johnson et al. (1980) estimated that Brown-headed Cowbirds constituted roughly 40% of a winter roost with an estimated 900,000 individuals.

Mortality, Survival Rates, and Longevity

High female mortality probably occurs after the first breeding season; Darley (1971) found that the male:female sex ratio among SY birds was 1.3:1, whereas the sex ratio among ASY birds was 2:1. Data from Arnold and Johnson (1983) suggest that higher female mortality does not occur during the winter. They found females had a greater survival rate (63%) than males (53%) in Texas during winter and attributed this to possible reduced competition for food due to their smaller size. Johnson et al. (1980) found in a Texas winter roost that ASY males were overrepresented among dead individuals and that within an age class among males, dead males had significantly longer wings, suggesting that as in Arnold and Johnson's (1983) study, mortality is related to depletion of food resources during winter. The time from banding to last recovery was six years for females and nine years for males (Fankhauser 1971).

Higher mortality of females has been explained by physiological demands of egg production (Darley 1971) and risks of host aggression. Arnold and Johnson's (1983) findings of lower mortality among females during winter lends support to this hypothesis; however, our knowledge of mortality is limited.

Diseases, Abnormalities, and Parasites

Stewart (1963) reported numerous external abnormalities of trapped Brown-headed Cowbirds, including bill deformities, cysts in the submucosa of cloacas, broken and deformed toes and legs, pox and mite nodules, scaly legs, claw hooked around tarsus, recurved claws, punctured eyeballs, and partial albinism. These abnormalities occurred in 3.9% of each sex (Stewart 1963). Cooper et al. (1973) recorded 20 species of internal parasites from the examination of 166 Brown-headed Cowbirds. Weatherhead and Bennett (1992) found that hematozoa parasitism depended on sex, age and year, but there were no differences between decoy-trapped and mist-netted birds. Percentage of infected individuals varied between 15% and 18% for SY males (n = 273), 26% and 31% for ASY males (n = 244), 14% and 31% for adult females (n = 311), and 0% and 26% for hatch-year birds (n = 136); in general, parasites did not appear to have a major effect on the health of infected individuals in terms of weight and recaptures in subsequent years (Weatherhead and Bennett 1992).

Vocalizations

Nestling and fledgling Brown-headed Cowbirds produce a loud warbling begging vocalization that is highly variable in length and frequency and is easily distinguishable to the human ear and on a sonogram (Broughton et al. 1987). Apparently, the intensity of begging vocalizations in comparison with most hosts serves as a superstimulus, eliciting guaranteed feeding responses by host parents. The wide frequency range of the begging vocalization may help trigger responses from a large number of hosts (Broughton et al. 1987). Juvenile Brown-headed Cowbirds also have a chirp call, variously described as *seer* (Friedmann 1929b), *yip* (Nice 1939), and *chip* or *chit* (Woodward 1983).

Distinctions between songs and calls can be somewhat arbitrary; however, male Brown-headed Cowbirds have two distinctly different vocalizations that functionally may be considered songs: perch songs and flight whistles. Other vocalizations used by males are the single syllable flight call and the "kek" note, emitted during copulation (Dufty and McChrystal 1992) or during foraging by both females and males (King and West 1983). Chatters are used primarily by females, but males occasionally chatter (Dufty 1981; Rothstein et al. 1988; pers. obs.). Males and females both may growl when agitated.

Chatter. Females emit a chatter which is used in a wide variety of contexts to communicate with both males and other females. The chatter is a loud harsh rattle (*chak-a-chak-a-chak-a-chak-a-chak*) and is the primary loud vocalization utilized by females. When chatter is emitted with a head-up display, it is usually interpreted as an aggressive display (Burnell and Rothstein

1994). When chatter is used in response to distant males, it is probably used to reveal location and is generally not considered aggressive (Burnell and Rothstein 1994). Although there is no apparent dialect in female chatter, individuals have recognizable chatters (Burnell and Rothstein 1994).

Perch Song. Perch songs consist of three phrases (King and West 1977; Johnsrude et al. 1994) and are described as a series of guttural bubbling notes—*glug glug gleeee* (Friedmann 1929b). As the term implies, perch song is almost invariably emitted while the male is perched. Perch songs are typically accompanied by a song spread or courtship.

Perch songs are directed at both males and females and are usually used to communicate less than a meter from the recipient (Rothstein et al. 1988). They function both inter- and intrasexually for courtship, in male-male interactions, and establishment of dominance hierarchies. Perch songs elicit copulation posture, or lordosis, in receptive females, and lordosis is used by researchers in evaluating song recognition by females and song potency.

Females can recognize phrase order, assuming lordosis more often with normal song than with altered phrases (Ratcliffe and Weisman 1987). Johnsrude et al. (1994), however, found that reversed order did not change the potency of the song, but females responded more to altered songs in which the initial or ending phrase was played first and responded the least to the middle phrase played first. Females also responded more to two-phrase songs than one-phrase songs. Females recognize and respond to precursors of song, suggesting that juvenile males possess species-specific signals early in song development.

Cowbirds are exposed to a wide range of host songs during their nestling phase; it might, therefore, be presumed that song development is largely innate and resistant to environmental influences (Mayr 1974). However, recent studies have shown a definite learned component to song, and exposure to adult song is necessary for normal perch song to develop (King et al. 1980). Ontogenetic forms of Brown-headed Cowbird perch song development include (1) subsong: variable, unstructured, low amplitude, unrecognizable to humans as cowbirdlike; (2) plastic song: poorly articulated and variable notes and whistles of both high and low frequencies; (3) formatted song: beginning with low frequency clustered notes and ending in high frequency whistles, with definite ordering and timing of the elements; and (4) stereotyped song: possessing consistent form and element. Brown-headed Cowbirds are, therefore, probably able to modify their songs according to different social stimuli.

In a captive setting, King and West (1988) determined that wild-captured young males sang 47% plastic song, 37% formatted song, and 16% stereo-typed song. They cautioned that their sample size was perhaps too small to come to definitive conclusions, but they suggested that the behavior of the audience probably has an effect on the ontogeny of perch song. Both perch song and flight whistles are learned and vary geographically. Females appear to have a role in the development of geographically distinct perch song (King and West 1983).

Flight Whistles. Flight whistles are frequently used by males just before or while taking flight, in flight, just before landing, and as another cow-bird approaches. Flight whistles may also be emitted while perched without approaching conspecifics, but this occurs less frequently. O'Loghlen and Rothstein (1995) claimed that flight whistles are usually not emitted when conspecifics are nearby. However, males that are perched with conspecifics may give the flight whistle just before they take flight. Flight whistles and single-syllable flight calls may also signal alarm (Rothstein et al. 1988). They contain two to three notes between 3 and 10 kHz (Rothstein and Fleischer 1987a; Rothstein et al. 1988) and are nearly pure tones or rapid frequency sweeps (Dufty 1988; Dufty and McChrystal 1992) with or without buzzes combined with the tones or trills (Rothstein et al. 1988). The function is probably for long distance communication, and males may be signalling to conspecifics whose location is unknown. Both males and females respond to flight whistles. Males respond by giving the flight whistle and flying toward the source; females respond with loud chatter with or without approach (Rothstein and Fleischer 1987b).

Flight whistles are also emitted immediately before copulation (Roth-stein et al. 1988), but this seems to be more prevalent in western cowbirds than eastern cowbirds. Often males and females are together less than five seconds before copulation occurs; whether flight whistles are emitted as a signal that copulation is about to take place or whether they are emitted in response to landing is unknown. Rothstein et al. (1988) suggested that flight whistles might function as a final check of individual identity. In general, flight whistles and single-syllable flight calls may be given less frequently by eastern males than by western males; however, the perceived difference may not be so much geographic as a difference among study methods. The studies of eastern cowbirds were conducted in captivity, where long distance communication of flight whistles is not relevant, and the studies of western cowbirds were conducted in the wild, where long distance communication is important.

Playbacks of flight whistles attract other cowbirds. In playback experiments, males responded to speakers with agitation and some head-up displays (Rothstein et al. 1988). Most emitted flight whistles or single-syllable flight calls as they approached the speaker. Rothstein et al. (1988) assumed that males responded because flight whistles represented new individuals. Females responded with less frequency (perhaps because they had other priorities), but some flew to the speakers and chattered in response.

Flight whistles are sometimes incorporated into male song (Dufty 1988; pers. obs.). Dufty suggested that this may represent a mistake in learning. However, the functional distinction between flight whistles and songs is not entirely clear. King and West (1977) reported that flight whistle playbacks did not elicit copulatory responses from captive females, but Rothstein (pers. comm. in Dufty and McChrystal 1992) observed that estradiol-implanted females responded to playbacks of flight whistles with copulatory postures. Flight whistles are also emitted more often than song just prior to copulation and during copulation attempts (Rothstein et al. 1988; Dufty and McChrystal 1992).

As with song, flight whistles have distinct dialects in some localities and are often separated by what may be unsuitable habitat for Brown-headed Cowbirds; where dialects do overlap, hybrid whistles are common (Rothstein and Fleischer 1987b). Females may show discrimination against males with foreign flight whistle dialects, and most SY males have flight whistles vocally distinct from ASY males (O'Loghlen and Rothstein 1995).[6] In some areas, the poor mating success of SY males might reflect female preference for the local dialect, and females may be able to assess male age and quality by the dialect males use. O'Loghlen and Rothstein (1995) suggested that most females probably do not acquire song preference until their first breeding season. It would be of tremendous interest to know the age and natal area of the test females; perhaps many of the females that responded to foreign dialects were SY females who had not yet learned the local dialect. Alternatively, females may learn their local dialect as juveniles and discriminate against males producing that dialect, thereby reducing the risks of inbreeding. If this were the case, females who returned to their natal area would respond more to foreign dialects, while females from other areas would prefer the local dialect.

Rothstein and Fleischer (1987b) hypothesized that flight whistle dialects are maintained because they are difficult to copy immediately (i.e., they must be learned) and, therefore, honestly signal social status and mate quality to females. This suggests that males who do not produce the local dialect are either newcomers or SY birds and may be inferior mate choices compared to ASY residents. Cheaters are unlikely because by the time they learn the whistle they are closer in quality to older residents. Rothstein and Fleisher found that ASY males produced significantly more flight whistles than SY males, and stated that this difference is a clear predictor of the honest convergence hypothesis. However, these whistles were elicited by walking up to the birds, and Rothstein et al. (1988:75) stated, "FWs [flight whistles] and SSs [single-syllable flight calls] are given as apparent alarm call. If males are approached by a human or another potential predator, they often give SSs and FWs . . . as they show signs of fright (looking up, cessation of feeding)." SY males may not emit alarm calls for reasons other than the fact that they do not conform to the local dialect, and just as dialect is learned, proper context might be learned. Nevertheless, this hypothesis was supported by the finding that SY males rarely sing the correct local dialect. Later, O'Loghlen and Rothstein (1995) provided evidence in further support of this hypothesis. Females appear to prefer males who produce the local flight whistle dialect.

The dialect of flight whistles is learned sometime after the juvenile period (Rothstein and Fleischer 1987a). In captivity, a male Brown-headed Cowbird even learned the more complicated flight whistle of a Shiny Cowbird kept in an adjacent cage. In the Mammoth area of California, O'Loghlen and Rothstein (1993) found that 9 of 11 (81.8%) ASY males produced complete versions of the local dialect, whereas only 2 of 16 (12.5%) SY males did.

The development of proper local dialect in some areas contributes to what is probably a myriad of mechanisms by which females select mates and by which males obtain status in dominance hierarchies. O'Loghlen and Rothstein (1993) found that testosterone-implanted SY males showed none of the vocal inhibitions of wild males (Rothstein and Fleischer 1987a), producing an equivalent number of flight whistles as ASY males.[7] In contrast, SY males in the wild appear to be more inhibited, perhaps because flight whistles elicit aggression from other males (Rothstein et al. 1988). Enlarged testes may, therefore, be important in song development and overcoming, to some degree, inhibitions caused by the presence of older males.

Single-Syllable Flight Call. Single-syllable flight calls are similar to flight whistles in both function and quality but consist of only one note with few frequency modulations; they are the loudest of male vocalizations (Rothstein et al. 1988). Single-syllable flight calls are rarely incorporated into flight whistles, but in some areas, elements of the flight whistle are indistinguishable from single-syllable flight calls (Rothstein et al. 1988).

Head-down or Invitation to Preen Displays

Many observations of head-down displays, earlier known as "invitation to preen displays," have been reported on both captive and wild Brown-headed Cowbirds. The display involves one bird approaching another with head bowed down and feathers of head and neck fluffed out. The bill is pointed straight down or toward the body, and the body is often held low or in a crouched position. After assuming this position, the displaying bird often moves closer to the recipient and may stop only a few centimeters away or stop when the head rests on the recipient. Responses that displaying birds receive from both other species and conspecifics vary greatly and include fleeing, no response, aggressive bill-tilt displays, supplanting the displayer, pecks, and gentle preening. Other cowbirds sometimes respond with the same display. Rothstein (1977b) reported that mutual displays may be maintained in a nearly motionless state for 5–10 minutes and that they occur between the same or different sexes. Nonresponsive recipients sometimes receive harsh treatment from displaying cowbirds, and occasionally recipients will mount the displayer in attempted copulation (Selander and La Rue 1961).

Selander and LaRue (1961) believed that the display serves an appeasement function, whereas Rothstein (1971a, 1977b) suggested that the display is aggressive and used to dominate other birds. More recently, Scott and Grumstrup-Scott (1983) suggested that the display functions in obtaining food, minimizing roosting energetics, and establishing flock order. Alternatively, Dow (1968) proposed that there may be no function. Many studies of this display were conducted on captive populations. Perhaps the display is given with higher frequency in captivity, but Rothstein (1977b) argued that these displays were so frequent in captivity (523 displays to conspecifics in 730 minutes) that they are unlikely not to have a function. Scott and Grumstrup-Scott (1983) observed 288 head-down displays in 59.2 hours of observation in a wild populations in Ohio and Pennsylvania.

Cowbirds may display to conspecifics or other species depending on flock composition and captive-wild status. In captivity, where cowbirds have ample interspecific opportunities, they display more to other species; in the absence of these opportunities, they display readily to each other. In wild populations, conspecific displays are observed less frequently (Rothstein 1971a; 1977b). In Scott and Grumstrup-Scott's (1983) study, only 4 of 288 (1.4%) displays were directed toward other cowbirds in wild populations, whereas in their captive populations, conspecific head-down displays were observed frequently (475 displays in 13.3 hours of observation).

Hypotheses Regarding Functions of the Display. Numerous hypotheses have been proposed regarding the functions of head-down displays.

1. The display is thought to reduce aggression by hosts, allowing the parasite to remain on the host's territory and approach the nest (Selander and La Rue 1961). In this hypothesis, it is assumed that the display is an invitation to preen. To directly support this hypothesis, it would be necessary to observe the behavior on a host's territory. The few reported interactions of cowbird behavior at host nests do not mention head-down displays. However, Scott and Grumstrup-Scott (1983) believed that the results of their study and results from Stevenson (in Scott and Grumstrup-Scott 1983), who found that displaying cowbirds were able to move closer to recipients than nondisplaying cowbirds, supported this hypothesis. As additional supporting evidence, they cited the fact that 25 of 288 displays observed in the wild resulted in preening. However, this is only 8.7% and might be considered as evidence against this hypothesis. Rothstein (1977b) felt that the term "invitation to preen display" was somewhat of a misnomer because these displays often do not result in preening; he subsequently referred to it as the "head-down display."

   Of 212 displays observed in Pennsylvania, 209 (98.6%) were directed toward House Sparrows; although the flocks studied consisted of only House Sparrows and Brown-headed Cowbirds, it nevertheless sheds some doubt that the display functions as an appeasement to hosts (Scott and Grumstrup-Scott 1983). Not only were these displays observed outside the breeding season, but House Sparrows are rarely hosts. Similarly, Hunter (1994) observed cowbirds directing the head-down display to Rusty Blackbirds (*Euphagus carolinus*) outside the breeding season; as with House Sparrows, Rusty Blackbirds are not common hosts.

2. The display enables cowbirds to assess the status or fighting potential of individuals (Rothstein 1980). For conspecific displays, this hypothesis is supported by dominant males displaying more often than subordinates (Rothstein 1977b; Scott and Grumstrup-Scott 1983). This does not necessarily imply that the behavior is aggressive; it could be an appeasement in order to get close to a subordinate (Scott and Grumstrup-Scott 1983). Interestingly, the head-down display is the opposite of the stereotypic

head-up or bill-point displays, which are clearly aggressive. Without the display, a subordinate might be supplanted by a dominant bird approaching so closely. Rothstein considered mutual displays to be tests of will, but Scott and Grumstrup-Scott observed that females and subordinate males perform this display with dominant males; so instead, the mutual display may serve to indicate acceptance of the close proximity and may replace a fleeing response by the recipient.

3. The display is deceitful mimicry, that is, it looks like an appeasement display, but is really aggressive (Rothstein 1980). Scott and Grumstrup-Scott (1983) suggested that the display cannot be mimicry because there is no model to mimic; that is, there are no similar behaviors of recipient species.

4. The display functions to obtain food, minimize roosting energetics, and establish flock order (Scott and Grumstrup-Scott 1983). This hypothesis consolidates portions of the first two hypotheses and may be most relevant for flocks. Individual spacing appears to play an important role, and it may not be coincidental that head-down displays are observed with higher frequency in captive populations than in the wild. These displays in the wild are most frequently observed during winter when close proximity may reduce energy demands. In captive populations and in winter flocks, individual spacing is closer than more loosely formed flocks during the breeding season, and head-down displays may function to reduce disorganization or fleeing responses, thereby allowing closer proximity. Therefore, I propose a fifth hypothesis that encompasses more fully the role of individual spacing.

5. The head-down display functions as a cue to the recipient that subsequent interactions will not be aggressive, allowing closer proximity to individuals in situations where space is at a premium (captive populations) or when close proximity is beneficial (winter roosting flocks).

Plumage and Molt

The medium gray natal down is replaced with a juvenile plumage similar to females but with more streaking and more yellow on the chin, breast, and belly. Postjuvenile molt begins as early as 34 days (Brackbill 1976). After the postjuvenile molt, females are virtually indistinguishable from older females, but males may be distinguished from older males by either retention of gray juvenile feathers in the underwing coverts or by browner primaries.

Brown-headed Cowbirds have an annual postnuptial molt that can occur from mid-July through October. Although there is overlap among individuals regarding the timing of breeding and molt (Lowther 1993), for the individual, breeding and molt are temporally separate (Payne 1973b). Friedmann (1929b) suggested that feathers on the head are molted in an irregular fashion, but Baird (1958) found a definite order in the progression of molt, which he described in detail for the entire body and head.

The *obscurus* race also has a partial prealternate molt, usually involving only the head, but occasionally the back (Payne 1973b). Apparently the *ater* race has no corresponding molt, and it is unknown whether *artemisiae* has a prealternate molt (Lowther 1993). Payne (1973b) suggested that partial spring molt in *obscurus* and lack of it in *ater* relates to differences in habitat.

Breeding Season

The breeding season of Brown-headed Cowbirds varies regionally and with latitude. The earliest eggs are laid in early to mid-April (Norris 1947), and the latest are laid in early August. More typically, egg laying occurs from early May through June. In many areas, several hosts begin breeding a few weeks earlier than cowbirds, and several hosts continue breeding after the cowbirds' breeding season ends. The peak of the cowbirds' breeding season peaks with that of most of their preferred hosts. Some hosts may avoid parasitism by breeding outside the season of the cowbird. American Goldfinches are known to be late breeders and, thus, largely avoid parasitism, but individuals breeding early may be parasitized (Berger 1948; Mariani et al. 1993).

Neither male (Payne 1965; Dufty and Wingfield 1986a) nor female (Payne 1965) reproduction is synchronized among individuals. Testicular regression in some individuals was observed as early as mid-June; in other individuals regression was not observed until July (Dufty and Wingfield 1986a). In Michigan, 65.5% (n = 29) of females had eggs in their oviducts between 20 June and 5 July, whereas between 6 and 10 July, only 33.3% (n = 6) of females had eggs in their oviducts (Payne 1965).

Increased daylight is probably responsible for onset of the breeding season. Females housed with 17 hours of artificial light for a month from 8 December to 8 January had ovaries enlarged to approximately the size of wild females' ovaries in April, whereas females kept under a natural light regime during the same period did not have enlarged ovaries (Payne 1967a).

Males kept under the same increased light regime also came into breeding condition, but once the refractory period begins, increased photoperiod has little to no apparent effect.

Circulating hormones in Brown-headed Cowbirds differ through the breeding season from nesting passerine species. In monogamous nesting species, luteinizing hormones (LH) generally are highest upon arrival of females and drop during incubation, with another peak if double clutching occurs. In male Brown-headed Cowbirds, however, plasma LH shows a plateau pattern from late April through late June (Dufty and Wingfield 1986a). Mean testosterone levels in Dufty and Wingfield's (1986a) study showed a double peak in mid-April and again in mid-May; by mid-June, the levels declined to prebreeding season levels. They attributed early overall decline to (1) testosterone levels of SY males declining earlier than ASY males, thereby depressing the overall mean; and (2) social relationships being well established by the time testosterone levels begin to decline, with less overt aggression required. High levels of testosterone apparently interfere with parental behavior; Pied Flycatchers (Ficedula hypoleuca) given exogenous testosterone treatments continued to be aggressive and court females, but they provided no parental care (Silverin 1980). Emancipated from parental care, male cowbirds can maintain high levels of testosterone, enhancing their reproductive success. This results in a rather extended period of male-male agonistic interactions. Dufty (1989) found that males implanted with testosterone during the breeding season suffered more injuries than nonimplanted males, and the return rate the following year was lower among implanted males than nonimplanted males.

Eggs and Egg Laying

Brown-headed Cowbird eggs are round with a shell considerably thicker (0.0125 mm, $n = 25$) relative to its volume than other (nonparasitic) icterines (Picman 1989). The background color varies from nearly white to various shades of blue or bluish green. Invariably, the eggs are maculated with light to dark brown spots. Spots are usually uniformly distributed throughout the egg, but they can be more concentrated on one end, or the spots may be larger at one end. Sometimes, concentrated brown pigments will form a dense circle or large blotch around one (or rarely both) ends. Egg size differs slightly depending on subspecies, with the smaller obscurus generally laying smaller eggs, but there is overlap in size between the subspecies (Schönwetter 1983).

Egg Variation. Brown-headed Cowbird egg morphology varies considerably among individuals, but variation within individuals is far less. Some studies have shown little intraindividual variation in egg morphology; morphological features among eggs of one individual are recognizable to humans to the extent that pairs of eggs can be matched to an individual with a reasonable degree of accuracy (Dufty 1983). Roby and Jackson (1992), however, noted that eggs from individuals vary considerably with regards to size, shape, and coloration. Mills (1987) found that among Brown-headed Cowbird eggs collected in Quebec and Ontario, the largest egg was 1.63 times larger than the smallest egg.

Compared to nonparasitic passerines, Brown-headed Cowbird eggs are small relative to female size, but the proportional composition of Brown-headed Cowbird eggs does not differ significantly from that of other altricial species, and in general, quality is not sacrificed for the large quantity (Ankney and Johnson 1985). It is not known how much egg weights are under genetic control, but Ankney and Johnson (1985) did find considerable variation in egg weights within the same female and suggested that such variation may represent fluctuations in dietary resources.

Rapid Laying. Average laying time of Brown-headed Cowbirds is only 41.0 (± 5.6, 3.5–119, $n = 28$) seconds (Sealy et al. 1995). This compares to laying time of other passerines ranging from 20.7 to 103 minutes and six other icterines ranging from 21.5 to 53.4 minutes (Sealy et al. 1995).

Time of Day Egg Laying Occurs. Brown-headed Cowbirds typically lay their eggs at dawn (Hann 1937; Harrison 1952, 1973). Neudorf and Sealy (1994) found the mean time of egg laying to be 35 ± 2.4 (SE, $n = 7$) minutes before sunrise in Manitoba. Norris (1947) reported laying to occur an average of 14.2 ($n = 5$) minutes before sunrise in Pennsylvania, and Scott (1991) reported laying to occur on average 9.1 ± 2.5 (SE, $n = 21$) minutes before sunrise in various localities. Captive birds, however, may retain an egg until nearly noon.

Number of Eggs Laid. An egg in the oviduct can usually be felt in the abdominal region by researchers. Fleischer et al. (1987) assessed the accuracy of these external examinations by retaining females overnight; the researchers were able to correctly assess an egg in the oviduct in 92% ($n = 36$) of the females tested.

Individuals have been known to lay as many as 77 eggs in a single breeding season (Holford and Roby 1993). Scott and Ankney (1980) reported an average of 40 eggs per breeding season, and Walkinshaw (1949) reported a cowbird (assumed to be a single individual) laying up to 25 eggs. Fleischer et al. (1987) reported an average of 35.2 eggs per season for SY females and 49.8 eggs per season for ASY females. The number of eggs laid by individuals varies greatly, and some females may not lay eggs at all (Payne 1965; Scott 1978). Holford and Roby (1993) found that the average fecundity of the same 12 females increased from 15.4 eggs in their first breeding season to 26.3 eggs in their second breeding season. Younger females may not lay as many eggs as older females for a couple reasons: (1) they often arrive on breeding grounds later than older females, and (2) younger birds may not be as adept at obtaining nutrients, such as calcium, necessary for egg production.

Clutch Size. Nice (1949) believed that Brown-headed Cowbirds lay their eggs in clutches of three or four with intervals of 6–12 days. Payne (1965) suggested that eggs were laid in clutches ranging from one to six eggs with periods of rest in between. Scott (1978) estimated clutch size to be 4.4 with an interval of 1.4–2.3 days, and Scott and Ankney (1983) suggested that the laying cycle is similar to domestic chickens. The results of more recent studies suggest that eggs are not laid in sets or clutches (Jackson and Roby 1992; Roby and Jackson 1992; Holford and Roby 1993). Females may lay one egg per day (Payne 1965; Jackson and Roby 1992) for up to 67 consecutive days (Holford and Roby 1993).

Limiting Factors of Egg Laying. Payne (1965) suggested that the apparent variability in the number of eggs laid by Brown-headed Cowbirds may reflect differences in the ability of individuals to find available host nests, with those finding fewer nests laying fewer eggs. This implies that females are able to regulate the number of eggs according to how many nests they find. However, the number of eggs laid may not be limited to the number of nests available, as Brown-headed Cowbirds in captivity commonly lay eggs without nests (Fleischer et al. 1987). Holford and Roby (1993) observed no difference in the fecundity of females held in captivity with and without nests available in their pens.

Dietary Requirements. Rothstein (1976a) believed that the energy demanded by large egg production is great enough that parasitizing rejector species is maladaptive. Ankney and Scott (1980) disagreed, and King (1973a) suggested that egg production is only a slight energetic demand. In a study on changes in nutrient reserves and diet, Ankney and Scott (1980) noted that the diet of females changed from primarily seeds prior to egg laying to insects during the egg-laying period, and back to seeds after egg laying, and nearly all females had mollusc shell in their gizzards during the egg-laying period. Intestine length increased during egg laying, which suggests an increase in food consumption (Ankney and Scott 1980). Females generally do not use protein reserves but probably use some calcium and fat reserves. Ankney and Scott (1980) felt that declines in fat reserves reflected a reduced need for large reserves rather than energy deficits due to egg laying. Although there may not be immediately recognizable costs to high egg production, as Fleischer et al. (1987) pointed out, reproductive organs may suffer from extensive use, and indirect costs of extra foraging needed probably include increased risks of predation. Later, Holford and Roby (1993) conducted experiments on wild birds captured and kept in captivity. They found that calcium-denied females averaged 61% fewer eggs than the females supplied with unlimited oyster shell.

Incubation. Incubation is somewhat variable but typically shorter than the incubation period of most hosts. Ten days is the usual incubation period for Brown-headed Cowbirds (Friedmann 1929b; pers. obs.). Norris (1947) and Hann (1937) reported a rather long average incubation period of 11.6 days.

Young

Brown-headed Cowbird nestlings may not have any specific adaptations for a brood parasitic mode of life, at least not to the extent that some cuckoos and honeyguides do. They generally do not eject host eggs or nestlings from nests (but see Dearborn 1996). Instead, when they become sole occupants of host nests, it is usually because they hatch earlier and develop more rapidly. Cowbird nestlings may become so large they fall through the bottom of some host nests (Brandt 1947). Cowbird nestlings have such enthusiastic begging responses, they may attract predators (Hofslund 1957). Brown-headed Cowbirds are often fed at a higher average feeding rate per hour than host young (Woodward 1983).

Brown-headed Cowbird nestlings do not mimic vocalizations of their hosts (Broughton et al. 1987), but fledglings may possess some subtle behavior, in addition to their enthusiastic begging behavior, that enables them to coax a wide variety of hosts to feed them once they leave the nest. They can be indiscriminate in this behavior, begging from any bird. Woodward (1983) observed fledglings being fed by different individuals of their host species and suggested that fledglings are indiscriminate in terms of soliciting care from individuals of their host species, but he never observed fledglings begging from other species even though opportunities to do so were abundant (but see Neff 1926; Klein and Rosenberg 1986).

Dolan and Wright (1984) reported possible pecking of host eggs and nestlings by Brown-headed Cowbird nestlings. In a study of Western Flycatchers (*Empidonax difficilis*), they observed damaged host eggs a couple days prior to hatching (but after the cowbird hatched), the severity of which was correlated with the parasite nestling age. They found no similar damage in nonparasitized Western Flycatcher nests. They also suggested that cowbirds may peck host nestlings and attributed the lack of previous reports to the long incubation period of Western Flycatchers (14–15 days, Harrison 1978).

Based on observations of 21 Brown-headed Cowbirds, Woodward (1983) described five stages of development from leaving the nest to independence.

1. Leaving the nest. This varies from 8 to 13 days (usually 10–11) and occurs any time of day.
2. Inactive. During this first 2–3 day postfledging stage, fledglings perch quietly and move mainly for comfort and self-maintenance.
3. Active. This period is usually from 3 to 13 days after fledging. During this period, fledglings fly 2–9 times an hour, sometimes following the host parent but usually independent of the host. Most behaviors necessary for independence have been observed during this period: flying, walking, exploratory pecking, preening, capturing prey, and self feeding.
4. Superactive. This stage is usually from 12 to 23 days postfledging but shortly before independence. Fledglings follow their host parents, relentlessly begging with characteristic loud warbling and wing quivering.
5. Independence. This occurs from 16 to 28 days postfledging (25–39 days old) and usually coincides with foster parents ignoring cowbird fledglings.

If more than one Brown-headed Cowbird fledges from the same nest, they may not associate with each other, as do fledglings of some host species (Woodward 1983), but cowbirds do associate with conspecifics by the end of the breeding season. Whether recognition of conspecifics is innate or learned is not understood as very little work has been conducted in this area. In a study using mounted Brown-headed Cowbirds, Red-winged Blackbirds, and Common Grackles (*Quisculus quiscula*), Graham and Middleton (1989) tested 26 Brown-headed Cowbirds at the ages of 25 days and again at 35 days for which mounts they preferred to associate with (perch near). The 25-day-old birds showed no preferential association with the cowbird mount, but some 35-day-old birds did. Graham and Middleton also conducted auditory tests with tapes of the same species to determine if 22- and 32-day-old cowbirds preferentially associated with speakers. Neither 22- nor 32-day-old cowbirds showed preferential associations with speakers broadcasting vocalizations of the three test species. Interpretation of such studies is tenuous at best; their data suggested that either the young cowbirds did not recognize cowbird vocalizations or they had no motivation to orient themselves toward the speakers. Perhaps the cowbirds also had no motivation to approach cowbird mounts; cowbirds at that age are still very much dependent on their hosts for care, and approaching adult cowbirds would probably result in no immediate rewards. Also, these birds were raised by humans and may not have performed behaviors of wild birds. It would be very difficult to conduct a similar study on a wild population; however, conducting a similar study with live birds that can provide rewards to young cowbirds when they are approached might at least reveal motivation behind associations.

Association between Adults and Young

Associations between nestling Brown-headed Cowbirds and adults are almost unheard of, but Bonwell (1895) observed a female Brown-headed Cowbird regularly feed a cowbird nestling in a Rose-breasted Grosbeak (*Pheucticus ludovicianus*) nest in his Nebraska orchard. The cowbird fed only the nestling cowbird and pecked the grosbeak nestlings on their heads when they begged for food. Bonwell remarked that he had never heard of a similar occurrence, and I am unaware of any other reports. However, Fletcher (1925) reported a female cowbird repeatedly feeding a fledgling. Whether or not cowbirds recognize and associate with independent offspring is not known. Using eight walk-in traps and DNA fingerprinting, Hann and Fleischer (1995) reported that females and their own juvenile offspring associated together. However, of 11 pair-wise trappings on common foraging grounds, only four pairs were probably related, which might be expected on a common feeding ground (see Mason 1987 for statistical judgement of pair formations).

How Brown-headed Cowbirds Find and Select Nests

Brown-headed Cowbirds appear to be adept at finding host nests. When they find nests that are well concealed from every angle, it is likely that they cue in on host activity. Norman and Robertson (1975) identified three types of nest searches used by female cowbirds: (1) silent watching of host activity, (2) silent systematic searching, and (3) noisily searching, apparently attempting to flush hosts. Thompson and Gottfried (1976, 1981) suggested that cowbirds may rely heavily on host activity to locate suitable nests, and several experimental studies have supported this.

Among the large number of predation studies in which artificial nests with eggs were used, few incidences of parasitism have been reported. For example, out of 1,532 artificial nests, only one cowbird egg was found (Jobin and Picman 1994). In some cases, nests have been set out specifically to look for incidences of parasitism; Laskey (1950) found none, and Thompson and Gottfried (1976) found no parasitism among 20 nests with Japanese Quail (*Coturnix coturnix*) eggs. Lowther (1979) found slightly higher rates of parasitism by adding one egg per day to real nests (6.1%, n = 33 nests). Two major differences between Lowther's study and other studies were that he added one egg per day, and perhaps more importantly, he used Barn Swallow and House Sparrow eggs, which are considerably smaller than Japanese Quail eggs and closer to the size of the eggs of typical cowbird hosts. In Pennsylvania, Yahner and DeLong (1992) also used (artificial) eggs that were the size of host eggs (15 × 22 mm) adding one per day for three days. Although many of the 300 nests were disturbed, presumably by predators, none were parasitized.

Brown-headed Cowbird eggs are often found in inactive nests (Buech 1982; Freeman et al. 1990; Ortega 1991), which may appear to shed some doubt on cowbirds using host activity to locate available nests. However, the history of many of these inactive nests is not known, and in some cases, the cowbird may have watched the nest owner construct the nest, or the host may have abandoned the nest after it was parasitized and before host eggs were laid. Captive Brown-headed Cowbirds appear to prefer nests with smaller eggs than their own, nests with only two host eggs, and nests that have not been previously parasitized (King 1979), but they apparently do not select certain nests based on the size of host eggs within a given host species (Murphy 1986).

Brown-headed Cowbirds watch nest building activities by potential hosts with unusual interest. They may watch the host for extended periods and inspect the nest numerous times during the construction process, suggesting more interest than merely obtaining information on location and construction stage. Mayfield (1961b) suggested that this behavior represents vestiges of a proprietary interest. Others have reported continued interest during the incubation phase or nest construction phase (Hann 1937; Norris 1947;

Payne 1973b; Norman and Robertson 1975), and Balda and Carothers (1968) reported behavior by two cowbirds in two separate incidences that could be interpreted as protective behavior of parasitized nests.

Hosts

The list of species known to be parasitized by Brown-headed Cowbirds increased from 157 in 1929 to 220 in 1985 (Friedmann 1929b; Friedmann and Kiff 1985). There are two primary reasons why the list has grown so dramatically: (1) far more research has been conducted on passerines in general, and (2) Brown-headed Cowbirds have come into contact with some new hosts in their range expansion. To this already expanded list may be added Marsh Wrens (Picman 1986), although parasitism on Marsh Wrens is undoubtedly rare. Also, several hosts have recently been split into two species. The Northern Oriole (*Icterus galbula*) was split into two species: Baltimore and Bullock's Orioles; the Brown Towhee (*Pipilo fuscus*) was split into two species: Canyon Towhee (*P. fuscus*) and California Towhee (*P. crissalis*); the Rufous-sided Towhee (*P. erythrophthalmus*) was split into two species: Eastern Towhee (*P. erythrophthalmus*) and Spotted Towhee (*P. maculatus*), and the Solitary Vireo (*Vireo solitarius*) was split into three species: Blue-headed Vireo (*V. solitarius*), Plumbeous Vireo (*V. plumbeus*), and Cassin's Vireo (*V. cassinii*), bringing the total up to 226 species (appendix B).

The list of 226 species known to have been parasitized is impressive, and this (previous total of 220) number of victims is commonly cited in papers and articles about Brown-headed Cowbirds to emphasize the widespread effect that they have on avian communities. However, many species on the list can be considered accidentally or rarely parasitized, and several species that have been listed as hosts "known to raise cowbirds" were presumed to do so based on observations of birds attending cowbirds outside the nest. Additionally, 17 species are ejectors, which may not be a final total because many hosts and potential hosts have not been tested for ejector-acceptance status. The list is, therefore, inflated and misleading, particularly when it is used to emphasize how many species are affected by cowbirds. It has become more like a bird-watcher's life list in which the object is to tick off as many species as possible. At this point, any species added to the list could probably be considered accidental. It is, nevertheless, useful to catalog newly parasitized species for future reference, but a list of hosts excluding accidentals and rarely parasitized species would be more reflective of the true impact Brown-headed Cowbirds have. For example, the fact that a cowbird egg was found in a Ferruginous Hawk *(Buteo regalis)* nest is irrelevant to our overall knowledge of the biology of either species. While the list is impressive and serves a purpose, it would be far more informative to also provide information on how many species actually raise Brown-headed Cowbirds.

Friedmann and Kiff (1985) compiled a list of 144 species known or presumed to have raised Brown-headed Cowbirds. With splitting Northern Orioles, Rufous-sided Towhees, Brown Towhees, and Solitary Vireos, the list increases to 149. Of these, eight are ejectors, and another nine are rarely parasitized (but well studied enough for us to know that parasitism is truly low). Subtracting ejector species and accidentals from the list of hosts known to raise Brown-headed Cowbirds whittles the list down to 132 biological hosts (species regularly parasitized that also raise cowbirds). I have retained Baltimore and Bullock's Orioles in the list even though they are ejector species because several innovative studies have demonstrated that orioles are not efficient at ejecting cowbird eggs, and cowbird parasitism has more of an impact on their nesting success than other ejector species that have been known on rare occasion to raise cowbirds. Of these 132 biological hosts, five hosts have only been presumed to have raised cowbirds because they have been observed feeding fledglings. Very little is known of some species known to raise cowbirds, particularly many wood warblers that nest in the canopy of forests.

Parasitism rates vary markedly both regionally and for different host species (appendix C). Mayfield (1965b) tentatively identified categories of parasitism rates: < 10% light, 10–30% moderate, and > 30% heavy. I would consider < 10% low, 10–20 low to moderate, 20–40 moderate, 40–50 moderate to heavy, and > 50 heavy.

Multiple Parasitism and Dump Nesting

Brown-headed Cowbird egg distribution and incidences of multiple parasitism have been the focus of many studies. Patterns of egg distribution can provide important information on territorial behavior, nest selection, and the number of eggs laid by individuals. Geographic variation exists in reports of both multiple parasitism and dump nesting. In Manitoba, Briskie et al. (1990) found a low level of multiple parasitism and no dump nesting. Zimmerman (1983), on the other hand, found multiple parasitism to be the rule among Dickcissel *(Spiza americana)* nests in prairie habitat and common in old field habitats in Kansas. Similarly, in Illinois, multiple parasitism appears to be the rule rather than the exception (Robinson 1992; Trine 1993).

Overall, about 70% of parasitized nests contain one Brown-headed Cowbird egg (appendix D).

To determine if eggs are laid at random, the observed distribution is compared to a Poisson series (Mayfield 1965a). Mayfield suggested that cowbird egg distribution among some hosts is random, and deviations from randomness are due to sampling errors. There are numerous complications with analyses of cowbird egg distribution: (1) Nests that were abandoned are not as easily found by researchers due to the lack of activity. Although these nests will sometimes be multiply parasitized, it is more likely to affect the category of one cowbird egg (Mayfield 1965a). (2) If the breeding season of host and parasite are not synchronized, the nests outside of the cowbird's breeding season are outside the realm of selection by the cowbird, and the category of 0 eggs will be overrepresented (Mayfield 1965a). (3) Ejector species' (e.g., Gray Catbird) effects on the analyses are unknown.

In some areas, Brown-headed Cowbirds even appear to avoid laying eggs in nests that are already parasitized (Ortega et al. 1994; Smith and Arcese 1994). Brown-headed Cowbirds in captivity also avoided laying in nests already parasitized (King 1979). Elliott (1977) found a nonrandom distribution of cowbird eggs in Kansas among pooled nests of Grasshopper Sparrows (*Ammodramus savannarum*), Dickcissels, and Eastern Meadowlarks (*Sturnella magna*). He suggested that this implies a certain amount of avoidance and that some nests may not have been discovered by the cowbirds. Orians et al. (1989) found a random distribution of cowbird eggs and suggested there was no evidence that cowbirds avoided already parasitized nests.

Reports on whether one or more females are responsible for multiply parasitized nests vary and may reflect geographic variation. Elliott (1977) felt that in his Kansas study, most multiply parasitized nests resulted from more than one female because most eggs in multiply parasitized nests were morphologically dissimilar. Relatively few reports exist for parasite success in singly versus multiply parasitized nests, perhaps because the emphasis is often on effects on hosts rather than effects on parasites. Smith and Arcese (1994) reported lower cowbird success in doubly parasitized nests (43%, $n = 30$ eggs) than singly parasitized nests (53%, $n = 294$ eggs).

Reasons for multiple parasitism are not clear; perhaps a cowbird could not find a suitable nest, or perhaps some nests are particularly attractive. I know of no study that specifically compares nest site characteristics of multiply parasitized nests with nests containing only one cowbird egg. Smith and Arcese (1994) suggested that multiple parasitism might be more prevalent in disturbed habitat where densities of cowbirds are greater and where they cannot monopolize host nests. The low multiple parasitism on Mandarte Island (off Vancouver) might be explained by the typical presence of only one laying cowbird (Smith and Arcese 1994). Multiple parasitism and dump nests are very common in Robinson's Illinois study, where cowbirds are abundant. Dump nests generally seem to be more common in some regions than others (Stoner 1919; Hann 1947; Byers 1950; Nickell 1955; Payne 1965; Klaas 1975; Scott 1977; Linz and Bolin 1982; Hatch 1983; Lowther 1977, 1983; Robinson 1992), but most dump nests are in the areas with the highest cowbird abundance.

Effects of Parasitism on Hosts

In general, Brown-headed Cowbird parasitism has a negative effect on the nesting success of hosts (table 7.6). Reasons for reduced success have been variously attributed to cowbirds removing eggs, the abandonment of parasitized clutches, costs of egg ejection, reduced hatching success for hosts, cowbirds gaining an advantage with a shorter incubation period, interference with heat exchange during incubation, and competition from larger cowbirds.

Egg Removal, Egg Puncturing, and Egg Eating. Brown-headed Cowbirds have been known to puncture eggs or to remove them from nests whole (Hann 1941; Norris 1947; Hofslund 1957), but not all nests that they remove eggs from are parasitized (Blincoe 1935). Most observations have been of females, but males will also remove and eat host eggs (Sealy 1994). Grzybowski et al. (1986) claimed that nestling cowbirds ejected vireo host eggs, but they provided no substantiating details regarding whether this behavior was actually observed or presumed. Later, Dearborn (1996) caught on videotape a nestling cowbird ejecting its Indigo Bunting host mate and suggested that such events may be more common than we assume. Vireo eggs hatch out considerably later than cowbirds (fig. 7.3), and the host nestlings may die very rapidly. Therefore, even a day between visits may yield the results of a host egg on one visit and no egg or nestling on the next visit. Mayfield (1961b) claimed that Brown-headed Cowbirds never entirely empty a nest of host eggs.

Brown-headed Cowbirds remove host eggs at all times of day: dawn (Hann 1941; Harrison 1952; Sealy 1994), midmorning (Sealy 1992, 1994), midday (Mengel and Jenkinson 1970), and late afternoon or sunset (Blincoe 1935; Olson 1943; Earley 1991; Sealy 1992). Sealy (1992) reported that approximately 13% of Yellow Warbler eggs were removed the day before parasitism occurred, 21–54% were removed on the same day of parasitism, and at least 33% were removed one or more days after parasitism.

Reports of host egg removal by Brown-headed Cowbirds vary. Sealy (1992) suggested that one-third of Yellow Warbler eggs are removed from parasitized nests; cowbirds may remove a Song Sparrow egg from about one-half (Smith 1981) to two-thirds (Smith and Arcese 1994) of parasitized nests. Slightly more than one Dark-eyed Junco (*Junco hyemalis*) egg is lost for every cowbird egg laid (Wolf 1987). For many hosts, the main cause of nesting loss among parasitized nests is clutch reduction (Zimmerman 1983). In other studies, few host eggs are removed from parasitized nests (Burgman and Picman 1989). The variation in egg removal patterns is not well understood. King (1979) found that cowbirds kept in captivity remove small host eggs more often than large eggs. Sealy (1992) suggested that some females remove host eggs all the time, while some females do not remove eggs at all, which, as he pointed out, would also explain annual differences that are sometimes observed on the same host species.

Many investigators have reported lower clutch size among parasitized nests than nonparasitized nests (Mayfield 1961a; Klaas 1975; Middleton 1977; Scott 1977; Finch 1983; Zimmerman 1983; Fleischer 1986; Wolf 1987; Petit 1991; Sealy 1992), but these figures can be confounded by other breeding characteristics. For example, Smith (1981) found that older Song Sparrows were parasitized significantly more often than yearlings. In contrast, I suspected that younger Red-winged Blackbirds were parasitized more often than older birds (Ortega 1991). Clutch size of younger birds is apt to be smaller than that of adult birds (Smith 1981), and therefore, unless eggs are known to have disappeared, one cannot safely assume that smaller clutch size in parasitized nests is necessarily a consequence of host egg removal by cowbirds. However, when eggs are known to disappear from parasitized nests, one can assume that cowbirds removed them, particularly if it is also established that eggs do not disappear from nonparasitized nests.

Brown-headed Cowbirds sometimes eat host eggs. Scott et al. (1992) reported that of 18 observations, only seven eggs (38.9%) were eaten, and concluded that cowbirds do not eat a large proportion of eggs they remove, which is surprising given the nutritional value. Scott et al. (1992) calculated that half the energy required to produce a cowbird egg could be obtained by consuming a 2-g host egg. Eggs that were consumed were in various stages of incubation, including eggs that were nearly ready to hatch (Scott et al. 1992), but Sealy (1992) reported that most (80%) removals of Yellow Warbler eggs were during the egg-laying phase. Data on the failure to eat

Table 7.6

Success (Sample Size in Parentheses) of Nests Parasitized and Nests Not Parasitized by the Brown-headed Cowbird

Host	Mean Hosts Fledged per Nest		Mean Hosts Fledged per Egg		Percent Successful Nests		Year	Location	Source
	Nonpara.	Para.	Nonpara.	Para.	Nonpara.	Para.			
Acadian Flycatcher	1.68 (50)	0.37 (16)	0.61 (138)	0.20 (30)	70.0 (50)	25.0 (16)	1945	Michigan	Walkinshaw 1961
Willow Flycatcher					56.3 (16)	18.2 (11)	1985–86	Colorado	Sedgwick and Knopf 1988
Eastern Phoebe	2.00 (3)				100.0 (3)	15.0 (13)	1987	California	Harris 1991
	3.77 (9)	0.75 (4)	0.89 (38)	0.17 (18)			1951	Michigan	Berger 1951
	2.25 (61)	0.14 (7)					1973–74	Kansas	Hill 1976a
			0.46 (1203)	0.09 (323)			1962–65	Kansas	Klaas 1975
Say's Phoebe	1.42 (36)	0.00 (1)					1973–74	Kansas	Hill 1976a
Horned Lark	0.25 (16)	0.00 (14)					1973–74	Kansas	Hill 1976a
Veery	2.50 (2)[1]	2.00 (4)[1]					1944–45	Pennsylvania	Norris 1947
Wood Thrush	3.33 (3)[1]	2.00 (1)[1]					1944–45	Pennsylvania	Norris 1947
Brown Thrasher			0.17 (18)	0.33 (6)			1974–75	Kansas	Elliott 1978
Bell's Vireo	0.50 (14)	0.00 (1)					1973–74	Kansas	Hill 1976a
	0.00 (1)	0.00 (1)					1973–74	Kansas	Hill 1976a
Plumbeous Vireo	2.35 (37)	0.50 (30)	0.69 (126)	18.1 (83)			1984–86	Colorado	Marvil and Cruz 1989
Yellow Warbler	3.81 (21)[1]	2.67 (9)[1]						Ontario	Burgham and Picman 1989
			0.80 (64)	0.44 (45)			1900	Ontario	Clark and Robertson 1981
Prothonotary Warbler	2.28 (75)	0.88 (43)			64.0 (75)	46.5 (43)	1983–84	Manitoba	Weatherhead 1989
	3.56 (48)[1]	1.90 (20)[1]					1983–84	Manitoba	Weatherhead 1989
	4.5 (42)[1]	3.5 (14)[1] early					1985	Tennessee	Petit 1991
	3.9 (42)[1]	2.9 (7)[1] late					1985	Tennessee	Petit 1991
Common Yellowthroat	0.60 (10)	0.11 (9)				13.6 (22)	1948–51	Michigan	Hofslund 1957
							1938	Michigan	Stewart 1953

Host	Mean Hosts Fledged per Nest		Mean Hosts Fledged per Egg		Percent Successful Nests		Year	Location	Source
	Nonpara.	Para.	Nonpara.	Para.	Nonpara.	Para.			
Northern Cardinal	0.44 (9)	0.00 (7)	0.19 (21)	0.00 (11)	25% (60)	36% (53)	1993–94	Ohio	Eckerle and Breitwisch 1997
	3.00 (1)[1]	1.50 (2)[1]					1951	Michigan	Berger 1951
	1.50 (4)	0.40 (5)					1944–45	Pennsylvania	Norris 1947
Indigo Bunting	0.42 (12)	0.00 (13)	0.75 (8)	0.29 (7)			1951	Michigan	Berger 1951
Dickcissel	3.7 (9)	1.8 (41)	0.25 (8)	0.11 (47)			1974–75	Kansas	Elliott 1978
	3.2 (54)	2.0 (57)	0.52 (63)	0.18 (419) prairie habitat			1973–74	Kansas	Hill 1976a
			0.31 (550)	0.13 (891) old field			1965–79	Kansas	Zimmerman 1983
							1965–79	Kansas	Zimmerman 1983
Eastern Towhee	3.50 (2)	2.14 (7)					1944–45	Pennsylvania	Norris 1947
Clay-colored Sparrow	2.38 (13)	0.60 (5)	0.49 (90)	0.23 (24)	49% (27)	31% (13)	1976	Minnesota	Buech 1982
Field Sparrow	2.76 (21)[1]	2.50 (2)[1]	0.66 (47)	0.19 (16)			1951	Michigan	Berger 1951
	1.88 (17)	0.64 (14)					1944–45	Pennsylvania	Norris 1947
Lark Sparrow	0.39 (49)	0.00 (13)	0.49 (65)	0.20 (46)	59% (17)	29% (14)	1968	Oklahoma	Newman 1970
Lark Bunting	3.2 (?)	1.3 (?)					1973–74	Kansas	Hill 1976a
Savannah Sparrow	0.50 (8)	0.50 (4)					1983–84	Minnesota	Johnson and Temple 1986
Grasshopper Sparrow	1.55 (18)	0.64 (28)	0.19 (32)	0.06 (17)	33% (9)	22% (9)	1974–75	Kansas	Elliott 1978
	3.33 (6)[1]	2.00 (3)[1]					1973–74	Kansas	Hill 1976a
Song Sparrow	2.7 (12)	1.5 (8)	0.41 (69)	0.20 (92)			1951	Michigan	Berger 1951
	0.99 (580)	1.13 (47)					1944–45	Pennsylvania	Norris 1947
	2.82 (203)[1]	2.04 (26)[1]	0.75 (179)	0.79 (98)			1975–79	Vancouver	Smith 1981
Dark-eyed Junco							1983–84	Virginia	Wolf 1987
Red-winged Blackbird	3.7 (654)[1]	2.8 (55)[1]	0.26 (2186)	0.32 (166)	35% (580)	55% (47)	1984–86	Colorado	Ortega 1991
			0.74 (774)	0.57 (93)			1984–86	Colorado	Ortega 1991
			0.73 (654)	0.65 (55)			1977–83	Washington	Røskaft et al. 1990
	0.1 (63)	0.4 (17)					1973–74	Kansas	Hill 1976a

Table 7.6 continued

Host	Mean Hosts Fledged per Nest		Mean Hosts Fledged per Egg		Percent Successful Nests		Year	Location	Source
	Nonpara.	Para.	Nonpara.	Para.	Nonpara.	Para.			
	1.24 (118)	1.33 (57)[2]			40.0 (118)	45.6 (57)	1983–84	Manitoba	Weatherhead 1989
	1.24 (118)	1.23 (13)[3]					1983–84	Manitoba	Weatherhead 1989
	3.15 (46)	2.92 (24)[2]					1983–84	Manitoba	Weatherhead 1989
	3.15 (46)	2.28 (7)[3]					1983–84	Manitoba	Weatherhead 1989
Eastern Meadowlark			0.24 (49)	0.12 (58)			1974–75	Kansas	Elliott 1978
Western Meadowlark	0.69 (16)	0.00 (1)					1973–74	Kansas	Hill 1976a
Yellow-headed Blackbird	1.1 (247)	0.9 (22)[4]					1985–86	Colorado	Ortega and Cruz 1988
Orchard Oriole	3.00 (3)	1.50 (2)					1973–74	Kansas	Hill 1976a
Pine Siskin	2.00 (7)	0.19 (16)					1973–74	Kansas	Hill 1976a

[1] Successful nests only.
[2] One cowbird egg present.
[3] More than one cowbird egg present.
[4] From artificially parasitized nests.

Fig. 7.3. Brown-headed Cowbirds typically hatch out first in the nests of Plumbeous Vireos.

removed eggs remains nebulous, but apparently consumption is not the primary reason for cowbirds removing eggs.

Removing Host Young from Nests. Adult females occasionally attack and/or remove host nestlings from both parasitized and nonparasitized nests (Du Bois 1956; Tate 1967; Beane and Alford 1990; Scott and McKinney 1994; Sheppard 1996). In nonparasitized nests, cowbirds may gain an advantage by this predatory behavior only if it causes the potential host to renest. If the host renests, the cowbird may then synchronize her own laying with that of the renesting bird. Arcese et al. (1996) found that in 5 of 19 years, nest failure of Song Sparrows was relatively low and coincided with the absence of cowbirds on their island study site. Cowbirds have not been observed eating nestlings (Scott et al. 1992).

Cost of Rejection

Hosts that reject Brown-headed Cowbird eggs, either through abandonment or ejection, often incur a high cost. This is particularly evident for hosts that abandon parasitized clutches late during the breeding season when there is

Table 7.7

Percent Abandonment (Sample Size in Parentheses) of Nests Unparasitized and Parasitized by Brown-headed Cowbirds

	Non-parasitized		Parasitized	Location	Source
Willow Flycatcher	50	(6)	53.8 (13)	California	Harris 1991
			81.9 (11)	Colorado	Sedgwick and Knopf 1988
Blue-gray Gnatcatcher			39.6 (48)	New Mexico	Goguen and Mathews 1996
Cedar Waxwing			43.3 (104)[1]	Michigan	Rothstein 1976b
			50.0 (16)	Michigan	Rothstein 1976b
Bell's Vireo			60.0 (5)	Illinois	Pitelka and Koestner 1942
Red-eyed Vireo	12	(41)	34 (44)	Ontario	Graham 1988
Yellow Warbler	3.1	(64)	23.8 (42)	Ontario	Clark and Robertson 1981
	1	(146)	23 (110)	Ontario	Graham 1988
			12.7 (236)[2]	Manitoba	Sealy 1995
Prothonotary Warbler	5.9	(136)	19.4 (36)	Tennessee	Petit 1991
Common Yellowthroat			10.5 (38)	Michigan	Hofslund 1957
Northern Cardinal	4	(70)	52 (27)	Ontario	Graham 1988
Dickcissel	0	(18)	16.7 (18)	Kansas	Elliott 1978
Chipping Sparrow	7	(61)	52 (100)	Ontario	Graham 1988
Clay-colored Sparrow			59.1 (22)	Manitoba	Hill and Sealy 1994
Field Sparrow			62.5 (16)	Illinois	Best 1978
Grasshopper Sparrow	22.2	(9)	22.2 (9)	Kansas	Elliott 1978
Song Sparrow	2	(61)	27 (78)	Ontario	Graham 1988
White-crowned Sparrow			18.5 (65)	California	Trail and Baptista 1993
Dark-eyed Junco	0.9	(107)	16.2 (68)	Virginia	Wolf 1987
Red-winged Blackbird	5.2	(580)	17.0 (47)	Colorado	Ortega 1991
Eastern Meadowlark	45.5	(12)	46.4 (28)	Kansas	Elliott 1978

[1] Nests abandoned in response to artificial parasitism.

[2] This does not include 20 parasitized clutches that were buried.

little opportunity to renest and for species that eject parasitic eggs by means of puncture ejection.

Røskaft et al. (1993) found that for Bullock's Orioles, the cost of ejection (0.26 host eggs per ejection) is higher than the cost of raising a cowbird (no cost) when the host brood includes only two nestlings, but the cost of ejection is lower than the cost of raising a cowbird (0.2 and 0.4 oriole chicks per brood) when the host brood is three or four.

Some hosts often abandon their parasitized nests, and among small hosts, this may be the most common defense (table 7.7). In some cases, it may not be the presence of a cowbird egg that elicits the desertion response but rather the reduction of the host's clutch. In a study of Clay-colored Sparrows, Hill and Sealy (1994) found compelling evidence in support of this: in experimentally parasitized nests without clutch reduction no nests were abandoned (n = 48 nests), whereas 66.7% (n = 15 nests) of nests with clutch reduction and no parasitism were abandoned, and 7.1% (n = 15 nests) of nests with both clutch reduction and addition of a cowbird egg were abandoned.

Desertion is profitable when the host renests and raises an unparasitized brood (Trail and Baptista 1993; Goguen and Mathews 1996), and some hosts may be able to assess their options. Petit (1991) found that among Prothonotary Warblers, females mated to males who defended areas with more available nest boxes were more likely to abandon parasitized clutches than females in territories with fewer available nest sites. For Prothonotary Warblers without opportunities to renest, acceptance of cowbird eggs may be adaptive because they can often raise their own (albeit fewer) nestlings along with the cowbird.

Other hosts desert by covering the parasitized clutches with a new nest floor. Many hosts have been reported to bury parasitized clutches (table 7.8); however, only Yellow Warblers respond to parasitized clutches with clutch burial on a consistent basis. Sealy (1995) found that Yellow Warblers were more likely to bury parasitized clutches when they contained fewer host eggs than parasite eggs. For some hosts, it is not entirely clear whether parasitism occurred before the finishing touches had been put on the nest.

Host Size and Growth. Brown-headed Cowbird parasitism generally has greater detrimental effects on hosts weighing less than 20 grams, both because smaller hosts seem more prone to abandon parasitized nests and because they suffer greater competition from larger cowbirds (Best and Stauffer 1980). Smaller species, particularly those with long incubation periods, suffer reduced growth and often raise no young of their own in parasitized nests (Marvil and Cruz 1989; fig. 7.4). Yellow Warblers have nearly equal nesting success and growth in parasitized and nonparasitized nests in southwestern Colorado (figs. 7.5 and 7.6) because they have an unusually short incubation period (10.5 ± 1.3 days, n = 16) and nestling period in southwestern Colorado (8–11 days).

Smith (1981) found that breeding success of Song Sparrows was reduced by 0.5–0.6 young per nest, but the weight of nestling Song Sparrows did not differ between parasitized and nonparasitized nests (Smith and Arcese 1994). Similarly, although parasitized Prothonotary Warbler nests fledged fewer hosts, the presence of a cowbird nestling had no apparent effect on growth of host nestlings (Petit 1991). Wolf (1987) observed that Dark-eyed Juncos were heavier on days six and seven in nonparasitized nests than in

Table 7.8

Brown-headed Cowbird Hosts Reported to Have Buried Cowbird Eggs in Their Nests

Host Species	Number of Eggs or Nests	Source
Acadian Flycatcher		Friedmann 1963
Willow Flycatcher	1/10 parasitized nests	Berger and Parmelee 1952
	2/11 parasitized nests	Sedgwick and Knopf 1988
Eastern Phoebe		Friedmann 1963
Eastern Kingbird		Friedmann 1963
Bushtit		Friedmann 1963
Blue-gray Gnatcatcher		Friedmann 1963
	2/63 parasitized nests	Goguen and Mathews 1996
Gray Catbird		Friedmann 1963
Bell's Vireo	1/24 parasitized nests	Barlow 1962
	1/7 parasitized nests	Mumford 1952
Solitary Vireo spp.		Friedmann 1963
Yellow-throated Vireo		Friedmann 1963
Warbling Vireo	2/14 parasitized nests	Ortega and Ortega, pers. obs.
Red-eyed Vireo		Friedmann 1963
Yellow Warbler	11/12 parasitized nests	Schrantz 1943
	1 five-storied nest	Johnson 1950
	9/11 parasitized nests	Berger 1951
	20/42 parasitized nests	Clark and Robertson 1981
	3/7 parasitized nests	Lowther 1983
	55/110 parasitized nests	Graham 1988
	32/94 parasitized nests	Weatherhead 1989
	23/96 parasitized nests	Briskie et al. 1990
	10/69 experimentally parasitized nests	Briskie et al. 1990
	85/236 parasitized nests	Sealy 1995
Chestnut-sided Warbler		Friedmann 1963
Yellow-rumped Warbler		Friedmann 1963
Prairie Warbler		Friedmann 1963
American Redstart		Friedmann 1963
Prothonotary Warbler		Friedmann 1963
Kentucky Warbler		Friedmann 1963
Common Yellowthroat		Friedmann 1963
Northern Cardinal		Friedmann 1963
Indigo Bunting		Friedmann 1963
Dickcissel		Lowther 1983
Clay-colored Sparrow		Friedmann 1963
Song Sparrow		Friedmann 1963
White-crowned Sparrow		Friedmann 1963
Red-winged Blackbird		Friedmann 1963
Eastern Meadowlark		Friedmann 1963
Hooded Oriole	1 nest	Hardy 1970
Baltimore Oriole		Friedmann 1963
	1/136 total nests	Hobson and Sealy 1987
American Goldfinch		Friedmann 1963; Berger 1948

Fig. 7.4. Brown-headed Cowbird nestlings outcompete Warbling Vireo foster siblings, which almost always results in death of the vireos within a few days of hatching.

parasitized nests, but most reduced success in parasitized nests was due to cowbirds removing eggs. Weatherhead (1989) found that in the beginning of the nestling period, Yellow Warbler nestlings in parasitized nests were consistently lighter than in nonparasitized nests, but the difference disappeared by the seventh day. He found no differences in the growth of Red-winged Blackbird nestlings between parasitized and nonparasitized nests. Eckerle and Breitwisch (1997) found no difference in the growth of Northern Cardinals between parasitized and unparasitized nests, and I found no difference in the growth of Red-winged Blackbirds and Yellow-headed Blackbirds between parasitized and unparasitized nests (Ortega 1991).

Interference with Heat Exchange During Incubation. When cowbird eggs are much larger than host eggs, host eggs may have a low probability of hatching. Zimmerman (1983) found that among Dickcissels, incubation time increased with the number of eggs in a clutch. Hann (1947) calculated, by water displacement, that cowbird eggs had 1.8 times the volume of Ovenbird *(Seiurus aurocapillus)* eggs. He suggested that this, together with host eggs located beneath the cowbird eggs, resulted in reduced heat to host eggs and contributed to reduced hatchability. Hofslund (1957) estimated that the heat received by cowbird eggs prevented two Common Yellowthroat *(Geothlypis trichas)* eggs from hatching per nest, and if a nest was parasitized with

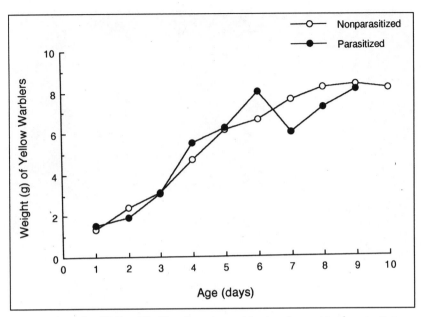

Fig. 7.5. Growth of Yellow Warblers in nonparasitized and parasitized nests. Data from Hesperus, Colorado, 1995 (Ortega and Ortega).

more than one cowbird egg, no host eggs hatched. Klaas (1975) observed decreased hatching success with multiple parasitism among Eastern Phoebes: 80% with one cowbird egg, 65.1% with two cowbird eggs, and 57.1% with three cowbird eggs. Other investigators who have reported lack of development of host eggs in parasitized nests include Middleton (1977) and Mariani et al. (1993) for American Goldfinches, and Petit (1991) for Prothonotary Warblers. In nests of larger hosts, cowbird eggs may not interfere with heat exchange. Lowther (1983) reported successful incubation in a Red-winged Blackbird nest containing three host eggs and five cowbird eggs.

Rothstein (1973) found an interesting association between hatching success and habitat. Cedar Waxwings, American Robins, Eastern Kingbirds, and Black-billed Cuckoos all experienced reduced hatching success in nests in orchards located adjacent to farms compared to nests located away from roads, farms, and other human activity. The reasons for the reduced hatching success were not identified, but some possibilities included pesticides, reduced incubation time, and different-aged individuals occupying different habitats. Rothstein felt the latter two possibilities were unlikely, but he did not rule out the possibility of pesticides having a negative effect on hatch-

Fig. 7.6. Yellow Warblers often successfully fledge in parasitized nests because they have a relatively short incubation period and rapid development.

ing. However, he did not notice any adult or nestling mortality, nor did he observe the eggshell thinning or flaking that often accompanies hatching failure due to pesticides in nonpasserines. These are the habitats that Brown-headed Cowbirds frequently use, and an interesting possibility for future studies exists on potential associations in host egg hatching success between parasitized nests adjacent to farms and parasitized nests in the interior of forests.

Host Suitability

The ability of hosts to raise cowbirds is sometimes expressed in terms of tolerance. Tolerance, as defined by Mayfield (1965b:19), is "the ratio of cowbirds fledged to the number of cowbird eggs laid in the nests of the host." Any host species that fledges more than 20% of cowbird eggs laid is considered tolerant. Overall tolerance is a reflection of how suitable a host is. Intolerant hosts do not necessarily escape parasitism, and some birds that appear to be suitable hosts are infrequently parasitized. Reasons for apparent lack of parasitism on some hosts that seem to be suitable have been investigated in numerous studies. Sometimes, single explanations for lack of parasitism can be readily identified: for example, the host has an ejection response (table 2.1), it provides an unsuitable diet (Middleton 1991; Mariani et al. 1993), or it breeds primarily outside the breeding season of the cowbird (Middleton

1991; Mariani et al. 1993; Brown 1994). Breeding parameters that can be investigated to explain low parasitism include: (1) egg acceptance/rejection status, (2) host breeding season, (3) incubation period, (4) proper incubation by host, (5) host care (feeding rates and diet), (6) nest type, (7) habitat, (8) nest site characteristics, (9) nest attentiveness, (10) host size, (11) host density, and (12) alternative host choices in the community.

Often, definitive conclusions cannot be drawn because a myriad of host breeding parameters that are marginally suitable may be identified, yet no factor by itself can fully explain low rates of parasitism for these hosts. Other birds that rarely serve as hosts are probably not suitable hosts; for example, Mourning Doves (*Zenaida macroura*) are rarely parasitized, yet in general they appear to accept foreign eggs (Holcomb 1968). Differences in feeding behavior might preclude them from being successful hosts. Also, Holcomb (1968) and McNicholl (1968) pointed out that the flimsy platform nest of Mourning Doves may not be suitable for holding a cowbird egg, which is small in comparison to their own; however, if this were a major obstacle, experimental additions of cowbird eggs should reveal disappearance. Indeed, experiments have resulted in some peculiar patterns, such as experimental eggs disappearing long after the standard five-day acceptance criterion. Holcomb and McNicholl also suggested that the sudden flushing of Mourning Doves from their nests may cause lighter cowbird eggs to fall from the nest.

While some potential hosts that are rarely parasitized appear to make good hosts (Eastzer et al. 1980; Ortega and Cruz 1991), other hosts that are moderately or heavily parasitized do not seem to do well at raising cowbirds or their own in parasitized nests (e.g., Abert's Towhee, *Pipilo aberti*, Finch 1983). High rates of parasitism in these studies may be due to local conditions, such as a shortage of available hosts.

Egg Acceptance/Rejection Status. This is one of the most simple parasite constraint parameters to study. Many experimental studies have revealed that low acceptance rates of parasitic eggs is the principal reason for low to zero rates of parasitism (chapter 2).

Host Breeding Season. Some birds may escape heavy parasitism and may escape parasitism entirely because they nest before or after the egg-laying period of Brown-headed Cowbirds. For example, House Finches (*Carpodacus mexicanus*) and Lucy's Warblers (*Vermivora luciae*) in Arizona nest before Brown-headed Cowbirds start laying (Brown 1994), and in at least some areas of Illinois, the egg-laying seasons of Common Grackles and cowbirds barely overlap (Peer and Bollinger 1997). In general, American Goldfinches

breed late, sometimes not even beginning until July or August, thereby es-
caping parasitism to a large degree (Middleton 1991; Mariani et al. 1993).
Individual American Goldfinches that nest early are often parasitized (Mariani et al. 1993), but their late breeding may not be in response to cowbird
parasitism. Instead, it may coincide with late-blooming composites from
which they feed. In addition to the late breeding season, American Goldfinches are poor hosts because they provide an inadequate diet high in seeds;
there are no recorded instances of Brown-headed Cowbirds unquestionably
fledging from American Goldfinch nests.

In Ontario, Northern Cardinals have a long breeding season that extends
through August, and individuals nesting in August experience no cowbird
parasitism. In Scott's (1963) study, rates of cowbird parasitism were considerably reduced after the first week of July; although overall parasitism
was 59.9% in his study, rates of parasitism varied between 71% and 100%
during the peak of the cowbirds' breeding season. Others have also reported
that late-breeding individuals escape cowbird parasitism (Best 1978).

Incubation Period. Host incubation periods are usually not a constraint because Brown-headed Cowbirds have an incubation period that is among
the shortest known. However, several hosts have equally short incubation
periods.

Proper Incubation. Middleton (1977) found low hatching success of Brown-
headed Cowbird eggs in nests of American Goldfinches and suggested that
it might be attributable to infertile eggs at the end of the cowbird's breeding
season. In a cross-fostering experiment, Mills (1988) found low hatching
success of Brown-headed Cowbird eggs (50%, $n = 14$) compared to Tree Swallows (*Tachycineta bicolor*, 95.5%, $n = 66$). Reasons for the comparatively low
hatching success were not apparent, but he suggested that Tree Swallows are
often away from their nests for extended periods during cold weather, and
cowbird eggs may be less cold tolerant. Larger host eggs do not appear to
have the effect of reducing hatching success of Brown-headed Cowbird eggs
(Ortega 1991; Ortega and Cruz 1991) to the same degree that Brown-headed
Cowbird eggs have on some smaller host eggs; however, Peer and Bollinger (1997) suggested that among cowbird eggs cross-fostered to Common
Grackle nests, half did not hatch and appeared insufficiently incubated.[8]

Host Care: Diet and Feeding Rates. Brood parasites are also constrained by
the diet provided by their hosts, and not all hosts provide a sufficient diet
for Brown-headed Cowbirds. American Goldfinches feed their young a diet

high in seeds, and they have not been known to successfully raise a cow-bird (Middleton 1991; Mariani et al. 1993; Friedmann et al. 1977; but see Friedmann 1963). Mariani et al. (1993) observed a cowbird in the nest of an American Goldfinch for ten days, but its development was clearly retarded, and it died in the nest. Mariani et al. (1993) attributed the lack of devel-opment to the content of the diet (highly granivorous) rather than to the quantity of food, as the cowbird was the sole occupant of the nest. Inter-estingly, Purple Finches *(Carpodacus purpureus)* that also provide nestlings a diet high in seeds appear to be somewhat successful in raising cowbirds and are parasitized at a higher frequency than American Goldfinches (Fried-mann et al. 1977).

In a study of Dickcissels, Hatch (1983) found that cowbirds are more affected by nest competitors than are hosts; he suggested that hosts may preferentially feed their own young. I have several times observed retarded growth and the death of Brown-headed Cowbirds in the nests of Warbling Vireos, even though they were the sole occupants of the nest. Hofslund (1957) observed that Common Yellowthroats provisioned cowbirds and host nestlings equally.

With generalist parasites, often the question is raised whether some hosts are better than others at raising the parasite (table 7.9). In a cross-fostering experiment, Eastzer et al. (1980) placed eggs of seven species, including Brown-headed Cowbirds, in nests of Barn Swallows and House Sparrows to compare the success of cowbird nestlings with nonparasitic species. Among House Sparrow nests, only cross-fostered House Sparrows fledged. In Barn Swallow nests, three of five nonparasitic species did better than cowbirds. The experiment's most crucial finding was that after fledging, cowbirds were fed by Barn Swallow hosts, whereas Red-winged Blackbirds and Gray Cat-birds were not, in spite of their loud, persistent begging vocalizations. Ap-parently, Barn Swallows are reluctant to feed fledglings of another species, but cowbirds were somehow able to break that code. They concluded that cowbird nestlings may not be more adapted to be reared by less-commonly used hosts, but they may be well adapted to coax reluctant hosts to feeding them as fledglings.

Proper Nest Type. In general, Brown-headed Cowbirds prefer to parasit-ize open cup nests and shy away from cavity nests and domed nests.[9] One notable exception to this is the Prothonotary Warbler (Petit 1991). Why Brown-headed Cowbirds shy away from cavity nests when other cowbirds species appear to use them without hesitation is not clear and warrants fur-

Table 7.9

Success of Brown-headed Cowbirds among Some Hosts

Host Species	Number of Cowbird Eggs	Percent Fledged per Egg	Location	Source
Alder Flycatcher	4	25.0	Michigan	Berger 1951
Willow Flycatcher	15	20.0	California	Harris 1991
Eastern Phoebe	139	41.7	Kansas	Klaas 1975
	10	40.0	Kansas	Hill 1976a
Say's Phoebe	1	0.0	Kansas	Hill 1976a
	8	62.5	Michigan	Berger 1951
Eastern Kingbird	19	5.3	Kansas	Murphy 1986
Horned Lark	24	29.2	Kansas	Hill 1976a
Tree Swallow	14[1]	0.0	Ontario	Mills 1988
Veery	8	50.0	Pennsylvania	Norris 1947
Wood Thrush	2	0.0	Michigan	Berger 1951
	2	50.0	Pennsylvania	Norris 1947
Gray Catbird	2	0.0	Michigan	Berger 1951
	1	0.0	Pennsylvania	Norris 1947
Brown Thrasher	6	16.7	Kansas	Elliott 1978
	3	0.0	Kansas	Hill 1976a
	2	0.0	Pennsylvania	Norris 1947
Bell's Vireo	5	0.0	Illinois	Pitelka and Koestner 1942
	1	0.0	Kansas	Hill 1976a
	15	20.0	Oklahoma	Ely in Wiens 1963
	15	6.7	Oklahoma	Wiens 1963
Plumbeous Vireo	49	44.9	Colorado	Marvil and Cruz 1989
Red-eyed Vireo	18	0.0	Pennsylvania	Norris 1947
Yellow Warbler	51	3.9	Manitoba	Weatherhead 1989
	33	6.1	Michigan	Berger 1951
Prothonotary Warbler	43	55.8	Tennessee	Petit 1991
Ovenbird	5	0.0	Michigan	Berger 1951
	12	25.0	Pennsylvania	Norris 1947
Louisiana Waterthrush	18	66.7	New York	Eaton 1958
Common Yellowthroat	40	30.0	Michigan	Hofslund 1957
	4	0.0	Pennsylvania	Norris 1947
Yellow-breasted Chat	21	14.3	Michigan	Nickell 1955
Scarlet Tanager	5	40.0	Pennsylvania	Norris 1947
Northern Cardinal	19	21.1	Michigan	Berger 1951
	4	25.0	Oklahoma	Wiens 1963
	2	0.0	Oklahoma	Ely in Wiens 1963
	4	50.0	Pennsylvania	Norris 1947
Blue Grosbeak	2	0.0	Oklahoma	Wiens 1963
	13	30.8	Oklahoma	Ely in Wiens 1963
Indigo Bunting	9	44.4	Michigan	Berger 1951
	7	14.3	Ohio	Phillips 1951
Painted Bunting	1	100.0	Oklahoma	Wiens 1963
	5	0.0	Oklahoma	Ely in Wiens 1963

Table 7.9 *Continued*

Host Species	Number of Cowbird Eggs	Percent Fledged per Egg	Location	Source
Dickcissel	44	11.4	Kansas	Elliott 1978
	14	14.3	Kansas	Hill 1976a
	620	24.7	Kansas	Zimmerman 1983
	5	0.0	Oklahoma	Wiens 1963
	1	0.0	Oklahoma	Ely in Wiens 1963
Eastern Towhee	3	0.0	Michigan	Berger 1951
	22	54.5	Pennsylvania	Norris 1947
Chipping Sparrow	1	0.0	Pennsylvania	Norris 1947
	7	14.3	Michigan	Berger 1951
Clay-colored Sparrow	27	0.0	Manitoba	Hill and Sealy 1994
	9	0.0	Saskatchewan	Fox 1961
Field Sparrow	7	14.3	Michigan	Berger 1951
	2	0.0	Oklahoma	Wiens 1963
	4	25.0	Oklahoma	Ely in Wiens 1963
	13	7.7	Pennsylvania	Norris 1947
Vesper Sparrow	2	0.0	Michigan	Berger 1951
Lark Sparrow	16	68.7	Kansas	Hill 1976a
	4	0.0	Oklahoma	Wiens 1963
	1	100.0	Oklahoma	Ely in Wiens 1963
Lark Bunting	15	6.7	Kansas	Hill 1976a
Grasshopper Sparrow	14	7.1	Kansas	Elliott 1978
	5	60.0	Kansas	Hill 1976a
Song Sparrow	82	30.5	Michigan	Berger 1951
	13	30.8	Pennsylvania	Norris 1947
	324	52.2	Vancouver	Smith and Arcese 1994
Red-winged Blackbird	7	14.3	Kansas	Elliott 1978
	19	10.5	Kansas	Hill 1976a
	96	27.1	Manitoba	Weatherhead 1989
	5	0.0	Michigan	Berger 1951
Eastern Meadowlark	86	5.8	Kansas	Elliott 1978
Western Meadowlark	1	0.0	Kansas	Hill 1976a
Yellow-headed Blackbird	23[1]	30.4	Colorado	Ortega and Cruz 1991
	1	100.0	Wisconsin	Young 1963
Orchard Oriole	3	33.3	Kansas	Hill 1976a
	9	44.4	Oklahoma	Ely in Wiens 1963
	4	100.0	Oklahoma	Wiens 1963
Pine Siskin	20	5.0	Kansas	Hill 1976a
American Goldfinch	4	0.0	Michigan	Berger 1951
	26	0.0	Ontario	Middleton 1977
	47	0.0	Ontario	Middleton 1991

[1] Data are from experimentally cross-fostered nests.

ther investigation. Undoubtedly, it is more difficult to determine if a nest owner is on the nest, and perhaps this makes parasitism of cavity nests more risky in terms of direct confrontation between host and parasite.

Habitat and Nest Site Characteristics. Habitat can have a major influence on rates of parasitism, and some species seem to escape parasitism because of habitat. For example, Holmes et al. (1992) found no parasitism among 125 Black-throated Blue Warbler (*Dendroica caerulescens*) nests located in an unfragmented forest in New Hampshire. Birds that nest in desert scrub may also experience less parasitism. Hergenrader (1962) and Linz and Bolin (1982) found high rates of parasitism in roadside ditches.

Buech (1982) hypothesized that Brown-headed Cowbirds might most easily locate less concealed nests in tall trees with low crown density. His observations, however, did not support this; parasitized nests were more concealed than nonparasitized nests, suggesting a high probability that cowbirds cued in on host activity. Fleischer (1986) found that parasitized Red-winged Blackbird and Dickcissel nests were located higher than nonparasitized nests; however, Wiens (1963) found that nest height did not differ between parasitized and nonparasitized nests.

Nest Attentiveness. Nest attentiveness may influence parasitism. Neudorf and Sealy (1994) observed no coordination between male and female hosts in nest guarding during the critical period when Brown-headed Cowbirds are apt to lay their eggs. Slack (1976) reported that female Gray Catbirds called their mates to relieve them of nest-guarding duties when they left to forage, and he speculated that this behavior might be an adaptation to prevent brood parasitism or predation. Similarly, Scott (1977) observed that Gray Catbirds were more attentive to their nests during egg laying than were more heavily parasitized Northern Cardinals. Smith (in Neudorf and Sealy 1994) also found similar nest-guarding behavior in Eastern Kingbirds. However, the relatively low rates of parasitism that we observe on Gray Catbirds and Eastern Kingbirds may reflect their egg ejection habits more than attentiveness to their nests.

Differences in territory size of hosts may play a role in frequency of parasitism, particularly for birds whose primary defense is aggression. Birds with smaller territories have an increased likelihood of detecting intruding Brown-headed Cowbirds and may come to the defense of their nests more rapidly. Scott (1977) noted that the size of Gray Catbird territories are roughly 20% the size of the more heavily parasitized Northern Cardinals. Similarly, the territory size of Yellow-headed Blackbirds that are rarely parasitized is much smaller than territories of more heavily parasitized Red-winged Blackbirds; however, for colonial-nesting species, host density may also have an effect.

Highly aggressive species do not necessarily escape parasitism, but aggression may lower the overall rate of parasitism for certain species. Red-winged Blackbirds are an excellent example of this (Ortega and Cruz 1991), but it may not be their aggression alone that influences parasitism; they are also larger than Brown-headed Cowbirds, they have a relatively short incubation period which does not give the cowbird a competitive advantage, and they are colonial nesters, which increases vigilance but also increases the chances of cowbirds finding nests (Ortega 1991; Ortega and Cruz 1991). In general, trends arising from aggression experiments with mounted Brown-headed Cowbirds are that levels of host aggression are proportional to intensity of parasitism (chapter 2).

Birds that are attentive to their nests at the time of cowbird egg laying may experience lower rates of parasitism than hosts that are generally absent from their nests during the early morning hours because individuals roosting on their nests overnight could intercept cowbirds. Neudorf and Sealy (1994) suggested that accepter species might be more vigilant during sunrise than rejecter species and that accepter species with low rates of parasitism would be more vigilant than those more heavily parasitized. However, among six accepter species (Least Flycatcher, Yellow Warbler, Red-winged Blackbird, Yellow-headed Blackbird, Brewer's Blackbird [Euphagus cyanocephalus], and Common Grackle), they found no significant correlation between roosting in nests overnight and rates of parasitism, nor did they find any significant correlation in time spent at the nest during the critical hours of 03:35–04:05, the time during which all parasitism events occurred. Females that roosted on their nests overnight spent significantly more time on their nests between 03:35 and 04:05 than conspecifics who had not roosted. However, Neudorf and Sealy (1994) also found a great deal of variation in roosting behavior within species. In agreement with their prediction, accepters were more likely than rejecters (Eastern Kingbird, American Robin, Gray Catbird, and Baltimore Oriole) to roost on their nests.

Host Size. At some point, there are limits to the size of appropriate hosts, and at the upper limits, some parasitic individuals or populations may be more daunted by host size and harm potentially caused by hosts of larger size. The size of large thrashers and Common Grackles appear to be the upper limit as rates of parasitism are fairly low, and no birds larger are parasitized.

Host Density. Host density may affect rates of parasitism (Freeman et al. 1990). Hosts may experience heavy parasitism when nesting density is reduced (Fretwell 1977). As Red-winged Blackbird density increases, rates of parasitism decrease (Freeman et al. 1990; but see Fleischer 1986). Zimmerman (1983) found that the percentage and intensity of Brown-headed Cowbird parasitism was inversely associated with Dickcissel nest density.

Colonial nesting increases host vigilance and may reduce risks to individuals through the swamping effect, but colonial nesting also provides a large number of nests for those cowbirds that can maneuver through a colony. Colony size and habitat may play important roles in rates of parasitism of colonial hosts. Among Red-winged Blackbirds, smaller colonies may be parasitized more frequently than large colonies; edges or roadside ditches are often heavily parasitized, and upland colonies are more heavily parasitized than marshes. Similarly, flooded tree stands are parasitized at a higher rate than higher density cattail marshes (Ortega 1991). Other potential hosts that nest in dense colonies are Tri-colored Blackbirds and Yellow-headed Blackbirds. These colonies are usually more dense than Red-winged Blackbird colonies and experience little parasitism. Out of tens of thousands of Tri-colored Blackbird nests, only one has been found to have been parasitized, yet experiments indicate that they are an accepter species (Friedmann et al. 1977).

Alternative Host Choices in the Community. Another influence on nest selection by Brown-headed Cowbirds is host availability in a given avian community, and this, more than any other reason, may explain the tremendous variation we observe in rates of parasitism within a specific host species. For example, in a comparison between my study (Ortega and Cruz 1991) of Yellow-headed Blackbirds that were not parasitized and Dufty's (1994) study in which they were parasitized, Dufty suggested that Brown-headed Cowbirds had few alternate choices in the avian community; that is, alternative host availability was low. In local areas surrounding Yellow-headed Blackbird colonies in my study, alternative hosts were abundant.

Suggestions for Future Research

Ample opportunities abound for expansion of Brown-headed Cowbird research. Below are only a few suggestions of future research topics. Other topics that would contribute to management are addressed in chapter 9.

Mating System

The mating system of Brown-headed Cowbirds may be influenced by sex ratio, habitat, home range, territoriality, host density, and cowbird density. Some trends are apparent, but research needs to be conducted in other regions and other habitats; in particular, more studies need to be conducted where feeding and breeding sites are not disjunct. The mating system of cowbirds in disturbed habitats could also be compared with the mating system of cowbirds under conditions closer to those previously experienced by cowbirds; Yellowstone and Grand Teton National Parks would provide an opportunity to study the mating system of cowbirds that associate with Bison rather than domestic cattle.

Dominance Hierarchies

Dufty (1986) suggested that singing is important in establishment of dominance hierarchies, but once ranks are established, they may be maintained through nonvocal means. In other studies of social dominance, plumage characteristics are important. During song spread, Brown-headed Cowbirds may be able to evaluate male age by the presence or absence of juvenile feathers in the underwing coverts. It would be worthwhile to evaluate this characteristic that varies somewhat with age (see aging). Manipulative experiments could reveal whether underwing coverts are used in the establishment and/or maintenance of dominance hierarchies.

Multiple Parasitism

Studies in spatial patterns of multiple parasitism (and also incidences of punctured cowbird eggs) and female territorial behavior might elucidate the connection between territoriality and multiple parasitism.

Hosts

Several studies have been conducted on hosts that are rarely parasitized in an attempt to understand host selection. Many other infrequently parasitized species that appear to be suitable hosts have not yet been investigated for breeding parameters that influence parasitism rates.

Summary

The Brown-headed Cowbird is a short grass/edge species; they forage mainly on the ground, often in association with grazing mammals. With deforestation, planting trees where none previously existed, and widespread introduction of domestic livestock, Brown-headed Cowbirds have expanded their range and abundance; they are now found throughout the United States and much of Canada.

The sex ratio of adult Brown-headed Cowbirds varies regionally but is almost always male-biased, suggesting higher mortality among females. Home range sizes of Brown-headed Cowbirds vary from 0.4 to 68 ha and appear to be correlated with the density of host nests and abundance of cowbirds. Brown-headed Cowbirds spend mornings in breeding habitat, laying eggs and searching for nests, and afternoons foraging. Foraging sites are often distant and disjunct from breeding sites; commuting between the two site types is common.

Males do not show clear behavior of defending territories, but female territoriality has been observed in a few studies. Males form dominance hierarchies; in general, dominant males have greater access to females, and older males dominate younger males. The mating system is largely monogamous, but some polygyny and promiscuity have also been noted. Males often guard and follow their mated females. Variation in the mating system of Brown-headed Cowbirds has often been attributed to differences in habitat, host density, cowbird density, and home range size. More reports of monogamy are from the north and northeast where host abundance is higher, where cowbird densities are lower, where the sex ratio is less male biased, and where habitat is less open.

Brown-headed Cowbirds are site faithful and have a strong homing instinct. Both males and females return to their natal site to breed, but the mechanisms of avoiding inbreeding are unexplored. They generally migrate between wintering and breeding grounds. Populations that winter in the same areas disperse widely to breeding grounds, and individuals that share breeding grounds disperse widely to wintering grounds.

Adults have several vocalizations that are used in different contexts. Females emit a loud harsh chatter that is used in a wide variety of contexts to communicate with both males and other females. Males have two songs: perch song and flight whistles. They are used to communicate with both males and females. Perch songs are used to communicate in close proximity, whereas flight whistles may be used more to communicate over longer distances. Recent studies have shown that there is a definite learned component to perch song and flight whistles, and exposure to adult song is necessary for normal perch song to develop. Both perch song and flight whistles have distinct dialects.

Brown-headed Cowbirds typically breed from May through June. They lay their eggs rapidly relative to other passerines, usually before sunrise. The number of eggs laid by individual females varies; up to 77 eggs have been laid by a single female during one breeding season. Egg laying appears to be restricted by dietary requirements, primarily calcium, which is acquired mainly through mollusc shells. Incubation ranges from 10 to 12 days. Young Brown-headed Cowbirds usually leave the nest when they are 10–11 days old and gain independence from their foster parents when they are 16–28 days old.

Brown-headed Cowbirds are known to parasitize 226 species but may regularly parasitize and be raised by 132 hosts. In general, Brown-headed Cowbird parasitism has a negative effect on the nesting success of hosts through egg removal, abandonment, costs of egg ejection, reduced hatching, and competition from larger cowbirds.

Some birds that appear to be suitable hosts are not parasitized or parasitized at a very low level. Breeding parameters that can explain low parasitism rates include egg acceptance/rejection status, host breeding season, incubation period, proper incubation by host, host care, nest type, habitat, nest site characteristics, nest attentiveness, host size, host density, and alternative host choices in the community.

Shiny Cowbirds

Shiny Cowbirds are the most widespread of all cowbirds, yet they are not as studied as Brown-headed Cowbirds. As they continue to spread into North America, they elicit fear from many concerned about potential hosts. A photograph of a trapped Shiny Cowbird in Texas was featured in the "Pictorial highlights" of *American Birds* (1992:1195); the caption read, "Perhaps a 'lowlight,' not a highlight, was the second record of Shiny Cowbird for Texas."

When brood parasitism in Shiny Cowbirds was first discovered is not known, but their parasitic habits were described by Azara in 1802, eight years before the first published account of Brown-headed Cowbird parasitism (Friedmann 1964). Friedmann found no evidence of knowledge by natives prior to Azara's description, but he doubted that locals were unaware of brood parasitism by Shiny Cowbirds. Common names are listed in table 8.1.

Subspecies

Geographic variation of Shiny Cowbirds has not been well studied but exists even amongst the *M. b. mimimus* range through the West Indies. For example, female Shiny Cowbirds from St. Lucia are morphologically distinct from those inhabiting Puerto Rico. There have been numerous subspecies suggested for the Shiny Cowbird. Friedmann (1929b) recognized eight subspecies; Blake (1968) recognized seven subspecies (table 8.2).

Description

Shiny Cowbirds are sexually dimorphic in both size and plumage. Males are entirely black with purple iridescence, but the wings and tail have green iridescence. Their feet and bills are shiny black. Females are ash brown above and paler beneath with a lighter chin; they may or may not have slight streaking on the chest and belly. Females may have a white post-

Table 8.1
Common Names of Shiny Cowbirds

English	Kuihrahu (Paraguay)
Argentine Cowbird	Murajú
Blackbird	Murajú Negro
Bluey-black	
Brazilian Starling	*Portuguese*
Common Cowbird	Corichó
Corn-bird	Papa (de) Arroz
Cowbird	Parasita
Glossy Cowbird [1,2]	Tordo Chopím
Glossy Starling	Virabosta
Guiana Cowbird	
Lazy-bird	*Spanish*
Shining Cowbird	Bovero
Silky Cowbird	Gauderio
	Mirlo
French	Pajaro Negro
Braunt-noir	Patero
	Renegrido
German	Tordo
Kuhvogel	Tordo Argentina
Viehstaar	Tordo Azul
	Tordo Comun
Native	Tordo Lustroso [1]
Burrumichi (Bolivia)	Todro Mulato(a)
El Guyra-hú	Tordo Negro
Gwihra-h (Paraguay)	Tordo Renegrido
Kosoleka	Verondica

Source: Friedmann 1929b.
[1] Raffaele 1983.
[2] Belcher and Smooker 1937.

ocular streak, which is most pronounced in *M. b. aequatorialis* and *M. b. occidentalis*. Their feet and bills are brownish black. The iris of both males and females is dark brown. Size varies geographically. For example, females in Suriname weighed an average of 31.5 g (Haverschmidt 1965); females in Argentina weighed an average of 44.9 g ($n = 78$, Mason 1987). In Colombia, females averaged 54 g (Kattan 1993). In Argentina males averaged 55.5 g ($n = 69$, Mason 1987).

Distribution and Abundance

Distribution and Range Expansion

Shiny Cowbirds were originally confined to South America, Trinidad, and Tobago, but the *M. b. minimus* subspecies began a northward expansion

Table 8.2
Type Locality and Range for Subspecies of the Shiny Cowbird

M. b. bonariensis, Buenos Aires, "Eastern and southern Brazil, north to Mato Grosso, Maranhão, Piauí, and Ceará; eastern Bolivia, Paraguay, Uruguay, and Argentina south to Chubut. Introduced into Chile and now established from Coquimbo south to Valdivia."

M. b. cabanisii, Santa Marta, Colombia, "Eastern Panama (Tuyra River, Darien), tropical and lower subtropical zones of Colombia west of eastern Andes (except Nariño), and east slope of eastern Andes in Ocaña region."

M. b. aequatorialis, Barbacoas, Colombia, "Tropical zone of southwestern Colombia, south of Río Patia, and western Ecuador south to Guayaquil and Puná Island."

M. b. occidentalis, Lima, Peru, "Extreme southwestern Ecuador (Casanga Valley of Loja) and western Peru, east to Province of Jaén in Cajamarca and south to Lima."

M. b. venezuelensis, Lake Valencia, Venezuela, "Tropical zone of eastern Colombia from Zulia Valley south to eastern llanos; northern Venezuela southward in the llanos to Apure and Orinoco rivers, and south of the Orinoco in northwestern Amazonas (San Fernando de Atabapo) and northern Bolívar."

M. b. minimus, British Guiana, "Lesser Antilles north to Martinique; Tobago; Trinidad; the Guianas and extreme northern Brazil in region of upper Rio Branco (Rio Causamé)." Also in West Indies and southeastern United States.

M. b. riparius, Pinhy, Rio Tapajóz, Brazil, "Lower Amazon Valley (to Obidos on north bank) west to Río Ucayali (Yarina Cocha), eastern Peru."

Source: Blake 1968.

through the West Indies in the early 1900s (Bond 1956). Expansion is thought to have been facilitated by deforestation of the islands and the introduction of domestic cattle. Also, caged birds may have been inadvertently introduced to some islands, but this is speculative. In addition to their range expansion being aided by human activities, Shiny Cowbirds possess certain characteristics that make them good colonizers: (1) they are flexible in their habitat selection; (2) they are able to subsist on a wide array of food items; (3) their host selection strategy is general; (4) they are gregarious, so more than a solitary individual is likely to invade new areas; and (5) they have high fecundity, allowing them to rapidly exploit new areas (Cruz et al. 1989).

Cruz et al. (1985) charted the northward spread of the Shiny Cowbird up through the West Indies (fig. 8.1). Shiny Cowbirds arrived at various locations on the following dates: 1899, Carriacou; 1901, Grenada; 1916, Barbados; 1924, St. Vincent; 1931, St. Lucia; 1934, St. Croix; 1948, Martinique; 1955, St. John and Puerto Rico;[1] 1959, Antigua; 1972, Hispaniola; 1982, Cuba. Since publication of Cruz et al. (1985), Shiny Cowbirds have been reported to have arrived in Curaçao in 1991 (Debrot and Prins 1992), in the Florida Keys in 1985, and on the mainland in 1987 (Smith and Sprunt 1987).

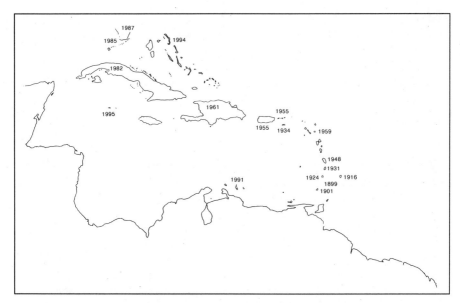

Fig. 8.1. Range expansion of Shiny Cowbirds through the West Indies.

Post et al. (1993) charted the first arrivals in the United States (fig. 8.2): in 1989, Shiny Cowbirds arrived in Georgia, Louisiana, North Carolina, and South Carolina; in 1990 they arrived in Alabama, Oklahoma, and Texas; in 1991 they were observed in Mississippi and Maine. During the summer of 1995, Shiny Cowbirds were seen in the Cayman Islands (T. Nakamura, pers. comm.). Most sightings have involved only one or two individuals, but a flock of 52 individuals was seen in Florida (Post et al. 1993). Recently, Shiny Cowbirds, including immatures, have been seen regularly in small numbers in Florida and Alabama (Jackson 1992, 1993; Langridge 1992, 1993; Paul and Schnapf 1993), and Davis (1993) noted that they continue to increase in the south Atlantic region of the United States. Although intensive searches were conducted in the Bahamas earlier, Shiny Cowbirds were not reported there until 1994 (Baltz 1995).

Post and Wiley (1977b) calculated an expansion rate of 8.3 km per year through the Lesser Antilles during the first 60 years of the expansion and a rate of 30.6 km per year through the Greater Antilles between 1955 and 1973. They suggested that the more rapid expansion through the Greater Antilles was facilitated by a greater diversity of hosts and foraging opportunities.

Shiny Cowbirds also invaded Chile and Panama in recent times. Friedmann (1929b) indicated that extension into eastern Chile had been recent,

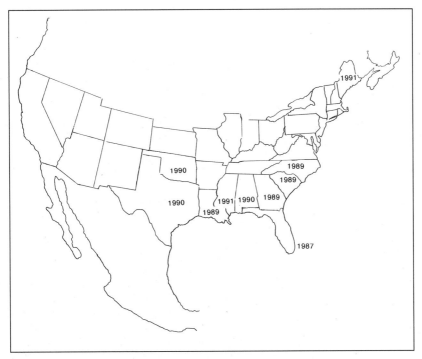

Fig. 8.2. Range expansion of Shiny Cowbirds into the United States.

but at the time his book was written, Shiny Cowbirds apparently did not inhabit western Chile. Johnson (1967) and Blake (1968) suggested that Shiny
Cowbirds were introduced to western Chile. Presently, they are common
throughout western Chile and in the foothills of the Andes (Johnson 1967).
Davis (1972) mentioned their presence in Panama, but Willis and Eisenmann (1979) did not. According to Ridgely and Gwynne (1989), they are
locally common or uncommon in central and eastern Panama; they appear
to be expanding west, as there are numerous reports of their presence in
western Panama. Howell and Webb (1995) suggested that Shiny Cowbirds
should also be looked for in Cancún and Isla Cozumel. Presently, Shiny
Cowbirds are widespread and abundant throughout much of South America,
with a major gap in their range through the Amazon rainforest (Cruz et al.
1990; fig. 8.3).

Fig. 8.3. Range of Shiny Cowbirds in South and Central America.

Natural History

Habitat

Shiny Cowbirds inhabit open and semiopen areas, from coasts to high elevations. They are common in farmland, mangroves, marshes, agricultural fields, plantations, thornscrub, and forest edges (Arendt and Vargas Mora 1984). They are attracted to disturbed habitat. Cavalcanti and Pimentel (1988) noted that Shiny Cowbirds are uncommon in undisturbed cerrado.

Feeding

Shiny Cowbirds eat a wide variety of arthropods and seeds. They often feed in flocks, sometimes with other species. Usually, they feed on the ground, often in association with cattle and other large grazing mammals, but they may also pick insects from the backs of large grazing mammals. Rice and grains set out for cattle and poultry also contribute significantly to their diet (Cruz et al. 1990). Friedmann (1929b) found that animal material predominated in stomach contents of 75 birds, but at times they may feed heavily on rice. Shiny Cowbirds do tremendous damage to rice fields, but Miller (1917) found that stomach analysis throughout the year of 60 birds revealed 90% animal matter. The variation in proportions of animal and vegetable matter may reflect seasonal variations.

Sex Ratios

The sex ratio of Shiny Cowbirds is generally male biased. In Chile, however, Johnson (1967) suggested that females outnumber males. Azara (in Bendire 1895) reported that the sex ratio was extremely male biased at 9:1; Bendire doubted this but, without quantifying the ratio, noted that males are more numerous than females. Barrows (in Friedmann 1929b) reported a male-biased sex ratio of 3:2 in Argentina. In the Dominican Republic, Arendt and Vargas Mora (1984) reported a sex ratio of 1.5:1 ($n = 96$). In Argentina, Mason (1987) observed a slightly male-biased sex ratio of 1.1:1 ($n = 150$).

Home Ranges and Territories

Very little work has been conducted on home range size and territoriality of Shiny Cowbirds. What little we do know suggests that home ranges are large and territories are not defended in the traditional sense. As with

Brown-headed Cowbirds, Shiny Cowbirds use different sites for foraging and breeding. On Puerto Rico, radio-tagged females traveled approximately 4 km between breeding and feeding sites (Woodworth 1993). Female home ranges overlap in Argentina (Fraga 1985) and Puerto Rico (Woodworth 1993). Woodworth (1993) found that radio-tagged females on Puerto Rico did not maintain exclusive territories; their ranges were large, but this would be expected from the low host density in the area where she conducted her research. Friedmann (1929b) suggested that territories were weak and easily modified, becoming less defined as cowbird density increased. He suggested that some females laid within an undefended home range or domain, and other females probably wandered from one domain to another. Bendire (1895) noted a high degree of sociability throughout the year, suggesting lack of territorial behavior.

Fraga (1985) did not find territorial females in Argentina; however, Shiny Cowbirds commonly punctured eggs of other cowbirds in nests of Rufous-collared Sparrows *(Zonotrichia capensis)*.[2] Hoy and Ottow (1964) mentioned a Chalk-browed Mockingbird nest in Argentina with 14 Shiny Cowbird eggs, all apparently laid by different females — yet from the photograph, it appears that not one was punctured. Friedmann. (1929b) noted that female Shiny Cowbirds often apparently removed each other's eggs from nests they laid in. The high incidence of multiple parasitism by Shiny Cowbirds suggests that exclusive spatial territories may not be defended at boundaries by females.

Courtship

Male Shiny Cowbirds have three types of courtship displays: terrestrial, arboreal, and aerial. The terrestrial display is performed most frequently; during the display, the male ruffles up his head and body feathers, bends forward with his head bowed down, raises and lowers his tail, holds his wings out horizontally (often with quivers, and sometimes the wings beat against the ground), and sings three guttural notes followed by three high glassy notes that Friedmann (1929b) described as *purrr-purrr-purrr-pe-tsss-tseeeee.* Females characteristically appear indifferent to the courtship displays. Arboreal displays are similar to terrestrial displays and are given fairly often. Friedmann (1929b) noted that during arboreal displays, Shiny Cowbirds are restless, changing perches frequently, whereas Brown-headed Cowbirds may remain planted in one spot displaying over and over again. The aerial display is given less frequently. It is described by Friedmann (1929b:70): "On several occasions I noted males fluffing out their body feathers while flying, and beginning to sing while ruffling their feathers. On one occasion a male even attempted to arch its wings with the result that it flew very stiffly for a few wing beats when it once more resumed its normal flight."

Mating System

After observing numerous pairs of Shiny Cowbirds during the breeding season and a only few cases of females mated to several males, Friedmann (1929b) concluded that Shiny Cowbirds are generally monogamous. The monogamy, however, may not last throughout the breeding season, especially where the cowbird population is dense. Friedmann (1929b) believed that where the population is dense, females mate with a male until they have exhausted the nest supply of his domain; they then move on to another male and another domain. The males, meanwhile, take what comes their way. Mason (1987) found no significant pairwise interactions in a trapping program in Argentina; in other words, individual males and females did not occur together more often than by chance. Mason also observed no obvious pair formation in the field; he suggested that while these findings would be consistent with a promiscuous mating system, conclusions remain tentative. Johnson (1967) suggested that in Chile, where females outnumber males, the mating system is polygamous.

Migration

Shiny Cowbirds are considered resident birds throughout much of their range, but in Argentina, a definite migration pattern can be observed. They migrate in large flocks consisting of up to thousands of individuals, often with other species. In general, migrants precede residents, older birds precede younger ones, and males precede females in each age class.

Site Fidelity

Mason (1987) suggested that Shiny Cowbirds are not as site faithful as Screaming Cowbirds, based on fewer recaptures and no individual observed in his study site for longer than 54 days. Fraga (in Mason 1987) felt that three weeks was a typical time females spent in an area. In the Dominican Republic, Arendt and Vargas Mora (1984) captured only 4 of 196 (2.0%) Shiny Cowbirds within 5 km of their banding locations a year and a half after banding.

Flocking Behavior

In some areas, Shiny Cowbirds roost at night with other species (Cruz et al. 1990; Post et al. 1993). On Puerto Rico, Shiny Cowbirds roost nocturnally and diurnally throughout the year in feeding areas with Yellow-shouldered Blackbirds (*Agelaius xanthomus*) that they parasitize heavily (Post and Post 1987). During the breeding season, they spread out more or less evenly over a given area; outside the breeding season, they congregate in large flocks (Miller 1917). Flocks are larger outside the breeding season in Argentina (Fraga 1978) and may contain up to tens of thousands of individuals (Hudson 1874, 1920). Hudson reported the flocks to be so dense as to make the ground appear to be carpeted in black and the trees foliaged in black.

Vocalizations

Nestling Shiny Cowbirds utter short loud peeps for the first four or five days, after which the easily recognizable calls become loud warbling or tremulous calls with a hissing quality (Fraga 1986). Fraga (1985) reported the frequency range of these calls to be 4–9 kHz.

Friedmann (1929b) described five vocalizations: male song, male twitter song (flight whistle), rattle (chatter), a "chuck" note, and a battle call. The male song lasts for approximately three seconds and consists of three guttural bubbling notes followed by another three high glassy notes. Friedmann (1929b) wrote the song as *purrr-purrr-purrr-pe-tsss-tseeeee*. Incomplete songs become common as the breeding season wanes, with the penultimate note often being omitted (Friedmann 1929b). The flight whistle of Shiny Cowbirds is similar to that of Brown-headed Cowbirds, but it is longer (Å 2 sec.) and more complex (Rothstein and Fleischer 1987a). Friedmann (1929b) noted that the flight whistle is often preceded by a chuck note. The flight whistle is emitted just prior to copulation (Fraga, pers. comm. in Rothstein et al. 1988).

Chatter is a loud harsh rattling vocalization emitted primarily by females, rarely by males. It is similar to the chatter of Brown-headed Cowbirds but more metallic, with a clicking quality (Friedmann 1929b). Chatter is often emitted just prior to flight. The chuck note is emitted by both sexes frequently. Its function is not understood, but it appears to be used in a general context (Friedmann 1929b). Friedmann (1929b) also described what he referred to as a battle call used by fighting birds. He described it as harsher and more grating than the chuck note and suggested that it may actually be the chuck note given during excitement and physical exertion.

Head-Down Displays

The head-down display of Shiny Cowbirds is similar to that of Brown-headed Cowbirds. Post and Wiley (1992) found that head-down displays resulted in preening by Yellow-shouldered Blackbirds in 34.1% of 170 displays and by conspecifics in 28.6% of 35 displays. Shiny Cowbirds performed the display to Yellow-shouldered Blackbirds far more often than to conspecifics. Post and Wiley (1992) suggested that Shiny Cowbirds may quickly learn or already know the identity of conspecific individuals, whereas they may be less familiar with individuals of another species and may not be able to interpret their motivations. Post and Wiley (1992) observed this display only in roosting areas and feeding situations, not in breeding areas; therefore, they believed that the display did not function to appease hosts in order to gain access to nests (chapter 7). Instead, they concluded that the display functioned to dominate individuals, which intraspecifically could increase mating success and which interspecifically could facilitate integration into flocks of potential hosts, increasing the chance of obtaining nest location information.

Plumage and Molt

The mousey gray natal down is replaced with juvenile plumage by a complete postnatal molt. The juvenile plumage varies with subspecies, but females are lighter than males, varying above in shades of brown or gray; the under parts are lighter, sometimes more yellow, and may be heavily streaked. The feathers on the upper surfaces and wings are emarginated in lighter shades; emargination is most pronounced on the wings. The bill and feet are medium pinkish brown to medium dark brown. The general pattern of males is similar, but they are noticeably darker than females and have very dark brown bills and feet.

The juvenile plumage is replaced in a few months with the first winter plumage, acquired by a nearly complete postjuvenile molt. Friedmann (1929b) suggested that more first-year Shiny Cowbirds may retain juvenile feathers than Brown-headed Cowbirds. In this plumage, young birds are very similar to older adults. Shiny Cowbirds have an annual complete postnuptial molt.

Breeding Season

In Argentina, the breeding season extends from September through late January (Fraga 1978, 1985) or February (Friedmann 1963; Salvador 1983; Fraga 1986). North of the equator, on Puerto Rico, the breeding season generally extends from early May through mid-September (Post and Wiley 1977a), but Wiley (1988) reported breeding to extend from March through August on Puerto Rico. Nakamura and Cruz (1993) reported the nesting season of hosts to extend from March through November. Arendt and Vargas Mora (1984) also suggested an extended breeding season in the Dominican Republic, from March through August. In Colombia, the breeding season is nine months long (Kattan 1993).

Eggs and Egg Laying

Shiny Cowbird eggs are round and thick shelled. Shell thickness and other measurements vary with subspecies (Schönwetter 1983; Rahn et al. 1988).

Egg Polymorphism. Among cowbirds, Shiny Cowbirds have the highest degree of egg polymorphism (fig. 8.4). The eggs vary from immaculate to heavily maculated. Fraga (1985) found that immaculate eggs in Argentina were more rounded than maculated eggs, but he did not elaborate on the significance of his findings. Eggs intermediate between immaculate and spotted are rare (Fraga 1985; Mason 1986a). Salvador (1983) found only 11 of 226 (4.9%) Shiny Cowbird eggs of the intermediate form in Argentina. However, Hudson (1874) suggested that many eggs appearing immaculate were actually sparsely speckled with pale pink or gray. Fraga (1978) found that some eggs had pale gray spots, resembling water marks, on a white background; he considered these to be immaculate. Background colors come in white, pale blue, pale green, gray buff, cream, and light brown. Colors of spots or blotches also vary: pale gray, yellowish brown, reddish brown, deep brown, and purplish gray. Underlying colors of grays and lilacs may also be present. Immaculate eggs are usually white or off-white but may be entirely brick red (Hudson 1874). Eggs can and do come in almost all combinations of backgrounds and patterns. Friedmann (1929b) remarked that to describe all possible combinations would be to describe all the eggs in his collection. It is assumed that each female probably lays eggs of similar morphology, but this has not been quantified (Hoy and Ottow 1964; Fraga 1985).

The frequency of immaculate eggs varies geographically and might be

Fig. 8.4. Egg polymorphism among Shiny Cowbirds. (Photo courtesy of
Alexander Cruz)

associated with acceptance status of local hosts. In Argentina, immaculate
Shiny Cowbird eggs may be common (Fraga 1985; but see Mason 1986b), but
they may not be as common elsewhere; Fraga (1985) found the frequency
of immaculate Shiny Cowbird eggs was only about 5% in coastal Ecuador
and Peru. Among Bay-winged Cowbird nests in Argentina, 35.0% (7 of 20)
of Shiny Cowbird eggs were immaculate (Fraga 1986). Salvador (1983) reported that 12.2% (n = 226 eggs) of Shiny Cowbird eggs in Argentina were
immaculate. King (1973b) found no immaculate eggs in Rufous-collared
Sparrow nests in northwestern Argentina, but in Buenos Aires, Fraga (1978)
found 32 of 59 (54.2%) Shiny Cowbird eggs in Rufous-collared Sparrow
nests were immaculate.[3] Rufous-collared Sparrows accept both immaculate
and maculated Shiny Cowbird eggs (Mason 1986a).

In contrast, perhaps resulting from preferential acceptance of spotted eggs

(Mason 1986a; but see Salvador 1984), Fraga (1985) found only 4.9% (5 of 102) of Shiny Cowbird eggs in Chalk-browed Mockingbird nests were immaculate. Fraga (1985) suspected that many other hosts behave similarly, rejecting immaculate eggs. Hudson (1874) suspected that Brown-and-Yellow Marshbirds ejected immaculate eggs, as he found many immaculate eggs beneath their nests. More than a century later, Mermoz and Reboreda (1994) determined that Brown-and-Yellow Marshbirds, indeed, ejected immaculate eggs (in five of six nests) but accepted spotted eggs. Fraga (1985) knew of only one host, the Yellow-winged Blackbird (*Agelaius thilius*), that preferentially accepts immaculate eggs over spotted eggs. Yellow-winged Blackbirds nested near Fraga's study area, and the prevalence of immaculate eggs might suggest that on a local level, they are not selected against.

Mason and Rothstein (1987) suggested that parasite egg morphology can be influenced by either host discrimination (which should favor mimicry) or by predation (which should favor crypsis). Predation was similar among immaculate and spotted eggs experimentally placed in Rufous-collared Sparrow nests; Mason (1987) concluded that maculation of Shiny Cowbird eggs is most easily explained by selection for mimicry. However, both immaculate and spotted Shiny Cowbird eggs contrast sharply with the eggs of most hosts. Mason (1986b) suggested that host selection does not vary between individuals. The high parasitism rate of rejecter species supports his conclusion because it is unlikely that females that lay in nests of rejecters were raised by the same, and females laying immaculate eggs do not appear to preferentially parasitize species that generally accept them. The possibility should not be necessarily ruled out, but it is unlikely that females laying immaculate eggs hatch from spotted eggs; both types are found in the same nests (Mason 1986b).

Time of Day Eggs Are Laid. Shiny Cowbirds lay their eggs in the morning (Hoy and Ottow 1964). As with Brown-headed Cowbirds, Shiny Cowbirds lay their eggs with extreme rapidity, in approximately 30 seconds (Wiley and Wiley 1980).

Fecundity. The numbers of Shiny Cowbird eggs laid in a season and in clutches are not well understood, and no captive studies have confirmed estimates of annual fecundity. Hudson (1874) remarked on the large number of Shiny Cowbird eggs he observed on the ground and in abandoned nests and suggested that they lay 60–100 eggs per season; Miller (1917) agreed, but Friedmann (1929b) suggested only six to ten eggs per season.

Friedmann and Kiff (1985) suspected the fecundity of Shiny Cowbirds to be similar to that of Brown-headed Cowbirds. More recently, Kattan (1993, 1995) suggested that their fecundity is remarkable; most females collected in Colombia during each month were reproductively active. He estimated the annual fecundity to be 120 eggs per year. Nakamura (1995), on the other hand, found that relatively few females were in breeding condition throughout the breeding season on Puerto Rico; the greatest proportion in breeding condition during any month was 28.2% in April and 41.6% in May.

Incubation. Shiny Cowbird eggs hatch in 11–13 days in the field and 11.2–12.1 days under constant laboratory conditions (Kattan 1995). Normally, incubation is 11 days (Cruz et al. 1990). Shiny Cowbirds attain this short incubation period by laying eggs with reduced energy content that are small relative to female size (Kattan 1993, 1995). Selection for short incubation may be opposed by the disadvantages of a smaller hatchling, but Kattan pointed out that in nests of smaller hosts the disadvantage of a smaller nestling is irrelevant. However, King (1973b) found that cowbirds often suffered from brood reduction if they hatched out several days after their Rufous-collared Sparrow hosts.

Young

Shiny Cowbirds hatch out with a small amount of down and with their eyes closed. Their weight can double in the first day. By three or four days, nestlings open their eyes, pin feathers emerge, and they begin to take notice of their surroundings, reacting to moving objects. By eight days, most feather sheaths are open. By nine days, they may fledge if alarmed (Friedmann 1929b), but usually they remain in the nest for 12–13 days (Fraga 1978). Several morphological features vary more in Shiny Cowbirds than in other cowbirds, but these differences do not necessarily correspond to subspecies. Fraga (1978) found considerable variation in skin color, rictal flanges, and mouth linings at the same study site in Argentina. The skin varies from pink to yellowish or orange-pink, and rictal flanges may be either yellow or white (Salvador 1984; Fraga 1986). Mouth linings vary from light pink to deep red or orangish red. Fraga (1978) did not find any correlation between variation in skin color, rictal flanges, and mouth linings with color or pattern of eggs.

The first call of Shiny Cowbirds is a soft peep, which is given more often than in Brown-headed Cowbirds (Friedmann 1929b). By the time they are four or five days old, Shiny Cowbirds will thrust their necks up and climb over host siblings. By seven or eight days, begging may be accompanied by wing fluttering. Gochfeld (1979a) found that removal of host nestlings had little effect on either the intensity or duration of begging vocalizations by Shiny Cowbird nestlings.

Friedmann (1963) noted a general lack of understanding between Shiny Cowbird nestlings and the distress calls of foster parents; instead of heeding warnings of their foster parents, young nestlings clamored for food with their loud begging vocalizations, often resulting in predation. Hudson (1920) rarely saw fledgling Shiny Cowbirds with Correndera Pipits (*Anthus correndera*) even though they were heavily parasitized; he attributed the lack of Shiny Cowbird fledglings to Chimangos picking off noisy fledglings. In addition to lack of attunement between parasite and host, Shiny Cowbirds may beg indiscriminately from inappropriate sources (Hudson 1920).

Shiny Cowbirds remain with their hosts for at least 20 days (Fraga 1985). Fraga (1986) observed that Shiny Cowbirds have an inactive postfledgling stage similar to that which Woodward (1983) described for Brown-headed Cowbirds. Shiny Cowbirds usually reach their maximum nestling weight (31–33 g) on day eight or nine (King 1973b; Fraga 1978, 1985, 1986; Kattan 1996). Post and Wiley (1977a) reported a fledging weight for *M. b. minimus* of 27.5 g (*n* = 23) on Puerto Rico, and Friedmann (1929b) noted that they typically weigh 35 g upon fledging in Argentina.

How Shiny Cowbirds Find and Select Nests

Wiley (1988) identified the same three nest-finding strategies that are found for Brown-headed Cowbirds: (1) silent watching of host activities, (2) active searches, and (3) flushing behavior. Shiny Cowbirds appear to find nests primarily by watching their hosts (Fraga 1985; Wiley 1988). Active searches, in which the female moved through nesting habitat in short flights and hops, accounted for only 14.7 % (*n* = 10) of nest-searching observations on Puerto Rico (Wiley 1988). Flushing behavior, consisting of noisy flights through nesting areas, accounted for 13.2% (*n* = 9) of Wiley's observations. He found that Shiny Cowbirds also visited nests of nonparasitized species but with less frequency than parasitized species. Shiny Cowbirds particularly cue in on nest construction of their potential hosts and closely monitor the progress in order to synchronize egg laying (Wiley 1988). Of 151 observations of nest visits to Stripe-backed Wren (*Campylorhychus nuchalis*) nests in Venezuela, Shiny Cowbirds successfully entered brood chambers only four times (2.6%, Piper 1994). Wiley and Wiley (1980) noted that at the time of egg laying, females flew swiftly and directly to host nests to lay their eggs, suggesting that nest selection occurs in advance of egg laying.

Fraga (1978) observed on 14 occasions females watching movements of potential hosts for up to 17 minutes. Fraga (1985) reported that often (16.7%, n = 42 observations) two or more Shiny Cowbirds watched nests together. Gochfeld (1979b) suggested that Shiny Cowbirds may use perches to watch the activities of potential hosts. In a comparative study between two closely related species of meadowlarks with similar breeding biology, Gochfeld (1979b) found that heavily parasitized Greater Red-breasted Meadowlarks rarely (1 in 24 nests) nest more than 3 m from elevated structures, usually a fence row, whereas unparasitized Pampas Meadowlarks (*Sturnella delfilippi*) rarely (1 in 12 nests) nest within 5 m of an elevated structure.

Hudson (1874) anecdotally relayed that he had set out several old nests in trees and later found Shiny Cowbird eggs in them. On Puerto Rico, Wiley (1988) set out host nests, adding one real gelatin-filled host egg per day, in alternating patterns with empty host nests. He found no Shiny Cowbird parasitism among empty nests (n = 81), whereas he found low parasitism among nests with host eggs (5.3%, n = 75). He also found a low incidence (5.5%) of parasitism among 55 additional nests with eggs and a dummy host placed next to the nest. His methods were similar to those of Lowther (1979), who found a low incidence of Brown-headed Cowbird parasitism (chapter 7). Given the high density of Shiny Cowbirds on Puerto Rico and the high rate of multiple parasitism, Wiley's results were somewhat surprising and strongly suggest that Shiny Cowbirds cue in on host activity.

Observations of males accompanying females during nest searches vary. In Argentina, Hudson (1920) often saw males close to females. Fraga (1978) never observed males accompany females and believed that Hudson's remarks were a consequence of mistaking Screaming Cowbirds for Shiny Cowbirds. Mason (1987) also noted that male Shiny Cowbirds do not accompany females on nest searches, but females may search together. On Puerto Rico, Wiley (1988) mentioned that males often accompanied nest-searching females. Young (in Friedmann 1963) suggested that in British Guiana, males search for nests more than do females.

Unusual Interest in Nests and Nest Building Activities

Friedmann (1963) felt that interest in host nests was stronger among Shiny Cowbirds than among Brown-headed Cowbirds and suggested that this interest, particularly in domed nests, may be a relict habit from a self-nesting mode. Friedmann (1963) also noted that interest in domed furnarid nests was greater than would be suggested by the rate of parasitism. For example, in Argentina, he found a parasitism rate of only 9.2% ($n = 217$) on Rufous Horneros. However, dump nests with the largest number of eggs had been constructed by Rufous Horneros. Friedmann (1929b:80) cited two examples of attempts to construct nests:

> [What Hudson] . . . has known [of] birds of the present species to attempt to build nests on two occasions is of real importance as it shows unmistakable evidence of a weak, inefficient recurrence of an ancestral habit. The only other pertinent evidence on this point comes from Page's observations on this species in captivity. He writes (*Avicultural Mag.* 1905 pp. 137–141) that he had a pair of these Cowbirds in his aviary and that the female "bullied" the male, pulling out feathers from the lores and forehead. "After about a month of this bickering, they patched up their differences and began courting again. Shortly after I found she had built a nest (I saw her finishing it, the male supplying material and looking on) in a large log provided for my Cockateels, putting in for a foundation small tufts of grass with earth attached, arranging on the top of this a loose structure of small sticks and bents similar to that of our Starling; no eggs were laid."

This is undoubtedly a very unusual observation and similar to that found in Steel-blue Whydahs (Nielsen 1956; chapter 1).

Hosts

Friedmann (1929b) reported 90 species known to be parasitized by Shiny Cowbirds. In 1963, Friedmann listed 146 species known to be parasitized and 26 known to have raised Shiny Cowbirds. Between 1971 and 1977 our knowledge of Shiny Cowbird hosts had increased rapidly; Friedmann et al. (1977) reported the list to be 176 species, with 36 species known to have raised Shiny Cowbirds. Friedmann and Kiff (1985) listed 201 species known to be parasitized and 53 host species known to have raised Shiny Cowbirds. Since 1985, Shiny Cowbirds have expanded their range, coming into contact with new hosts, and the list has increased by at least 13. This list will probably continue to increase as Shiny Cowbirds come into contact with new hosts in the United States. Presently, 214 species are known to be parasitized, and 68 hosts are known (or presumed) to have raised Shiny Cowbirds (appendix E).

Multiple Parasitism

Shiny Cowbirds appear to engage in multiple parasitism more than other cowbirds. Fraga (1985) commented that multiple parasitism is more common than single parasitism. On Puerto Rico, the mean number of Shiny Cowbird eggs in parasitized nests is higher than the mean number of host eggs for several species. Multiple parasitism may reflect a high density of Shiny Cowbirds. On Puerto Rico, where the density of Shiny Cowbirds is approximately 49 individuals per km^2, 74.3% of parasitized nests were multiply parasitized; on St. Lucia, where density of Shiny Cowbirds is approximately six individuals per km^2, 37.9% of parasitized nests were multiply parasitized (Post et al. 1990). Multiply parasitized nests are sometimes a result of one female, but often, more than one female parasitizes a single nest. Hoy and Ottow (1964) observed a nest with 14 Shiny Cowbird eggs and believed each egg was from a different female.

Wiley (1985) noted that the incidence of multiple parasitism was greater in nests of species parasitized at a heavy rate (2.7 Shiny Cowbird eggs per nest) than nests of species parasitized at a low rate (1.1 Shiny Cowbird eggs per nest; appendix A and table 8.3). Frequently used Fork-tailed Flycatchers (*Tyrannus savana*) received an average of 3.5 Shiny Cowbird eggs per parasitized nest (Hudson 1874). Interestingly, Cavalcanti and Pimentel (1988) found a very low rate of parasitism in Fork-tailed Flycatchers in Brazil and commented on Mason's (1986a) findings of ejection in Argentina. Hudson's work was conducted more than a century before Mason found Fork-tailed Flycatchers to be ejectors in the same general area; possibly, they had become ejectors in the interim.

Shiny Cowbirds also engage in dump nesting (Friedmann 1929b, 1963; Hoy and Ottow 1964; Fraga 1978; Williams 1981; Cruz et al. 1985), perhaps more frequently than Brown-headed Cowbirds. Friedmann (1963) noted that domed furnarid nests were often the object of dump nesting at the end of a breeding season and that first-year individuals may have been primarily responsible. Up to 37 Shiny Cowbird eggs have been found in individual nests (Miller 1917).

Effects of Parasitism on Hosts

Shiny Cowbirds usually have a detrimental effect on the nesting success of their hosts. Hoy and Ottow (1964) remarked on how uncommon it is to find nonparasitized nests of many species in Argentina once the Shiny Cowbird

Table 8.3

Shiny Cowbird Egg Distribution in Nests of Some Hosts from Argentina

Host	Number of Shiny Cowbird eggs							Source
	0	1	2	3	4	5	≥6	
Rufous Hornero	13	4	0	0	0	0	0	Mason 1986b
Freckle-breasted Thornbird	6	1	0	0	0	0	0	Mason 1986b
Yellow-browed Tyrant	6	4	0	0	0	0	0	Mason 1986b
Fork-tailed Flycatcher	2	0	1	1	3	2	1	Hudson 1874
	3	2	1	1	0	0	0	Mason 1986b
White-rumped Swallow	6	3	1	0	0	0	0	Mason 1986b
House Wren	7	3	3	1	2	0	1	Mason 1986b
Rufous-bellied Thrush	0	2	1	0	0	1	0	Mason 1986b
Chalk-browed Mockingbird	18	17	12	11	3	3	4	Mason 1986b
	19	22	11	10	7	0	0	Fraga 1985
Saffron Finch	44	1	0	0	0	0	0	Mason 1986b
Grassland Yellow-Finch	12	4	0	0	0	0	0	Salvador 1983
Chestnut-capped Blackbird	165	41	3	3	1	0	0	Salvador 1983
Brown-and-Yellow Marshbird	19	25	10	7	3	2	0	Mermoz and Reboreda 1994
Bay-winged Cowbird	11	4	0	0	0	0	0	Mason 1986b
Rufous-collared Sparrow	11	15	4	6	2	2	0	Fraga 1978
	40	8	1	2	0	0	0	Mason 1986b

breeding season commences. For a few species, Shiny Cowbirds appear to have little impact on nesting success; for example, Cruz et al. (1990) found that among Yellow-hooded Blackbirds (*Agelaius icterocephalus*), parasitized individuals were equally successful as nonparasitized individuals. For most species, however, nonparasitized nests are more successful than parasitized nests (table 8.4)

Egg Removal, Egg Puncturing, Egg Eating. Host egg destruction appears to be more prevalent among Shiny Cowbirds than other cowbirds. While Mayfield (1961b) suggested that Brown-headed Cowbirds never entirely empty a nest of host eggs, Fraga (1985) reported that Shiny Cowbirds removed all host eggs in 8 of 50 (16.0%) parasitized Chalk-browed Mockingbird nests. Females and males both participate in the behavior (Hoy and Ottow 1964), and Shiny Cowbirds also puncture each other's eggs (Hudson 1874; King 1973b). Egg puncturing extends beyond the Shiny Cowbirds' own breeding season with continued high frequency (Friedmann 1963). Hoy and Ottow (1964) commonly found punctured eggs in March, beyond the breeding season of Shiny Cowbirds in Argentina. Egg puncturing by Shiny Cowbirds takes place at any time during the egg phase—before, during or after egg laying.

Shiny Cowbirds may eat host eggs (Hudson 1874; Hoy and Ottow 1964), but the high observed frequency of punctured host eggs suggests that consumption is not as common as puncturing without consumption, and it is doubtful that the purpose of the egg-puncturing habit is for nutritional value. Shiny Cowbirds also often remove host eggs from nests, as is evidenced from the smaller clutch size of parasitized nests compared with nonparasitized nests (Fraga 1978, Mason 1980, 1985, 1986; Salvador 1984; Cruz et al. 1985; Wiley 1985).

The behavior is not necessarily universal, and patterns in the variation are not readily apparent, but egg puncturing may occur more frequently in areas of recent colonization. Cruz et al. (1990) reported no significant differences in clutch size between parasitized and nonparasitized Yellow-hooded Blackbird nests in Trinidad; they found only ten instances of suspected host egg damage by Shiny Cowbirds among 153 parasitized nests. Similarly, Wiley and Wiley (1980) found almost no evidence of Shiny Cowbirds puncturing eggs of Yellow-hooded Blackbirds in Trinidad, Venezuela, and Suriname.

On recently colonized Puerto Rico, Shiny Cowbirds may indiscriminately puncture a large percentage of host eggs. Shiny Cowbird eggs were as likely to be punctured as host eggs, and egg puncturing was significantly lower in singly parasitized nests (7.4%, n = 476) than in multiply parasitized nests (46.7%, n = 152, Nakamura and Cruz, in press). Nakamura and Cruz (1993) observed that approximately 88% of nests which contained one or more punctured eggs failed, primarily as a consequence of abandonment. What maintains this destructive and apparently maladaptive behavior is not well understood, but Nakamura and Cruz suggested that Shiny Cowbirds may puncture host eggs in order to manipulate the host to renest. Nakamura and Cruz observed that banded females monitored renesting attempts and subsequently may have parasitized new clutches. They further suggested that egg puncturing by Shiny Cowbirds appears to be more frequent in the tropics than in temperate zones, and the difference may reflect an extended breeding season in the tropics, with ample opportunities for hosts to renest. However, puncturing is also very common in temperate Argentina (Friedmann 1963; Hoy and Ottow 1964; Fraga 1978, 1986), where the breeding season extends only from September through late January (Fraga 1978, 1985). Nakamura and Cruz found that the incidence of egg puncturing was highest under high Shiny Cowbird densities and lowest in areas of marginal

Table 8.4

Success (Mean Number Fledged; Sample Size in Parentheses) of Nests Parasitized and Nests Not Parasitized by the Shiny Cowbird

Host	Mean Hosts Fledged per Nest Nonpara.	Para.	Mean Hosts Fledged per Egg Nonpara.	Para.	Percent Successful Nests Nonpara.	Para.	Year	Location	Source
Caribbean Elaenia					27 (15)	50 (2)	1984-85	St. Lucia	Nakamura 1995
Puerto Rican Flycatcher	3.00 (2)	0.64 (11)			100 (2)	60 (11)	1973-82	Puerto Rico	Cruz et al. 1985
Gray Kingbird					70 (64)	64 (14)	1982-88	Puerto Rico	Nakamura 1995
Red-legged Thrush	1.21 (24)	0.00 (1)			80 (75)	0 (1)	1973-82	Puerto Rico	Cruz et al. 1985
Northern Mockingbird					50 (24)	0 (1)	1973-82	Puerto Rico	Cruz et al. 1985
					60 (43)	100 (1)	1973-82	Puerto Rico	Cruz et al. 1985
Chalk-browed Mockingbird			0.34 (50)	0.08 (39)			1979-84	Argentina	Salvador 1984
	1.2 (14)	0.38 (50)		0.15 (127)	50 (14)	20 (50)	1972-79	Argentina	Fraga 1985
Puerto Rican Vireo					40 (5)	21 (34)	1982-88	Puerto Rico	Nakamura 1995
Black-whiskered Vireo	2.00 (2)	0.33 (9)			100 (2)	60 (9)	1973-82	Puerto Rico	Cruz et al. 1985
					0 (3)	43 (7)	1982, 83, 86	Puerto Rico	Post et al. 1990
					46 (26)	29 (77)	1982-88	Puerto Rico	Nakamura 1995
					50 (2)	29 (21)	1982-84	St. Lucia	Post et al. 1990
					67 (3)	24 (21)	1984-85	St. Lucia	Nakamura 1995
Yellow Warbler	0.58 (26)	0.14 (81)			30 (26)	40 (81)	1973-82	Puerto Rico	Cruz et al. 1985
					20 (22)	50 (65)	1975-81	Puerto Rico	Wiley 1985
					none	62 (8)	1982, 83, 86	Puerto Rico	Post et al. 1990
					40 (139)	32 (123)	1982-88	Puerto Rico	Nakamura 1995
					54 (11)	19 (16)	1982-84	St. Lucia	Post et al. 1990
					50 (12)	19 (16)	1984-85	St. Lucia	Nakamura 1995
Adelaide's Warbler					39 (31)	100 (1)	1982-88	Puerto Rico	Nakamura 1995
Troupial					none	80 (5)	1973-82	Puerto Rico	Cruz et al. 1985
					18 (11)	0 (1)	1982-88	Puerto Rico	Nakamura 1995

Host	Mean Hosts Fledged per Nest		Mean Hosts Fledged per Egg		Percent Successful Nests		Year	Location	Source
	Nonpara.	Para.	Nonpara.	Para.	Nonpara.	Para.			
Black-cowled Oriole		0.33 (12)			none	80 (12)	1973–82	Puerto Rico	Cruz et al. 1985
Yellow-hooded Blackbird	0.13 (63)	0.18 (44)	0.06 (127)	0.08 (95)	11 (63)	16 (44)	1979–81, 84	Trinidad	Cruz et al. 1990
Yellow-shouldered Blackbird	0.75 (20)	0.40 (48)	0.38 (40)	0.18 (105)	50 (20)	25 (48)	1973–75, 76	Puerto Rico	Post and Wiley
	0.33 (12)	0.25 (152)			20 (12)	40 (152)	1973–82	Puerto Rico	Cruz et al. 1985
		0.34 (44)		12.0	none	29 (44)	1982	Puerto Rico	Cruz et al. 1985
					55 (20)	38 (163)	1982–88	Puerto Rico	Nakamura 1995
Greater Antillean Grackle	1.48 (195)	1.36 (22)			70 (195)	50 (23)	1973–82	Puerto Rico	Cruz et al. 1985
					46 (96)	0 (4)	1982–88	Puerto Rico	Nakamura 1995
Carib Grackle					37 (76)	100 (2)	1984–85	St. Lucia	Nakamura 1995
Rufous-collared Sparrow	0.55 (11)	0.24 (29)	0.17 (35)	0.08 (92)	18 (11)	10 (29)	1970–<78	Argentina	Fraga 1978
	1.00 (10)	0.47 (19)	0.43 (23)	0.28 (32)	40 (10)	31 (19)	1969–70	Argentina	King 1973b
Village Weaver	1.75 (126)	0.91 (12)	0.58 (126)	0.35 (11)	68 (126)	55 (11)	1974–77	DR	Cruz and Wiley 1989a
	2.00 (24)	0.67 (9)	0.68 (24)	0.23 (9)	71 (24)	44 (9)	1982	DR	Cruz and Wiley 1989a
Bronze Mannikin	4.00 (5)	0.00 (1)			80 (5)	0 (1)	1973–82	Puerto Rico	Cruz et al. 1985

Table 8.5
Some Species that Abandon Nests Parasitized by Shiny Cowbirds

Species	Location	Source
Vermillion Flycatcher	Argentina	Hudson 1920
Great Pampa-Finch	Argentina	Hoy and Ottow 1964
Many-colored Chaco-Finch	Argentina	Hoy and Ottow 1964
Red-crested Finch	Argentina	Hoy and Ottow 1964
Saffron Finch	Argentina	Hoy and Ottow 1964
Black-capped Warbling-Finch	Argentina	Hoy and Ottow 1964
Chestnut-capped Blackbird	Argentina	Hoy and Ottow 1964
Chalk-browed Mockingbird	Argentina	Fraga 1985
Rufous-collared Sparrow	Argentina	Fraga 1978
Brown-and-Yellow Marshbird	Argentina	Hudson 1874

habitat. Post and Wiley (1977a) also believed that their observed high incidence of egg-puncturing behavior is reflective of an imbalance between host
and parasite populations.

Removing Host Young. Reports of Shiny Cowbirds removing host young
from nests do not appear as frequently as they do for Brown-headed Cowbirds. Fraga (1985) suspected one case of a Shiny Cowbird removing a host
nestling from the nest of a Chalk-browed Mockingbird.

Cost of Abandonment. Some hosts regularly abandon their parasitized nests
(table 8.5). Desertion can be an adaptive response to Shiny Cowbird parasitism if renesting occurs and does not result in another parasitized clutch.
Whether a host responds with abandonment may depend on investment.
Wiley (1985) found that for four hosts on Puerto Rico, abandonment was
particularly high (91%) if the nest was parasitized before host eggs were
laid, and abandonment was less likely if the host had a completed clutch.

Host Size and Growth. In nests of smaller hosts, Shiny Cowbirds outcompete
host siblings for food and/or trample them. In nests of larger species, host
nestlings may not suffer as much mortality as do smaller hosts. On Puerto
Rico, Shiny Cowbirds reduced the nesting success of Yellow-shouldered
Blackbirds during the egg stage, but once hosts hatched, they successfully
competed with Shiny Cowbird nestlings (Post and Wiley 1977a). However,
Mermoz and Reboreda (1994) found when brood reduction occurred in

parasitized Brown-and-Yellow Marshbirds, host nestlings were affected more than Shiny Cowbird nestlings.

Interference with Incubation. Host eggs may have reduced hatching success in parasitized nests, and this may intensify as the number of parasitic eggs increases. The number of hatched Chalk-browed Mockingbird eggs in parasitized clutches was significantly lower (75.0%, n = 40) than in non-parasitized clutches (95.5%, n = 22), and most unhatched eggs occurred in clutches of six eggs or more (Fraga 1985). Similarly, Hudson (1874) noted that eggs in parasitized Fork-tailed Flycatcher nests were often addled while Shiny Cowbird eggs were viable. Mermoz and Reboreda (1994) reported that although the number of host eggs in Brown-and-Yellow Marshbird nests did not differ between parasitized and nonparasitized clutches, hatching failure was higher among parasitized nests.

Host Suitability

Not all birds are suitable hosts, and Shiny Cowbirds are limited by the same breeding parameters that constrain Brown-headed Cowbirds (chapter 7). Shiny Cowbirds show obvious preferences for certain hosts; for example, on Puerto Rico, Yellow-shouldered Blackbirds are parasitized at or nearly at a 100% rate, whereas other hosts are parasitized at much lower rates. Similar situations are found elsewhere. In Brazil, Sick (1958) found Rufous-collared Sparrows to be preferred and heavily parasitized. In Chile, the preferred host is the Common Diuca (*Diuca diuca*, Johnson 1967). White-headed Marsh Tyrants (*Arundinicola leucocephala*) are the most common host in British Guiana (Young in Friedmann 1963), and House Wrens (*Troglodytes aedon*) were the most preferred host in Suriname (Haverschmidt, in Friedmann 1963). In Cordoba, Argentina, Salvador (1984) suggested that Chalk-browed Mockingbirds are the preferred local host, and although it is parasitized at a heavy rate of 88%, only 6.5% of cowbird eggs in Chalk-browed Mockingbird nests produced fledglings. Variation in parasitism rates on certain species may reflect alternatives in the host community.

Egg Acceptance/Rejection. Brood parasites are constrained by whether or not a host or potential host accepts and incubates their eggs. Relatively few hosts or potential hosts are known to eject Shiny Cowbird eggs (table 2.1), but several experimental studies have revealed that some species that might otherwise be good hosts eject Shiny Cowbird eggs. Mason (1986b) noted that like North American ejector species, South American ejector species tend to be larger than accepter species. Acceptance-rejection status is more complex for South American and West Indian hosts than it is for North American hosts because Shiny Cowbird eggs are polymorphic. Polymorphism of Shiny Cowbird eggs confounds the possible dichotomy of accepters and ejectors because several species accept spotted eggs and eject immaculate eggs (Chalk-browed Mockingbird, Mason 1986a; Brown-and-Yellow Marshbird, Mermoz and Reboreda 1994), while another species accepts immaculate eggs and ejects spotted eggs (Yellow-winged Blackbird, Fraga 1985). Additionally, several species appear to be intermediate between accepters and rejecters. Hosts that eject all cowbird eggs are obviously not suitable hosts, but it is not so clear how suitable hosts with intermediate responses are. Fraga (1985) and Mason (pers. comm.) found that Chalk-browed Mockingbirds eject immaculate eggs but accept most spotted eggs, yet this host is considered highly preferred (Fraga 1985).

Rufous Horneros (*Furnarius rufus*) accept cowbird eggs depending on size; widths of less than 88% of their own egg width were ejected, and there was no overlap in egg widths of accepted and ejected Shiny Cowbird eggs (Mason and Rothstein 1986). Shiny Cowbird eggs in Uruguay are generally larger than in Argentina, and this allowed a larger number of eggs to be accepted by Rufous Horneros in Uruguay; 58% (n = 162 eggs) fell above the discrimination threshold in Uruguay, whereas only 15% (n = 167 eggs) did in Argentina (Mason and Rothstein 1986). The Rufous Hornero is an excellent host (Mason 1986a), and Mason and Rothstein (1986) explained the smaller Shiny Cowbird egg size in Argentina by suggesting that selection on Rufous Horneros by Shiny Cowbirds is a recent phenomenon in Buenos Aires Province. Hoy and Ottow (1964) remarked on the frequency with which Shiny Cowbird eggs are seen beneath the nests of Rufous Horneros. White Monjitas (*Xolmis irupero*) using old *Furnaris* nests may also eject Shiny Cowbird eggs before the onset of their own egg laying.

For some hosts, the number of Shiny Cowbird eggs may, in part, determine abandonment. Fraga (1978) reported that 7 of 29 Rufous-collared Sparrow nests were abandoned; the abandoned nests contained a higher average number of Shiny Cowbird eggs (2.9) than parasitized nests that were not abandoned (1.8).

Burial of Shiny Cowbird eggs is not common. Williams (1981) reported a buried Shiny Cowbird egg in a Cinereous Finch (*Piezorhina cinerea*) nest in

Peru. Yellow-browed Tyrants *(Satrapa icterophrys)* sometimes bury parasitized clutches. Hudson (1920:78) relayed a chilling account of a live burial:

> Finding a very thick nest one day, containing two half-fledged young birds besides three addled eggs, I opened it, removing the upper portion or additional nest intact, and discovered beneath it three buried *Molothrus* eggs, their shells encrusted with dirt and glued together with broken-egg matter spilt over them. In trying to get them out without pulling the nest to pieces I broke them all; two were quite rotten, but the third contained a living embryo, ready to be hatched, and very lively and hungry when I took it in my hand. The young Tyrant-birds were about a fortnight old, and as they hatch out only about twenty days after the parent-bird begins laying, this parasitical egg with a living chick in it must have been deeply buried in the nest for not less than five weeks. Probably after the young Tyrant-birds came out of their shells and began to grow, the little heat from their bodies, penetrating to the buried egg, served to bring the embryo in it to maturity; but when I saw it I felt (like a person who sees a ghost) strongly inclined to doubt the evidence of my own senses.

Breeding Season. Parasites and hosts may share the same cues that signal the onset of the breeding season. In temperate zones, increased photoperiod brings Brown-headed Cowbirds into breeding condition (chapter 7). Shiny Cowbirds inhabiting the tropics, where changes in photoperiod are more subtle, may have to rely more on host activity in order to synchronize their breeding activity with preferred hosts. Individual hosts nesting early and late during the Shiny Cowbird's breeding season may experience lower parasitism rates or escape parasitism altogether (King 1973b; Fraga 1985).

Incubation. The incubation period of Shiny Cowbirds is short relative to most of their hosts; therefore, they are usually not constrained by host incubation periods. When host incubation is equally short, Shiny Cowbirds may experience reduced fledging success. Incubation of Yellow-hooded Blackbirds is only 10–11 days, and only 4 of 28 (14.3%) Shiny Cowbird eggs in Yellow-hooded Blackbird nests successfully fledged (Wiley and Wiley 1980). Wiley and Wiley suggested that a short incubation period is the most important defense of Yellow-hooded Blackbirds against Shiny Cowbirds.

Host Care: Diet and Feeding Rates. As with most passerines, Shiny Cowbird nestlings require a diet high in animal protein. Hosts that provide a low animal protein diet might be considered unsuitable hosts. Mason (1986a) observed three seed specialists on his study site in Argentina, none of which have been known to raise Shiny Cowbirds. In cross-fostering experiments, Mason (1986a) found Saffron Finches *(Sicalis flaveola)* and House Sparrows feed Shiny Cowbird nestlings an unsuitable diet. Wiley (1988) found that most granivorous, nectivorous, and frugivorous passerines were not parasitized on Puerto Rico, nor were columbids and cuckoos. Additionally, Wiley found that species that required special feeding behavioral or structural adaptations as nestlings were not parasitized: doves, hummingbirds, and mannikins.

Fraga (1985) found that Shiny Cowbird nestlings in Chalk-browed Mockingbird nests often died of undernourishment while their host siblings did not. However, in the same area, Mason (1986a) observed a much lower rate of starvation among Shiny Cowbirds in Chalk-browed Mockingbird nests. Brood reduction, or adjustment of brood size according to the availability of food and favoring larger nestlings, has been observed in Chalk-browed Mockingbirds, and because Shiny Cowbird nestlings in Fraga's (1985) study were smaller than Chalk-browed Mockingbird nestlings, the higher mortality of Shiny Cowbirds might have been a consequence of brood reduction rather than specific discrimination against the brood parasites. King (1973b) suggested that the brood load of Rufous-collared Sparrows may be reduced to 3.4–3.5 sparrows or 1.75 Shiny Cowbirds; whether mortality occurred among the sparrows (which fledge at roughly half the weight of Shiny Cowbirds) or cowbirds appeared to depend on which eggs hatched first, with the last nestling hatched suffering mortality. Therefore, while Rufous-collared Sparrows, and perhaps other small passerines, may be suitable hosts for one cowbird, they may be unsuitable for multiple parasitism because they are generally incapable of raising more than one healthy cowbird. Interestingly, Mason (1986a) noted that a trend existed for multiple parasitism to be greater in the nests of larger hosts. Brood reduction also occurred among Brown-and-Yellow Marshbirds, and even though the host nestlings were larger (33–58 g, $n = 23$) than cowbird nestlings (31.5–47 g, $n = 14$), brood reduction affected more hosts than cowbirds (Mermoz and Reboreda 1994). However, Mermoz and Reboreda pointed out that brood reduction also occurred in nonparasitized nests and may have been affected more by the lack of helpers than by brood parasitism. Nevertheless, it is clear that Brown-and-Yellow Marshbirds are excellent hosts and can successfully raise four cowbirds in parasitized nests (Mermoz and Reboreda 1994).

Nest Types. Shiny Cowbirds are apparently not limited by nest type. They parasitize a wide variety of nests: open cups, natural cavities, nest boxes, domed nests, and pendulous nests. Hoy and Ottow (1964) reported that even Short-billed Canasteros *(Asthenes baeri)* nests, with entrances covered in thorns and spikes, are often parasitized. Shiny Cowbirds do not shy away from cavity nests as Brown-headed Cowbirds do. Many excellent and preferred hosts, such as Rufous Horneros, House Wrens, and Yellow-shouldered Blackbirds, nest in cavities. Mason (1986a) found Rufous Horneros to be excellent hosts and attributed this to their superior mud dome nests; other birds using their vacated nests also experience a high rate of nesting success.

The advantages of cavity nests can best be seen by comparing open versus cavity nests in the same host. Yellow-shouldered Blackbirds provide us with this unusual opportunity because they nest both in the open and in cavities. Post and Wiley (1977a) found that Shiny Cowbird eggs were more successful in cavity nests (0.52 fledglings/egg, $n = 21$) than in open nests (0.19 fledglings/egg, $n = 59$; see also Post 1981). Post and Wiley (1977a) suggested that the conspicuousness of some Yellow-shouldered Blackbird nests may have facilitated parasitism. Predation was greater among open nests than cavity nests (Wiley 1988).

Habitat. How much Shiny Cowbirds are limited by the habitat of their hosts is not well understood, but Shiny Cowbirds are generally not abundant in dense tropical rainforests. The large gap in their range through the Amazon Rainforest is probably an indication that this is unsuitable habitat, and birds nesting in dense rainforests, for the most part, escape Shiny Cowbird parasitism. Shiny Cowbirds also are not seen in some plantations, such as sugar cane, and birds nesting in these habitats may also escape parasitism (T. Nakamura, pers. comm.).

Nest Site Characteristics. Little work has been published on comparison of nest site characteristics between parasitized and nonparasitized nests. Gochfeld (1979b) suggested that Greater Red-breasted Meadowlarks were more heavily parasitized than Pampas Meadowlarks because they nest near fence rows where Shiny Cowbirds sat on perches. Shiny Cowbirds may also parasitize with low frequency individuals nesting in the central portions of marshes, but this may be confounded with host density.

Nest Attentiveness. Aggression toward intruding parasites does not necessarily prevent parasitism; in fact, aggression may provide cues to the location of nests and quality of the host. Host aggression may be circumvented by larger group size of parasites. Fraga (1985) observed that when Shiny Cowbirds approached nests of Chalk-browed Mockingbirds, who are generally extremely aggressive toward Shiny Cowbirds, cowbirds arrived in groups. However, Hoy and Ottow (1964) uncovered no parasitism on Common Thornbirds (*Phacellodomus rufifrons*) that effectively chase off all intruders.

Cruz et al. (1995) compared nest attentiveness between male Yellow-hooded Blackbirds in Trinidad and male Yellow-shouldered Blackbirds on Puerto Rico to determine if the hosts recognized Shiny Cowbirds as a threat. They found Yellow-hooded Blackbirds were more attentive (95.7%, $n = 47$ Shiny Cowbird intrusions) and more aggressive (36 chases of 47 Shiny Cowbird intrusions) than Yellow-shouldered Blackbirds (54.1%, $n = 85$ Shiny Cowbird intrusions and 13 chases of 85 Shiny Cowbird intrusions). Even though Yellow-hooded Blackbirds are not as heavily parasitized as Yellow-shouldered Blackbirds, they are parasitized at a moderately high rate (33.3%, $n = 15$, Wiley and Wiley 1980; 40.6%, $n = 377$, Cruz et al. 1990). However, as Neudorf and Sealy (1994) pointed out, attentiveness during early morning hours may be a critical determinant of host suitability. Mason (1980) reported low aggression levels and poor attentiveness of Chalk-browed Mockingbirds presented with Shiny Cowbird models near their nests, but Fraga (1985) suggested that Chalk-browed Mockingbirds are both attentive and aggressive toward Shiny Cowbirds.

Host Size. Shiny Cowbirds parasitize a wide size range of hosts, but trends of size preference have not been adequately explored to draw conclusions. Mason (1986b) suggested that larger hosts are preferred based on three lines of evidence: (1) larger hosts are parasitized more frequently than smaller hosts, (2) larger ejectors are parasitized more frequently than many accepters, and (3) when larger hosts are limited, Shiny Cowbirds switch to smaller hosts. He observed that parasitism on Rufous-collared Sparrows dropped from 72.5% to 43.5% when new agricultural conditions attracted larger White-browed Blackbirds (*Sturnella superciliaris*) to the area. In other areas, however, smaller hosts, such as House Wrens, seem to be preferred. When exploring size preferences, it would be useful to consider variations with cowbird densities to determine any possible associations between multiple parasitism, which appears to increase with density, and inability of smaller hosts to raise more than one cowbird.

Host Density. One might expect that common species would be parasitized at a higher frequency than rare species if cowbirds form search images similar to the search images of some predators (Friedmann et al. 1977). However, there is little evidence to support this, and other breeding parameters, such as nest concealment and aggression, may confound any association between host density and incidence of parasitism. On Puerto Rico, Wiley (1988) found no association between host density and rates of Shiny Cowbird parasitism.

Additionally, high host density can facilitate parasitism or make parasitism difficult, depending on host aggression and habitat. Shiny Cowbirds may find entering centers of dense colonies difficult if hosts are aggressive and/or if trees are unavailable to perch in. For example, Cruz et al. (1990) found that in Trinidad, Shiny Cowbird parasitism on Yellow-hooded Blackbirds nesting on the periphery of marshes (< 20 m from edge) was 50% (n = 26), for those nesting centrally parasitism was 24% (n = 21), and for those nesting in isolation from the colony parasitism was 76% (n = 21). On the other hand, for colonies that Shiny Cowbirds can easily travel through, high nest density may minimize nest-searching efforts. For example, Yellow-shouldered Blackbirds nest in colonies within mangroves, which provide perches, and Yellow-shouldered Blackbirds are not highly aggressive toward intruding cowbirds.

Alternative Host Choices in Community. The observed variation in local host preference may depend on host availability within a given avian community. If a preferred host is absent, cowbirds will likely select another preferred species or increase the number of hosts. Mermoz and Reboreda (1994) found Brown-and-Yellow Marshbirds to be heavily parasitized (74.3%) in Buenos Aires Province. Where they were absent at one site, Mason (1986b) noted a wider range in host selection.

Are Some Hosts More Suitable than Others as Foster Parents?

Some hosts are clearly more suitable than others, as evidenced from their success in nests of different hosts (table 8.6), and a suite of host breeding characteristics contribute to differential suitability. Success of cowbirds in the nests of different species may be complicated by their own activity (such as puncturing each other's eggs), and therefore, a host that fledges a low number of cowbirds may actually be a good host. Mason (1986a) provided a good example of this paradoxical problem. Chalk-browed Mockingbirds are good hosts; therefore, competition for their nests is high. Consequently, they are often multiply parasitized and subject to a high frequency of egg puncturing by Shiny Cowbirds. This trend is also observed for Yellow-shouldered Blackbirds on Puerto Rico.

Table 8.6

Success of Shiny Cowbirds among Some Hosts

Host Species	Number of Cowbird Eggs	Percent Fledged per Egg	Location	Source
Caribbean Elaenia	2	50.0	St. Lucia	Post et al. 1990
Puerto Rican Flycatcher	8	37.5	Puerto Rico	Post et al. 1990
	35	31.4	Puerto Rico	Wiley 1985
Fork-tailed Flycatcher	23	4.3	Argentina	Salvador 1983
Gray Kingbird	1	0.0	Puerto Rico	Wiley 1985
Northern Mockingbird	1	0.0	Puerto Rico	Wiley 1985
Chalk-browed Mockingbird	31	6.5	Argentina	Salvador 1983
	102	5.9	Argentina	Fraga 1985
Red-legged Thrush	1	0.0	Puerto Rico	Wiley 1985
Black-whiskered Vireo	9	22.2	Puerto Rico	Post et al. 1990
	23	34.8	Puerto Rico	Wiley 1985
	28	32.1	St. Lucia	Post et al. 1990
Yellow Warbler	8	75.0	Puerto Rico	Post et al. 1990
	153	25.5	Puerto Rico	Wiley 1985
	21	9.5	St. Lucia	Post et al. 1990
Adelaide's Warbler	1	0.0	Puerto Rico	Post et al. 1990
Rufous-collared Sparrow	59	6.8	Argentina	Fraga 1978
Troupial	8	6	Puerto Rico	Wiley 1985
Black-cowled Oriole	47	76.6	Puerto Rico	Wiley 1985
Yellow-hooded Blackbird	28	14.3	Venezuela	Wiley and Wiley 1980
Yellow-shouldered Blackbird	164	25.6	Puerto Rico	Post and Wiley 1977a
	374	14.2	Puerto Rico	Post et al. 1990
	481	20.2	Puerto Rico	Wiley 1985
Chestnut-capped Blackbird	48	4.2	Argentina	Salvador 1983
Red-breasted Blackbird	18	16.7	Venezuela	Ramo and Busto 1981
Greater Antillean Grackle	2	0.0	Puerto Rico	Post et al. 1990
	26	7.7	Puerto Rico	Wiley 1985
Bay-winged Cowbird	20	5.0	Argentina	Fraga 1986
Bronze Mannikin	1	0.0	Puerto Rico	Wiley 1985

Comparison of Historic with Recent Hosts

The range expansion that Shiny Cowbirds have undergone has brought them into contact with a new array of hosts, as reflected in the ever-growing list of hosts. In areas of longer contact and reasonable stability, host-parasite relationships may remain relatively stable. Cruz et al. (1995) compared historic records of host use by Shiny Cowbirds on Trinidad and Tobago and found that host use has been stable throughout the last century. The same birds that were heavily parasitized nearly a century ago remain heavily parasitized, and only one new host, the Tropical Kingbird (*Tyrannus melancholicus*), has been added to the list based on a sole observation.

Suggestions for Future Research

Research opportunities on Shiny Cowbirds are abundant, and the wealth of knowledge we possess on Brown-headed Cowbirds leaves ample room for some valuable comparative work. Research that would contribute to a better understanding of how to manage Shiny Cowbirds should take high priority. In addition to further information on basic breeding biology of Shiny Cowbirds and their hosts, the following topics may be of particular interest.

Movement Patterns

It is important to understand migration and patterns of movement within the breeding and nonbreeding seasons in order to predict how far the Shiny Cowbird will spread into North America.

Mating System

Almost no work has been published on the mating system of Shiny Cowbirds. Although more research is required for a clear understanding of Brown-headed Cowbird mating systems, interesting comparative work awaits.

Territoriality, Egg Puncturing and Egg Morphology

The relationship among egg puncturing, apparent lack of territorial behavior in females, nest searches by more than one female, social cohesiveness during the breeding season, and the tremendous variation in egg morphology has not been investigated. We need to first establish that females lay a consistent type of egg; so far this has only been assumed. If females recognize their own eggs and avoid puncturing them, perhaps egg polymorphism evolved from selection pressure to compete with other females. It may be more advantageous for females to maintain social cohesiveness during the breeding season than to segregate into exclusive territories. In this case, territorial behavior may be expressed through egg puncturing instead of overtly aggressive behavior at territory boundaries.

Summary

Shiny Cowbirds are sexually dimorphic in both size and plumage. Seven subspecies are currently recognized. Originally confined to South America, Trinidad, and Tobago, Shiny Cowbirds began a northward range expansion through the West Indies and are presently found throughout the southeastern United States. The range expansion was probably aided by deforestation and introduction of livestock. Shiny Cowbirds prefer disturbed habitat and commonly inhabit forest edge, farmlands, and agricultural fields, where they feed on arthropods and seeds.

Sex ratios are usually male biased. Shiny Cowbirds are not territorial in the traditional sense of defending exclusive ranges, but puncturing each other's eggs may express territorial behavior. The mating system of Shiny Cowbirds is not well understood, but both monogamy and polygamy have been suggested. They are partially migratory in some areas of their range.

Shiny Cowbird eggs are polymorphic and range from immaculate white to heavily spotted. Some hosts preferentially accept immaculate or spotted eggs. It is not known how many eggs Shiny Cowbirds lay in a season, but estimates have been as high as 120 eggs per season. Incubation takes 11–13 days, which is shorter than most of their hosts' incubation periods.

Three different nest searching strategies have been identified for Shiny Cowbirds: (1) silent watching of host activities, (2) active searches, and (3) flushing behavior. Silent watching is the most frequently observed strategy. Several experiments have revealed that host activity is an important cue for finding nests. However, laying eggs in inactive nests is not uncommon in some areas, particularly when cowbird density is high.

The list of known Shiny Cowbird hosts has been increasing rapidly due to additional field work and range expansion. They are now known to parasitize at least 214 species; 68 species are known to raise Shiny Cowbirds. Multiple parasitism is common and may be even more apparent in areas of recent colonization where cowbird populations are dense. Multiple parasitism also appears to be more frequent in heavily parasitized species, suggesting intense competition for these preferred hosts. Shiny Cowbirds also engage in dump nesting to a greater degree than other cowbirds.

Shiny Cowbirds usually have a detrimental effect on the nesting success of their hosts. The biggest impact they have is due to their habit of puncturing and removing host eggs. Both males and females puncture eggs, and the behavior is indiscriminate, lasting beyond their own breeding season. More often than not, nests with punctured eggs fail. As with multiple parasitism, egg puncturing appears to be most frequent in areas of high cowbird density. Shiny Cowbirds also puncture each other's eggs, and this behavior may in some regards express territoriality and negate the need for potentially harmful aggressive behavior.

Not all hosts are suitable for Shiny Cowbirds; cowbirds are constrained by whether the hosts accept parasitic eggs, the breeding season, host incubation period, host care and diet, nest site characteristics, host attentiveness, and host density. At a local level, Shiny Cowbirds show obvious preference for certain hosts, and the variation in the incidence of parasitism on certain species may reflect alternatives in the host community.

The Management Challenge

The fascinating behavior and ecology of Brown-headed Cowbirds is often overshadowed by the dramatic impact they seem to have on numerous species of migratory birds. With apparent declines in migratory songbirds, negative attention has been turned to cowbirds. They are often placed in the category of pests and blamed for the declines, but as Smith (1994) pointed out, they may be more accurately represented as convenient scapegoats for anthropogenic changes. Even admitting that cowbirds are scapegoats, we are left with the weighty ethical dilemma of our responsibility for alleviating problems we have created for many species of migratory songbirds. Unfortunately, we do not fully understand what solutions work for various problems. Additionally, landscape patterns and parasite-host dynamics seem to be changing faster than we can accumulate knowledge. Clearly, our knowledge of cowbird management is still in its infancy, yet cowbird researchers are often called upon to make management recommendations. Although much work must be conducted in the area of neotropical migrant management, some patterns have emerged that may be widely applicable, and for managing endangered species or populations, the efficacy of cowbird trapping programs is fairly well understood. In this chapter, I address only problems of Shiny Cowbirds and Brown-headed Cowbirds, as there is no indication that the other cowbird species need to be managed.

Cowbird Populations

Many papers on the management of cowbirds are based on an assumption set forth in the often cited Brittingham and Temple (1983) paper, which claimed that Brown-headed Cowbirds have increased dramatically throughout this century. Brittingham and Temple reported an alarming increase of Brown-headed Cowbirds from 1900 through 1980. However, their data were based upon Audubon Christmas counts and may not accurately reflect populations during the breeding season when cowbirds have the potential to reduce populations of other birds.

Furthermore, their data were based on the percentage of counts in which cowbirds were detected, not on actual abundance. Their data, therefore, reflect the use of more wintering areas, not necessarily increased abundance. However, they stated that "Cowbirds have been reported with increasing frequency from 1900 to 1980 . . . and we interpret this as an indication that cowbird numbers have also been steadily increasing" (1983:31). While cowbird populations appeared to increase dramatically through the first half of the twentieth century, the Breeding Bird Survey (BBS), conducted during the breeding season, indicates that overall, from 1966 through 1992, Brownheaded Cowbirds have declined at an average rate of 0.9% per year, and the area of greatest abundance (midcontinent from North Dakota through Oklahoma) has remained rather stable (Peterjohn and Sauer 1993).[1] From 1966 through 1976, they increased slightly, but Brittingham and Temple's data from winter counts showed a steady and rapid increase, even during years when data from the BBS indicated declines. In some of the same forests of Wisconsin where Brittingham and Temple (1983) conducted their research, Ambuel and Temple (1982) observed Brown-headed Cowbirds in fewer forests than Bond (1956) did in forests of the same region during 1954; data from the BBS indicate a statistically significant decline of Brown-headed Cowbirds in Wisconsin.

In areas where Brown-headed and Shiny Cowbirds are recent arrivals or have increased dramatically, disequilibrium conditions often occur. Hosts in these areas are often intensely parasitized and suffer greater consequences from cowbirds compared to hosts that have long been sympatric with cowbirds. In a review of old and new hosts, Mayfield (1965b) found that hosts are more heavily parasitized in areas of recent contact (e.g., eastern forests) than historic areas (short grass plains of the midcontinent). Cruz et al. (1985) predicted that the frequency of parasitism and the reduction in nesting success of parasitized nests would be related to length of exposure to cowbirds; specifically, they predicted that Shiny Cowbirds would have greater impact on hosts on Puerto Rico, where Shiny Cowbirds arrived around 1955, than on Trinidad and mainland South America where Shiny Cowbirds have coexisted with their hosts for a long time. On Puerto Rico, rates of parasitism on endemic Yellow-shouldered Blackbirds reached 100% in 1982, whereas parasitism rates on Yellow-hooded Blackbirds on Trinidad have been moderately high (approximately 40–45%) but stable through time. Host aggression, number of punctured eggs, and multiple parasitism are also higher, and nest success is lower among Yellow-shouldered Blackbirds than among Yellow-hooded Blackbirds.

Further, Cruz et al. (1985) found 61.1% of 18 resident passerines on Puerto Rico were parasitized, while 15.0% of 20 resident passerines on Trinidad were. This is also higher than the 21.4% of 28 resident passerines parasitized and 40.0% of 35 resident passerines parasitized that Fraga (1985) and Gochfeld (1979b), respectively, found in Argentina.

Conditions that cause disequilibrium between host and parasite are not well understood. Lack of antiparasitic defenses by hosts recently exposed to brood parasitism is often cited as the primary cause for disequilibrium (Cruz et al. 1985, 1989; Rothstein et al. 1987). However, *effective* antiparasite defenses are rare, even among hosts that have long coexisted with cowbirds. Only a couple dozen species are known to eject cowbird eggs on a regular basis (table 2.1). Several more species sometimes reject cowbird eggs through burial or abandonment, and only a handful of species are known to be *effectively* aggressive toward intruding cowbirds.

Rothstein (1982b, 1990) stated that the only reason for recognition and ejection of dissimilar eggs is in the context of brood parasitism; similarly, Hoover and Brittingham (1993) believed that egg ejection has no adaptive value in the absence of brood parasitism. While hosts in areas newly exposed to brood parasites often do not possess antiparasite defenses, this is not always the case. For example, Cruz et al. (1989) observed on Puerto Rico that six species of passerines never reported to be parasitized, or parasitized at a very low rate (appendix A), ejected artificial Shiny Cowbird eggs from their nests: Red-legged Thrush (*Turdus plumbeus*, 100%, $n = 4$), Pearly-eyed Thrasher (*Margarops fuscatus*, 81.0%, $n = 21$), Northern Mockingbird (77.8%, $n = 9$), Bananaquit (*Coereba flaveola*, 64.3%, $n = 14$), Stripe-headed Tanagers (*Spindalis zena*, 100%, $n = 2$), and Greater Antillean Grackles (*Quisculus niger*, 88.9%, $n = 36$). Post et al. (1990) similarly found Gray Kingbirds (*Tyrannus dominicensis*) on Puerto Rico, parasitized at only 0.0–1.1%, rejected 100% ($n = 11$) of experimentally added Shiny Cowbird eggs. It is possible that these species evolved an ejection response to CNP, but it is not known how these species react to conspecific eggs.

Brown-headed and Shiny Cowbirds are not tied to the population of one host; nevertheless their own population is totally dependent upon the community of hosts. Theoretically, an optimal cowbird:host community ratio exists. If the cowbird population exceeds this ratio, the number of available nests would decrease, resulting in the reduction of the number of cowbirds. Cowbird and host populations probably oscillate near an optimal cowbird:host community ratio, similar to predator-prey oscillations. In the process, some host species could be eliminated. As preferred hosts become rare, they may still be heavily parasitized (Wiley 1985), especially those nesting conspicuously, such as Yellow-shouldered Blackbirds. Wiley (1985) cited an example of the Sedge Warbler (*Acrocephalus schoenobaenus*) that became rare and then disappeared after years of increasing parasitism by European Cuckoos.

Lown (1980) used data from the Breeding Bird Census to determine the stability of the Brown-headed Cowbird:host community ratio over time. He found that the cowbird:host community ratio (three breeding females to 100 host pairs) remained stable over five decades. Lown assumed that females laid ten eggs per season, and this resulted in an overall parasitism rate of 30%. We have learned since this time that cowbirds may lay far more than ten eggs per season. In light of the fact that three cowbirds may easily lay a combined 100 eggs in a season, the overall parasitism rate may be somewhat alarming. However, many nests are multiply parasitized in some areas, and undoubtedly, many eggs are wasted on rejecter species and inability to find nests. Interestingly, Lown (1980) found that the abundance of cowbirds was not related to the abundance of rejecter species.

Host Populations

Concern regarding declines of migratory songbirds is widespread, yet whether declines are real or not is currently under debate (Hagan and Johnston 1992; Finch and Stangel 1993; Rappole and McDonald 1994). The origins of declines are not well understood, primarily because concern is relatively recent, and many factors both on breeding and wintering grounds probably contribute to declines. Furthermore, international aspects of songbird ranges and mobility confound logistics of studying their populations on breeding and wintering grounds; therefore, population dynamics and regulatory mechanisms are not well understood either. Caution must be exercised in making generalizations because closely related species and even populations of the same species may exhibit different migratory and foraging behavior, site fidelity, and responses to brood parasitism and predation (May and Robinson 1985). Reports and opinions vary regarding almost every aspect of migratory bird declines, yet there is pressing need to recommend management strategies based upon what little we do know. Therefore, we must look for trends and continue to conduct long-term studies that have management implications.

An overall steady decline during the 1900s of many neotropical migrants has been well documented by numerous investigators (Terborgh 1992; Hagan and Johnston 1992; Finch and Stangel 1993; Rappole and McDonald 1994). Declines have been reported for at least 109 neotropical migrants (Rappole and McDonald 1994), but whether the declines have exceeded what might be expected directly from the amount of habitat destruction is relatively unexplored. Other neotropical migrants have remained stable or increased over the same time period (Robbins et al. 1993). Some studies have indicated that the number of young produced may not be able to replace their parents (Laymon 1987), while other studies have reported that more young are produced than needed for parental replacement (Holmes et al. 1992). Why some neotropical migrants have declined and others have not is not readily apparent, but declines appear to be more common in the eastern United States than in the west (Robbins et al. 1993). Common threads among species that have undergone declines may be that these species occupy critical, declining habitat on both breeding and wintering grounds. For example, Wood Thrush (*Hylocichla mustelina*) and Kentucky Warbler (*Oporornis formosus*) populations appear to have declined. They are area sensitive forest dwellers on breeding grounds and require mature rainforests in Latin America during winter; both habitats have been undeniably encroached upon (Robbins et al. 1993).

In an analysis of 100 neotropical migrants based on data from the BBS, Sauer and Droege (1992) reported an overall long-term (1966–1988) increase throughout the United States but a more recent (1978–1988) decrease in the east. Declines in the east were primarily in species that nest in forested habitat; however, Sauer and Droege found no association between population trends and forest acreage change. Contrary to the view that migratory birds occupying forested habitat are declining, although they found geographic variation in population trends, James et al. (1992) noted that four of eight warbler species in the south central United States have increased during the first 22 years of the BBS, and these four (Prothonotary Warbler, Hooded Warbler, Northern Parula [*Parula americana*], and Yellow-throated Warbler [*Dendroica dominica*]) are all forest-dwelling species, whereas two species (Common Yellowthroat and Prairie Warbler, *Dendroica discolor*) occupying more open habitat have undergone a decline during the same time period. Roth and Johnson (1993) found that Wood Thrush abundance declined by 4% per year over one period in their 13-year study but recovered to predecline levels during 1990.

Ambuel and Temple (1982) documented a significant decline of several long-distance migrant hosts in southern Wisconsin forests from 1954 through 1979. Terborgh (1992) noted that as reports of declines accumulated, patterns of fragmentation became apparent. Most declines were from isolated forest tracts, and the smaller the tract, the more extreme were the declines. Terborgh (1992) cited Rock Creek Park in Washington, D.C., as an example where numbers of breeding birds in the 1970s had dropped to one-third the level of the 1940s. Rock Creek Park is a large urban park that has existed for more than a century. While Rock Creek Park itself may have changed little over the last century, it has become more isolated as a consequence of heavy development of lands surrounding Washington, D.C.

In an attempt to identify the source of neotropical migrant declines, Rappole and McDonald (1994) presented 14 predictions based on the hypothesis that declines are a result of breeding ground factors. Examination of their predictions did not support the hypothesis, and they tentatively concluded that declines were mostly due to alterations of wintering habitat; over 40% of tropical forests have been destroyed and converted to other land uses (Petit et al. 1993). One might expect that if wintering grounds were limiting populations, populations arriving on suitable wintering grounds would either remain stable or increase as individuals would be forced to pack together. This is not what Faaborg and Arendt (1992) found during a long-term winter banding program on Puerto Rico. However, as they pointed out, it is unknown where populations that winter on Puerto Rico disperse to breed; therefore, it is difficult to make generalizations. The largest declines corresponded to serious drought conditions on breeding grounds. Current evidence suggests that factors occurring on both breeding and wintering grounds, as well as on migratory routes, are responsible for declines (Sherry and Holmes 1992, 1993; papers in Hagan and Johnston 1992), and the management of these birds is a monumental international challenge. Even when rates of brood parasitism and predation have remained unchanged over time, some birds are experiencing declines in population (Rappole and McDonald 1994).

Terborgh (1992) suggested that because some species wintering under heavy deforestation conditions appear to be stable, whereas some species wintering in relatively unaltered habitat are in a decline, tropical deforestation was not to be blamed for the declines. He mentioned that during migration, thousands of birds pass through Rock Creek Park; some males stop and sing but move on, apparently rejecting the park as suitable habitat. He blamed predators and cowbirds, giving credit to songbirds for being able to assess their risks of being parasitized or preyed upon. Terborgh discussed the effect that predators and parasites could have on a population, but he did not mention the important ecological role of parasites and predators. Another useful way of identifying whether alterations on breeding or wintering grounds are responsible for declines is to compare migrants with residents. Ambuel and Temple (1983) found that populations of nonmigratory or short-distance migrants in the same forests either increased or remained stable. They suggested that destruction of wintering habitat and habitat along migratory routes may contribute to declines of long-distance migrants. Residents also showed fewer declines than neotropical migrants in various eastern states (Johnston and Hagan 1992). O'Connor (1992) reported that according to data from the BBS, increases of neotropical migrants and residents alike were larger than needed to compensate for years of decrease, but once neotropical migrants are lost from forest patches, recolonization may not take place. Expansion by resident species often follows such losses of migrants and may impede reestablishment of migrants (O'Connor 1992).

Whitcomb et al. (1981) suggested that destruction and fragmentation of forest are probable causes of migratory bird declines and that forest fragmentation leads to increased cowbird parasitism and competition from edge species. Indeed, parasitism by Brown-headed Cowbirds is often cited as a primary cause for declines (Mayfield 1977; Gates and Gysel 1978; Brittingham and Temple 1983; Terborgh 1992; Hoover and Brittingham 1993; Schram 1994). Historical effects of cowbird parasitism on host populations are relatively unknown, and general host selection by both Shiny and Brown-headed Cowbirds complicates the assessment of changes over time. Even though we lack consistent data on reproductive success, it is often assumed that parasitism rates have increased and contributed significantly to declines (Terborgh 1992; Brittingham and Temple 1983). The justification for this assumption is that in recent times, for some host populations, parasitism has become so intense that host populations may not be able to produce enough young to maintain their numbers (Brittingham and Temple 1983). Least Bell's Vireos (*Vireo bellii*) experienced a resurgence in local populations with an effective cowbird control program, strongly suggesting that, for some species, cowbirds can have a detrimental impact not only on an individual's reproductive success but on the entire population. However, parasitism is greater for some species in the midwest than the east (Robinson et al. 1993), and neotropical migrant declines appear to be most prevalent in eastern states, which does not support the view that cowbirds are the major contributor to declines of all hosts.

Data from other studies also do not support this view. Sherry and Holmes (1992) reported declines of American Redstarts *(Setophaga ruticilla)* in an unbroken forest of New Hampshire that coincided with trends throughout both New Hampshire and the rest of the United States, yet they reported a 0% rate of parasitism and noted that parasitism in their study site was almost nonexistent on other species as well. Predation, on the other hand, had a major effect on nesting success and yearling recruitment; Sherry and Holmes also cited starvation and adverse weather conditions to be critical factors during the nesting period in some years.

Increased predation may be the most important source of declines for some populations. Predation is still the major cause of nest failure for many open cup nesting birds in North America (Gates and Gysel 1978; Best and Stauffer 1980). Best and Stauffer (1980) found that parasitism was not the most important factor in nest failure; in decreasing order of importance were predation, desertion, parasitism, and natural disasters. Others have also reported that among cowbird hosts, predation is the major source of nesting failure (Goossen and Sealy 1982; Holcomb 1972). Effects of brood parasitism may be overestimated when losses due to predation are not fully recognized.

For other species, in some areas, cowbird parasitism is the major source of nesting mortality. In southwestern Colorado, cowbird parasitism contributes far more than predation toward nesting failure for Plumbeous Vireos and Warbling Vireos, whereas in the same area, predation, not parasitism, is the major source of nest failure for Yellow Warblers.

To evaluate the importance of cowbird parasitism on the population dynamics of migratory hosts, detailed long-term studies of reproductive success and demographic patterns will be needed over a broad range of habitats and human influences, including habitats intact and fragmented to varying degrees. Few studies of cowbird hosts have involved large numbers of marked individuals over a sufficient number of years to generate data on reproductive performance over a lifetime. The four-year study on Black-throated Blue Warblers in an unfragmented forest by Holmes et al. (1992), the nine-year study of American Redstarts in an unfragmented forest by Sherry and Holmes (1992), and the 13-year study on Wood Thrushes in Delaware by Roth and Johnson (1993) provide excellent examples of the kind of data that are needed for comparing host population dynamics with various sets of ecological parameters.

Such long-term detailed studies are few and far between and will probably continue to be slow to accumulate. However, historic nest records may be used in a simplistic manner to compare cowbird parasitism by regions over time. In a study of parasitism on Wood Thrushes in various regions, Hoover and Brittingham (1993) used nest records from the Cornell Laboratory of Ornithology Nest Record Program and BBS data. They found within two of three regions (midwest and northeast), rates of parasitism by decade did not vary, and in the mid-Atlantic region parasitism rates were lowest during the most recent decade. The BBS data indicated that the abundance of cowbirds was highest in the midwest region. Correspondingly, incidence of parasitism was highest and abundance of Wood Thrushes was lowest in the midwest region.

Endangered Species

Brown-headed Cowbirds have been implicated in the endangered status of several taxa, including Kirtland's Warbler (*Dendroica kirtlandii*), Least Bell's Vireo (*V. b. pusillus*), Black-capped Vireo (*Vireo atricapillus*), and Southwestern Willow Flycatcher (*E. t. extimus*). Similarly, Shiny Cowbirds have been implicated in the endangered status of Yellow-shouldered Blackbirds. Of particular concern are endemic hosts on islands that have recently been colonized by Shiny Cowbirds. Cruz et al. (1989) suggested that in addition to monitoring populations of Yellow-shouldered Blackbirds, particular attention should be paid to other parasitized endemics.

Kirtland's Warbler. During the breeding season, Kirtland's Warblers occupy dense Jack Pine (*Pinus banksiana*) forests of lower Michigan and are restricted to an area of approximately 120 by 160 km (Probst and Weinrich 1993). Depletion and fragmentation of Jack Pine habitat and organized firefighting were probably the primary reasons for the decline in Kirtland's Warblers, and the restricted population was at higher risk for further decrease by cowbird parasitism. Prior to a cowbird control program, they were a preferred host and parasitized at a much higher rate than other species in the area (Walkinshaw 1972). Most parasitized Kirtland's Warblers fail to raise young of their own.

In 1951 and 1961, the numbers of singing males were estimated at 432 and 502, respectively. By 1971, the population had decreased to 200 singing males (Mayfield 1973, 1977; Shake and Mattsson 1975). With parasitism rates between 60% and 70%, a cowbird trapping program was implemented in 1972, and Kirtland's Warblers were listed as endangered under the Endangered Species Act of 1973. Upon initiation of the trapping program in 1972, parasitism decreased to 6.5%, the average clutch size increased from 2.3 to 4.2 eggs per nest, the number of fledglings per nest increased from 0.8 to 2.8 (Shake and Mattsson 1975), and the birds often double brooded. Mayfield (1977) suspected that double brooding occurred less often among parasitized pairs, perhaps because of the burden of feeding voracious, demanding cowbirds. No parasitism was observed in 1973, and success was similar to the previous year; the total number of singing males rose by 9.2% (Shake and Mattsson 1975). However, even with continual trapping, the number of singing males declined again in 1974 to 167, suggesting that cowbird parasitism was not the only reason for the declines.

The trapping program and subsequent success of Kirtland's Warblers has been disappointing overall. After more than 15 years of trapping, the population remained no more than stable at approximately 200 singing males between 1972 and 1989, even though rates of parasitism were estimated to be low. In 1990, the population began to increase again, and in 1993, 488 singing males were located; DeCapita (1993) attributed the increase to an increase in suitable breeding habitat. In addition to being restricted to Jack Pine forests, Kirtland's Warbler populations are highest in stands from 5 to 12 years old, regenerated from wildfires, and from 1.4 to 5.0 m high (Probst and Weinrich 1993). The lack of living lower branches in older stands may contribute to the lack of suitability of older stands. The recovery plan called for periodic burning to regenerate the Jack Pine ecosystem so that about 1,120 ha per year would be regenerated (Probst and Weinrich 1993). Prescribed burning, however, has generally failed to provide regeneration at the density required by Kirtland's Warblers, primarily because regeneration has been deemphasized as an objective (Probst and Weinrich 1993).

While the trapping program has undoubtedly contributed to stabilizing the Kirtland's Warbler population, it has had virtually no effect on the local cowbird population. In 22 years of trapping, more than 89,000 cowbirds have been removed from the Kirtland's Warbler breeding habitat, averaging 4,071 per year (DeCapita 1993). DeCapita estimated that the population of Kirtland's Warblers could not be maintained at a sustained parasitism rate of 35%. Without a cowbird control program, the Kirtland's Warbler would likely become extinct in a short time.

Least Bell's Vireo. The Least Bell's Vireo was listed as endangered by the U.S. Fish and Wildlife Service in 1986. Populations are disjunct and widely dispersed because the riparian habitat that they depend exclusively upon has been severely depleted in California. Therefore, they are especially vulnerable to stochastic events. Prior to 1930, Least Bell's Vireos were regarded as common and extended into Shasta County in northern California (Goldwasser et al. 1980). By 1970, they were extirpated from northern California (Goldwasser et al. 1980), and in 1987, only 283 breeding pairs were known (Franzreb 1990). Clearly, the major cause of the decline was destruction of critical habitat; nevertheless, heavy cowbird parasitism contributed significantly to reproductive failure in the already stressed population, and the unexpectedly rapid growth of the population in the last eight years is undoubtedly attributable to a successful trapping program.

Habitat destruction may not entirely explain the decline of Least Bell's Vireos; Goldwasser et al. (1980) pointed out that considerable suitable habitat along the Sacramento and Salinas Rivers was not inhabited by the vireos. However, it is not known how destruction of suitable habitat along their migratory route may affect their migratory patterns. Destruction of their breeding habitat continued even as they were in the process of being listed as endangered, and the pressure to develop their habitat continues. For example, through an interagency cooperative agreement, the California Department of Transportation initiated a cowbird trapping program in 1986 as part of mitigation required for a 404 permit issued to the U.S. Army Corps of Engineers for construction of a bridge. While destruction of the Least Bell's Vireo's habitat is more or less permanent, mitigation may be unstable and very costly.

Least Bell's Vireos have an incubation period of 14–15 days, and one hatched cowbird is sufficient to cause total reproductive failure. Laymon (1987) calculated female Least Bell's Vireo population growth at four different levels of parasitism: 13%, 30%, 48%, and 69%. He assumed adult mortality at 40% and an average of 2.55 female eggs per female per year, but he did not include nesting mortality due to predation. At 13%, this is an annual increase of 33% per year, which is very close to the actual increase of 30–35% at the Camp Pendleton Marine Base from 1988 through 1995, where predation is 20–30% and parasitism is almost nonexistent (J. Griffith, pers. comm.). Laymon estimated the increase of the female population to be 7% annually, with a parasitism rate of 30%, which is much lower than the rate Griffith observed with nesting mortality (from predation) of 20–30% per year. The difference may be due to an underestimation in Laymon's estimate of eggs laid per breeding season. Griffith reported that Least Bell's Vireos are presently producing between six and eight eggs per year.

The recovery plan includes habitat restoration, cowbird trapping, and monitoring nests to remove cowbird eggs; the goal of the recovery plan is to increase the number of pairs to 5,000 (Franzreb 1988, 1990). A reintroduction program to northern parts of California has also been considered. Private lands with suitable riparian habitat in the Central Valley of California, owned by conservation organizations such as the Nature Conservancy and the Audubon Society, have been identified as ideal reintroduction locations (Franzreb 1990). Several reintroduction methods have been suggested: capture and release of first-time breeders and juveniles, capture of birds in autumn and release the following spring, captive breeding, and cross-fostering. Of all these proposed methods, capture and release appears to be the most feasible (Franzreb 1990). In addition to the challenges of reintroducing a short-lived migratory bird, Least Bell's Vireos are site tenacious. If it reinvades areas of its former range, it will likely take decades to do so (Franzreb 1990). Reintroduction programs have been met with varying success for game birds (Franzreb 1990). However, for small passerines limited studies have been conducted (Franzreb 1990), and for short-lived migratory species, such as Least Bell's Vireos, reintroduction may be especially difficult. To date, reintroduction has not been attempted (J. Griffith, pers. comm.).

While cowbird trapping is only a stop-gap measure, not a long-term solution, it can contribute to temporarily stabilizing a host population. After implementation of the trapping program, the estimated number of Least Bell's Vireo breeding pairs in the United States increased to 350 in 1988. The trapping program along the Santa Margarita River on the Camp Pendleton Marine Base has been particularly successful. While the number of cowbirds trapped per year has not decreased, parasitism has been almost nonexistent since 1988, and as of 1996, there were approximately 700 breeding pairs (J. Griffith, pers. comm.). The explanation is simply that cowbirds are removed during the end of March and early April before the vireo's breeding season. Griffith estimated that approximately 85–90% of these territories represent breeding pairs, and many of these pairs are double-brooding, producing six to eight young per breeding season. Presently, the population is increasing by 0–35% annually, and in 1995, the total population was estimated at 1,500–1,600 pairs (J. Griffith, pers. comm.). Effectiveness of the trapping program is also apparent with other host populations within the trapping area. Griffith and Griffith (in press) have also noted increases in populations of Warbling Vireos, Wilson's Warblers (*Wilsonia pusilla*), Hutton's Vireos (*Vireo huttoni*), Swainson's Thrushes (*Catharus ustulatus*),

Yellow-breasted Chats, Blue Grosbeaks *(Guiraca caerulea)*, and Lazuli Buntings *(Passerina amoena)*. The increase in the Least Bell's Vireo population can also be partially explained by a low rate of predation (20–30%).

Yellow-shouldered Blackbird. Yellow-shouldered Blackbirds, endemic to Puerto Rico, were described as widespread and abundant prior to 1940. Population data from 1940 through 1972 are largely lacking (Post and Wiley 1976, 1977b); therefore, it is difficult to determine if the population decline coincided with arrival of Shiny Cowbirds on the island around 1955. In 1972, Post and Wiley (1976) estimated the population to be approximately 2,400. Although they suggested that other factors may have contributed to their decline, Post and Wiley (1976) observed that Shiny Cowbird parasitism had a significantly negative effect on the population of Yellow-shouldered Blackbirds. Between 1974 and 1982, the population had declined by approximately 80% (Wiley et al. 1991). Yellow-shouldered Blackbirds are the preferred host of Shiny Cowbirds on Puerto Rico and are heavily parasitized. Nesting success is reduced by at least half in parasitized nests.

Yellow-shouldered Blackbirds nest in a variety of habitats, including mangroves, lowland pastures, suburbs, and palm plantations. On Mona Island, they nest on coastal cliffs. Shiny Cowbirds arrived on Mona Island by 1971 but apparently do not parasitize Yellow-shouldered Blackbirds there (E. Hernandez-Prieto, pers. comm.). In addition to nesting in different habitats, their nests may be open cups or constructed in natural cavities or nest boxes. Even though Yellow-shouldered Blackbirds occupy a wide variety of nesting habitats, reduction in nesting habitat may also have contributed to their initial decline (Post and Wiley 1976). Nest boxes were parasitized at an equally heavy rate (96%, $n = 103$) as open nests (91%, $n = 77$) and natural cavities (100%, $n = 20$, Wiley et al. 1991). However, predation is lower in cavities and boxes (18.2%) than in open nests (31.9%); therefore, even though box nests are parasitized, host production can be improved with their addition (Wiley et al. 1991). Additional advantages of providing boxes are: (1) they can be placed on trees or poles fitted with antipredator polyvinyl chloride pipes, (2) they can be placed in traditional nesting sites, (3) they can be monitored, and (4) cowbird eggs can be removed (Wiley et al. 1991). However, the microclimate of nest boxes also attracts nest mites, which can lead to abandonment; Wiley et al. (1991) suggested the use of insecticides, such as 2% Sevin powder, but he cautioned that insecticides are potentially hazardous.

Shiny Cowbird control programs were initiated in Boqueron Forest, in southwest Puerto Rico, in 1983 by the U.S. Fish and Wildlife Service and the Department of Natural Resources. The program has met with some success but will probably need to be continued for a indefinite period of time in order to ensure the continued existence of Yellow-shouldered Blackbirds.

What Can and Should Be Done to Manage Parasitism

Over the last century, we have altered landscape patterns and created conditions which favor cowbirds and predators alike—most notably by increasing edge and introducing domestic livestock. Total eradication of cowbirds is not a reasonable solution, nor is partial but widescale eradication. Therefore, management of conditions that favor cowbirds is the most reasonable long-term solution. Cowbird removal programs should be viewed as a stop-gap measure, not a long-term solution. These measures can be extremely useful, however, in buying time and a second chance for some endangered hosts when used in conjunction with managing conditions that favor cowbirds.

In setting priorities for managing cowbirds, the generalist host-selection strategy of both Shiny and Brown-headed Cowbirds must be considered. Possibly, neither Shiny nor Brown-headed Cowbirds select nests based upon specific host species but, instead, select nests based upon a set of nest site characteristics. Therefore, except in situations of endangered hosts, it may be most appropriate to manage cowbirds from a community approach rather than a single species approach. Even when cowbird management targets a particular threatened or endangered host species, effects on other community members should be considered. As we face lean times with less consideration given to threatened and endangered species, it will be critical to allocate limited funds in a judicious manner that maximizes the number of species targeted. A multivariate approach that allows species to be grouped together because of similar threats (e.g., cowbird parasitism) provides a powerful tool for setting management priorities (Given and Norton 1993) and maximizing funds.

The natural ecological role of cowbirds should be considered in all management plans, and assumptions should not be embraced as truths. It is often assumed in areas of recent range expansion of cowbirds that hosts have not evolved effective antiparasite defenses and that intense parasitism of naive hosts has contributed significantly to declines of migratory songbirds. It is also commonly assumed that cowbird abundance has increased to alarming proportions. These premises are often the basis of cowbird management, but as discussed earlier, these assumptions need reexamination. Furthermore, absolute abundance of cowbirds may not be as important as the cowbird:host ratio. Because we are still in the infancy stage of our knowledge of cowbird management, we need to remain flexible in our ideas and not implement plans without thorough consideration of benefits and consequences.

Habitat fragmentation and introduction of domestic livestock must be considered together because cowbirds are readily able to commute fairly long distances between breeding and feeding sites. This ability of the birds to commute is one of the major challenges facing cowbird managers. Even small feeding areas, such as local horse corrals, livestock pens, and birdfeeders, in relatively remote areas may enhance cowbird access to large expanses of breeding habitat that would, otherwise, be unsuitable for cowbirds. For example, forest inholdings, particularly those with livestock, within dense forests set up an inviting scenario for cowbirds. As we continue to fragment our forests through logging, residential development, and other human activities, we increase edge, allowing cowbirds to penetrate large expanses of forest. Grazing in national forest lands also increases risks of cowbird parasitism within the forests. In the Sierra National Forest in California, Verner and Ritter (1983) observed that Brown-headed Cowbirds prefer meadow edges for breeding habitat, but their abundance in these meadows was directly related to the location of human-based food resources, such as pack stations. Cowbird abundance declined with increasing distances from such food resources. Furthermore, they found that relative abundance of preferred hosts was negatively correlated with abundance of cowbirds. At the junction of the Sierra Nevada and Cascade ranges in California, Airola (1986) also found that evidence of parasitism was strongly associated with habitat disturbance.

Basic strategies for controlling cowbirds include: (1) preserve undisturbed forests with large core areas; (2) restore large expanses of native grasslands grazed on long-term rotation of perhaps up to 20–25 years; (3) reduce edge habitat within forests and tallgrass prairies; (4) reduce further development that creates islands within forests and tallgrass prairies; (5) follow principles of refuge shape and size: create compact shapes that reduce edges; (6) reduce or control foraging conditions for cowbirds, including grazing and mowing; (7) restructure grazing regimes; (8) eliminate grazing near areas of logging; (9) revegetate logging roads and clearcuts; (10) relocate dairies and stockyards from areas of high risk; (11) increase efficiency of harvesting techniques that would reduce winter food availability; (12) increase the amount of federal land that excludes grazing; (13) work with landowners to help define and fulfill their role in bird conservation within the context of stewardship; and (14) remove cowbirds only in areas where aggressive control measures are needed. Habitat restoration is the most appealing solution for numerous reasons. It is more or less permanent, it expands the reproductive potential of hosts, and continuation would require little additional funding or human-power.

Much land in need of restoration is in private hands, and landowners vary in their ownership objectives, interest in wildlife, and degree of cooperation. However, many individuals would respond to tax rebates offered for habitat restoration. Partnerships, such as Partners in Flight, can work with landowners to help define and fulfill their role in bird conservation within the context of stewardship (Wigley and Sweeney 1993). Activities that the private sector can be involved in to increase cooperative participation include research, survey, fund raising, management, and education (Johnson 1993). Resource managers also face continual opposing pressures from society and conservationists. As society, ecological views, and knowledge change, resource managers must adapt and work within a malleable framework that organizes relationships between people and the environment (Bonnicksen 1991).

Habitat

In 1965, Mayfield (1965b:24) stated that

> There are still large unbroken tracts of woodland in the East where nesting birds are untouched by the cowbird. It would be useful to know how far into a dense woodland the cowbird penetrates for egg-laying, and how the cowbird population varies according to the availability of open space in a region largely wooded. . . . For most species we will be slow to notice any general effects on their populations as long as the hosts continue to have breeding areas where the cowbird is scarce. But if the cowbird continues to find access to more nests of various woodland species, I believe the effect on some may become appreciable.

In the northeast, many birds nesting in large unbroken tracts are still more or less untouched by cowbirds. However, 85% of Maryland's forest tracts are less than 20 ha, and 59% are less than 8 ha; similarly 84% of Delaware's forest tracts are less than 20 ha (Roth and Johnson 1993). Clearly, further human development in forest tracts, particularly that which would increase human-based food resources for cowbirds, will likely increase cowbird abundance and parasitism rates and subsequently decrease host populations.

The preferred (disturbed) habitat of Shiny Cowbirds has also increased dramatically; some hosts, such as Rufous-collared Sparrows, also benefit from mildly disturbed habitat, but as they expand their range, Shiny Cowbirds follow. Habitat alteration has exposed new hosts to Shiny Cowbird parasitism, hosts that may also be poorly adapted to habitat disturbance; for example, in Brazil, White-banded Tanagers (*Neothraupis fasciata*) were recently added to the list of Shiny Cowbird hosts, and White-rumped Tanagers (*Cypsnagra hirundinacea*) may be suffering from increased parasitism as a consequence of habitat disturbance.

Habitat Quality. In order to effectively manage cowbird and host populations, we should have a good conception of host habitat quality, and assessment of habitat quality should be at the foundation of any wildlife management plan. Such assessment is often based on the assumption that density of a species is a direct measure of habitat quality. However, some species are found in high densities within habitat that can be considered suboptimal for reproduction. Nowhere is this more apparent than in edge habitat where densities are high and nesting success is low. Van Horne (1983) suggested that in addition to density, mean individual survival and fitness should be used in assessment of habitat quality. Johnson and Temple (1986) found that tallgrass prairies distant from forest edges and those recently burned were the most productive in terms of nesting success and, therefore, highest quality for five common grassland species. However, the density for all five species was lower in the habitat quantified as highest quality than in small fragments. Low nesting success in small fragments located close to edge could be attributed to both predators and Brown-headed Cowbirds. Johnson and Temple aptly pointed out that if they identified the highest quality habitat as that which had the highest density, management practices could have impaired the ability of these species to maintain their populations. Habitat quality assessment is complicated by the fact that cowbirds are generalists. Also, habitat quality differs depending on the host species; therefore, compromises have to be made.

Forests. Fragmentation has been identified as one of the major causes of migratory bird declines, and for some species, there appears to be a strong correlation between their numbers and the size of contiguous forests (Ambuel and Temple 1983). These species are identified as area sensitive or forest interior specialists; they are generally found only in large, unbroken tracts and are usually uncommon in smaller forest patches. O'Conner and Faaborg

(1992) suggested that sensitivity may be due, in part, to cowbird parasitism. However, this has not been adequately addressed, and other aspects of host biology, such as foraging ecology, would need to be investigated to more fully understand the role of cowbirds in host sensitivity to area.

Brown-headed Cowbirds are often found in disturbed forests and may be absent in less disturbed forests. Disturbance patches within forests may reduce the commuting distance of some individuals and allow a higher number of female cowbirds, even in areas where cowbirds are territorial because as they decrease food acquisition time allowances, they can devote more time to thorough nest searches. Coker and Capen (1995) found cowbirds in 46% of the 702 disturbance patches they surveyed within a Vermont forest. They also found that cowbirds were more likely in larger, less remote disturbance patches than in smaller, more remote disturbance patches. Logging operations within the forest, when remote from other human disturbance, particularly animal husbandry, did not appear to increase cowbird presence. Coker and Capen (1995) suggested that large unbroken forest tracts may be effective barriers to cowbird movements; they detected cowbirds in only 4 of 399 sample points in undisturbed forests. While they did not quantify cowbird parasitism in their study, Donovan et al. (in press) found both higher abundance and higher rates of parasitism in fragmented forests than unfragmented forests in Missouri. Similarly, in an unfragmented hardwood forest of New Hampshire, Sherry and Holmes (1993) never observed a parasitized nest in more than 20 years, and only twice did they observe hosts feeding fledgling cowbirds. Holmes et al. (1992) reported no cowbird parasitism and a low rate of predation (22%) on Black-throated Blue Warblers in an unfragmented forest in New Hampshire.

In Pennsylvania, Hoover et al. (1995) found that success of slightly area sensitive Wood Thrushes was 86% in continuous forest habitat of greater than 10,000 ha (two sites within one forest), 72% in forests of between 103 ha and 126 ha (two sites), and 43% in forest of less than 80 ha (seven sites). Success in different sized woodlots was due to predators, not Brown-headed Cowbirds. Predation was 10%, 22%, and 56% in forests of >10,000 ha, >100 ha, and <80 ha, respectively. Cowbird parasitism was fairly low (9%) and did not differ between woodlots of different sizes. Interestingly, Hoover et al. (1995) could not attribute total nest failure in any case to cowbird parasitism.

Although cowbirds penetrate large expanses of forests in Illinois, this may not be the case throughout their range, and careful consideration of increased edge should be given when facing decisions of interior clearcuts.

O'Conner and Faaborg (1992) recommended that forests should be at least 1,000 ha to reduce cowbird parasitism. Total area, however, may not accurately predict responses to fragmentation because it does not take into account shape and resulting amount of edge. A core area model provides a more accurate prediction of responses by area sensitive species. This model relies on a core area greater than 100 m from the edge (Temple 1986). Temple pointed out that two forest fragments of similar size (39 ha and 47 ha) may have 0 ha and 20 ha of core area, respectively, as a consequence of elongation and irregularity of the former fragment. Robbins et al. (1989) suggested that at least some reserves must be protected that are many thousands of acres, with the core area containing the richest habitat. They found that the minimum area that could be expected to contain all or most breeding birds of the mid-Atlantic states is 3,000 ha. In the midwest, where cowbirds penetrate deeply into forests, forests of 20,000–50,000 ha may be necessary to protect forest interior birds from cowbird parasitism (Robinson et al. 1993).

Ambuel and Temple (1983) suggested two explanations for increased bird diversity with increased woodlot size: (1) area-dependent changes in the avian community could be caused by area-dependent vegetational changes, and (2) area-dependent changes could be a consequence of changes in interactions with predators, brood parasites and competitors (biotic interactions model). Forest edge and farmland species are more common in smaller woodlots, presumably as a consequence of increased edge. Additionally, increased edge also attracts predators (Chasko and Gates 1982) and cowbirds. Ambuel and Temple pointed out that these edge-dependent or farmland species exclude area sensitive species, and these competitors may be a more important influence on migratory birds in small woodlots than are vegetational changes within forests. Very large woodlots tend to have a greater diversity because they attract area sensitive species in addition to edge species along the borders. All eight studies reviewed by Paton (1994) indicated a positive relationship between patch size and nest success, and Wilcove (1985) found that predation rates were higher in urban woodlots than in rural woodlots.

Fragmentation generally creates edge, and while edge is beneficial for some species, it can be a trap for others. The edge effect concept is based upon ecological relationships that exist at the boundaries of different plant communities. Density and diversity tend to increase in edge habitat because species characteristic of both adjacent communities, plus the edge specialists, inhabit the ecotone (Gates and Gysel 1978). Historically, edge was thought to be a positive landscape feature. While edge may increase diversity, it is a positive aspect for only certain species, among them predators and parasites. The positive value of edge is questionable for some nesting birds, and for these birds, edge habitat may be an ecological trap where density-dependent predation and parasitism occur.

Birds tend to be more abundant in edge habitat than interior forests. For example, Gates and Gysel (1978) found most of 194 nests of 21 passerines within 10–20 m of the edge between an oak-hickory forest and an herbaceous field. Predators and cowbirds are both abundant in edge habitat, and several investigators have found a significant correlation between fledging success and distance to edge (Gates and Gysel 1978). Host density is often highest in forest edges (Brittingham and Temple 1983), and these edges might also be considered the highest quality territories for cowbirds. In a review article, Paton (1994) noted that most studies support the hypothesis that nesting success is lower close to edges. The lower success in ecotones has been largely attributed to predators and parasites. Predation rates are highest within 50 m of edge in most studies (Paton 1994). Not all edges are equal, however, and Morgan and Gates (1982) commented that forest edges do not necessarily all have a well developed shrub layer that contributes to higher diversity and abundance (Ranney et al. 1981). Morgan and Gates (1982) suggested that wider shrubby borders may be preferable to narrow shrubby edges because they allow birds to disperse their nests.

In a heavily parasitized community in a Wisconsin forest, Brittingham and Temple (1983) found that cowbird density declined with distance to the nearest forest openings (>0.2 ha with >50% open canopy). They also found that parasitism rates decreased with increasing distance from forest openings, and that parasitism increased with increasing open habitat within 200 m of the nest. Additionally, they reported that nests within 35 m of smaller openings (0.01–0.2 ha) were parasitized more frequently than nests more than 35 m from such openings. Other studies have resulted in similar findings. Wolf (1987) found that parasitism was higher in open canopy (41–45%) than in closed canopy (20–32%).[2] In Missouri, cowbirds penetrated at least 400 m into forests, but their relative abundance decreased significantly with increasing distance from the edge of forests (O'Conner and Faaborg 1992). The linear relationship between increasing distance from forest edge and decreasing rates of parasitism and predation may reflect higher host densities in edge habitat.

Many studies have demonstrated that predation also decreases with increasing distance from edge (Andrén and Angelstam 1988; Small and Hunter 1988; Møller 1989). Artificial nest studies have similarly shown that predation is more frequent in small patches than large patches (Wilcove 1985; Small and Hunter 1988; Yahner and Scott 1988).

In moderately forested areas of southern Illinois, Robinson et al. (1993) observed no significant decrease in parasitism rates even at 800 m from forest edge. Similarly, parasitism was higher in a mature 1,300 ha forest than in adjacent old field and edge habitats in New York (Hann and Hatfield 1995). In a dry forest of Puerto Rico, Woodworth (1993) found no relationship between forest edge and either host abundance or Shiny Cowbird:host ratios. Small woodlots in Illinois may be population sinks for many neotropical birds nesting there because most nests fail due to extraordinarily high predation (89%) and parasitism (76%) rates, and return rates are low (15%, Robinson 1992). Host populations in these sinks may be maintained by source populations up to 200 km away (Robinson 1992). It is, therefore, important to identify locations of source populations in order to protect these areas.

Paton (1994) commented that inconsistent methods among edge effects studies may explain varying results. He suggested that future research on edge effects be concentrated within 100–200 m of the edge in 20–25 m increments; he further proposed that only forest openings with a diameter greater than three times the height of adjacent trees be included. While consistency of methods has obvious value for comparative purposes, the proposed methods may not be appropriate for all forest types and regions. For example, in forests with very high canopy, cowbirds may perceive openings less than three times the height of adjacent trees as edge, whereas openings greater than three times the height of adjacent trees in the Elfin Woods of Puerto Rico may not be perceived by cowbirds. Additionally, for some forest interior species, success may increase dramatically more than 200 meters from the edge; Temple and Cary (1988) found that nest success was only 18% within 100 m from the edge of a forest in southern Wisconsin, but success increased to 58% between 100 and 200 m, and to 70% more than 200 m from the edge. Distinctions between natural and anthropogenic edges are relatively unexplored, and studies of edge effects are lacking from the Pacific Northwest and southeast where much of the timber industry is located.

Although fragmentation generally increases edge, with consideration to the size and placement of fragments, edges can be minimized. The equilibrium model of island biogeography has been evoked in management of insular forest islands (Ambuel and Temple 1983; Blake and Karr 1984; Askins et al. 1987). Four general principles are: (1) a single large island is superior to several islands of equivalent cumulative size, (2) compact islands with reduced edge are superior to elongated islands with a high edge to area ratio, (3) islands in close proximity to one another are superior to islands widely separated, and (4) islands linked together by corridors are better than isolated islands (Diamond 1975; Wilson and Willis 1975). Ambuel and Temple's (1983) biotic interactions model (see above) supports the first two principles of the equilibrium model, neither supports nor disagrees with the third principle, and diverges dramatically from the fourth principle. In the equilibrium model, corridors serve as bridges facilitating bird movements among forests, reducing isolation and presumably increasing numbers. Although some investigators recommend corridors to connect forest habitats (Robbins et al. 1993), the need for corridors for birds has not been adequately demonstrated. In Ambuel and Temple's biotic interactions model, corridors act as a funnel, attracting predators, brood parasites, and competitors into the forest. Coker and Capen (1995) also suggested that Brown-headed Cowbirds may travel along disturbed corridors. In addition, interior clearcuts expose a greater area to penetration by cowbirds (O'Conner and Faaborg 1992). Protection of animals depending on connecting corridors needs to be evaluated, and compromises based on triage approach will undoubtedly have to be made on a case by case basis.

Robbins et al. (1993) recommended clustering snags near forest edges instead of scattering them throughout forests, implying that such action would reduce cowbird parasitism. For example, Kirtland's Warbler nests near dead snags may be at higher risk of cowbird parasitism (Anderson and Storer 1976). However, many other forest interior species, including mammals, may also depend on snags, and Robinson et al. (1993) suggested that we are not yet in a position to make management recommendations regarding placement of snags.

Gustafson and Crow (1994) constructed a timber harvest allocation model that runs within Geographic Information System software and incorporates location and relative quality of cowbird feeding sites and the relationship between distance from forest edge and parasitism rates. In their model, simulated on four 237 km^2 areas within Hoosier National Forest in southern Indiana, they found that for a goal of retaining 4,000 ha of forest interior with a mean harvest size of 1 ha, less than 1% of the timber base could be harvested per decade. The percentage of timber that could be cut with the same goal increased with increasing average size of cuts. Their model demonstrates that practices of allocating small areas throughout a forest for clearcutting makes management of contiguous tracts of undisturbed forest interior more difficult than allocating fewer but larger blocks. Models such as this could be extremely useful for management decisions regarding identification of areas of highest and least impact for forest nesting birds.

Open Grasslands. More than 99% of tallgrass prairies have been destroyed in Minnesota, and conversion to agriculture has made former prairies unsuitable for grassland nesting birds (Johnson and Temple 1990). Many management practices of tallgrass prairies focus on perpetuation of native grassland species without consideration for how various attributes of prairie fragments affect nesting birds (Johnson and Temple 1990). For example, Callenbach (1996) discussed the need for grassland corridors for genetic exchange among Bison herds; however, no consideration was given to the effects this may have on nesting success of birds.

Edge effect may be as apparent in unforested habitats as in forested habitats (Paton 1994). In a study of Field Sparrows (*Spizella pusilla*), Best (1978) reported that the average distance of parasitized nests from shrub-woodland was 13.4m, and no parasitism occurred farther than 26 m from shrub-woodland. Johnson and Temple (1986) found that nest productivity was highest in tallgrass prairies located far from forest edges and in those burned recently, and rates of predation and parasitism decreased with increased distance from the edge. They suggested that management of remaining fragments of tallgrass prairies should entail frequent burning and reducing edge. However, other factors, including regional differences, must be considered. For example, while recently burned areas were more productive in terms of nesting success in Johnson and Temple's study, large herbivores may graze on recently burned grasslands more intensely than on unburned grasslands (Anderson 1982), and this factor should be incorporated into management decisions. This may be particularly important in the management of future Bison reserves.

Mangroves. Mangroves are the preferred nesting habitat for endangered Yellow-shouldered Blackbirds on Puerto Rico; Post and Wiley (1976) reported that of 105 nests, 90 (85.7%) were constructed in mangroves. Large areas of mangroves have already been destroyed on Puerto Rico, and they may be particularly vulnerable to further exploitation and human disturbance by squatters and houseboats (Post and Wiley 1976; Wiley et al. 1991).

Riparian Zones. Riparian zones have been identified as one of the most important habitats for breeding birds in North America, and changes in riparian vegetation have been associated with increases or declines of various nesting birds that depend on riparian ecosystems (Kauffman and Krueger 1984). Riparian zones are particularly important in the southwestern United States, where approximately 77% of 166 nesting species are associated with riparian habitat, and 50% of nesting bird species are completely dependent on riparian ecosystems (Johnson et al. 1977). When riparian vegetation is removed, numbers and diversity decrease; 72% fewer species of birds and 93% fewer numbers nested after vegetation was removed (Henke and Stone 1978). In California, riparian areas have been altered severely by agriculture, livestock grazing, flood control, and urban development; overall, riparian areas in California have been reduced by 95% (Franzreb 1990). Concomitantly, birds depending exclusively on riparian areas for breeding have declined.

Although riparian zones vary in size, shape and complexity, particularly in the western United States, most have a high edge to area ratio and can be considered ecotonal. Therefore, they provide classic examples of edge effect, with a greater diversity of wildlife than most other habitats in North America (Kauffman and Krueger 1984). Indeed, cowbirds are common in riparian habitats, particularly in those that are grazed or have grazing nearby. It is, therefore, important to curtail further development in and around riparian zones. Riparian areas in unlogged and ungrazed forests may be especially important.

Association with Large Grazing Mammals

Before Europeans settled North America, Brown-headed Cowbirds presumably used to follow Bison. Grazing by hooved mammals was part of the natural ecosystem in North America long before the introduction of domestic livestock. However, Bison did not remain in one area as cattle do today; they grazed areas for a short time, promoting new growth (Williamson et al. 1989), and moved on, giving the recently grazed areas long periods of rest before the Bison returned, perhaps years later. Cowbirds probably followed the herds and also may not have returned to certain areas for long periods of time, allowing host populations to recover from the negative effects of parasitism.

Heavy grazing by domestic cattle has unequivocally altered and degraded many habitats, particularly in areas where native ungulates were absent or scarce (Bock et al. 1993). Cattle grazing has had a dramatic effect on grasslands by changing plant composition, reducing perennial grass cover, and converting grasslands into desert shrub; furthermore, grazing in riparian zones can dramatically change stream morphology and associated flora and fauna. Although some of the best examples of deforestation for grazing have occurred in the tropics, deforestation for grazing has also occurred in the United States. An estimated 167 million acres in ten western states administered by the Bureau of Land Management (BLM), 138 million acres administered by the U.S. Forest Service, and 212 million acres of private land are currently grazed, and 83% of federal rangeland is considered to be in fair to poor condition (BLM 1994; W. Werkmeister, BLM, pers. comm; J. Windell, pers. comm.).

Overgrazing in riparian areas has a negative impact on riparian ecosystems. When vegetation is removed, soils may become compacted, which decreases infiltration of precipitation and causes topsoil to erode into the aquatic zone. Loss of vegetation and canopy cover may result in loss of allochthonous detrital matter entering the stream. Many aquatic invertebrates rely primarily on this material, and the decrease in aquatic invertebrates, in turn, affects populations of birds depending on them.

The competing interests of conservationists and ranchers, whose economy, lifestyles, and history depend on cattle, make management of grazing issues contentious and difficult. While grazing by domestic cattle may be neither universally detrimental nor beneficial to native wildlife, cattle appear to be a keystone species in most of the habitats they occupy (Bock et al. 1993), and the cattle industry makes a substantial contribution to feeding North Americans. Bock et al. (1993) recommended that we increase the amount of federal land from which cattle are excluded and increase the recovery time allowed for rotational grazing to perhaps 20–25 years. Sharp (1995) pointed out that in addition to costs of administering the grazing program in National Forests far exceeding grazing revenues, the high number of cowbirds attracted to grazed forest lands should be considered an additional cost.

Laymon (1987) suggested that grazing near riparian areas should be entirely eliminated. While this may not have much effect on cowbird densities in some areas because of their ability to commute long distances between breeding and feeding sites and because riparian zones are essentially edge habitat, it would improve breeding habitat for hosts, easing the general stress on their populations, and denser growth may inhibit the cowbirds' ability to find nests. Laymon (1987) also pointed out that elimination of grazing from mountain meadows may be particularly effective if grasses subsequently grow to a level unsuitable for cowbird foraging. Relocation of dairies and stockyards more than 7 km from riparian areas might cut down on cowbird use of these areas; however, radiotelemetry studies would be needed to determine distances that cowbirds travel from stockyards and other feeding sites to riparian areas. Coker and Capen (1995) found that the number of livestock areas with 7-km commuting distance was an important indicator of cowbird occurrence in a forested landscape in Vermont.

As Bock et al. (1993) pointed out, the problem is not so much with the presence of cattle as with the ubiquity of them. New stockyards and pack stations should be located near already existing areas of human influence and as far from bird breeding sites as possible. Because cowbirds readily commute at least 7 km between breeding and feeding sites, stockyards and pack stations should be placed at least this far from breeding sites, preferably farther. Thompson (1994) observed cowbirds moving distances as great as 10 km, but most cowbirds commuted less than 2 km between breeding and feeding sites. Minimizing interspersion of cowbird feeding areas is an important step in conservation efforts (Thompson 1994).

Food Supply

Human-based food supplies may be directly associated with cowbird densities and parasitism rates. In the Sierra Nevada of California, areas with high availability of human-based food resources have a greater concentration of Brown-headed Cowbirds and higher parasitism rates than areas lacking these resources. Grain resources have allowed cowbirds to increase their over-winter survival. Robinson et al. (1993) suggested that more efficient harvesting methods could reduce this resource.

Removing Cowbird Eggs from Nests

Some management programs, such as the Least Bell's Vireo Recovery Plan, specify removal of cowbird eggs from parasitized nests, but it is important to understand the consequences of such removal before the practice has widespread implementation. If removing the egg promotes additional cowbird parasitism, especially if the cowbird removes or otherwise destroys host eggs, it might be worth considering adding cowbird eggs or replacing cowbird eggs with artificial cowbird eggs. In an experimental study on Red-winged Blackbirds, Brown-headed Cowbirds appeared to avoid parasitizing nests that contained artificial eggs (Ortega et al. 1994). Before implementing this practice on endangered or threatened species, a pilot study should be conducted on a closely related species to evaluate whether the practice lowers nesting success in any way, such as abandonment, or reduction in clutch size, hatching success, or fledging success.

Nest Boxes

In general, Brown-headed Cowbirds seem to shy away from cavity nests, but they regularly parasitize cavity-nesting Prothonotary Warblers. Prothonotary Warblers using nest boxes and natural cavities were parasitized with equal frequency (Petit 1991). Density of nest boxes did not affect parasitism rates among Prothonotary Warblers; however, females in territories of high nestbox density were more likely to abandon parasitized clutches and renest than parasitized females in territories with lower nestbox densities (Petit 1991). Holes in nest boxes (or milk cartons, Fleming and Petit 1986) might have to be less than 3 cm to prevent parasitism. Packard (1936) reported on a Black-capped Chickadee *(Parus atricapillus)* nestling in a box with a 3.8 cm hole that was parasitized by a Brown-headed Cowbird, and Petit (1991) reported parasitism in nest boxes with holes 3 cm in diameter.

If and when Shiny Cowbirds use more hosts in the north, caution should be taken for cavity-nesting hosts. Erection of nest boxes may not decrease the incidence of parasitism by Shiny Cowbirds. Mason (1986b) found that in Argentina, House Wrens nesting in boxes were more heavily parasitized than House Wrens nesting in natural cavities. Similarly, Kattan (1993) found that parasitism on House Wrens nesting in natural cavities was 50%, whereas it was nearly 100% for House Wrens nesting in boxes. Shiny Cowbird parasitism on Bay-winged Cowbirds was equivalent between natural sites and nest boxes (Fraga 1986). The advantages of erecting nest boxes for Yellow-shouldered Blackbirds on Puerto Rico are less clear because both host and parasite experience higher nesting success in cavities, but Wiley et al. (1991) recommended the use of nest boxes for Yellow-shouldered Blackbirds (see above, under endangered species).

Trapping and Removal

If it is determined that a cowbird eradication program must be initiated in order to buy time for an endangered species, a couple methods of cowbird removal can be considered: trapping and shooting. Shooting can be aided by the use of cowbird vocalizations to draw the birds to within firing distance. Rothstein et al. (1987) suggested that shooting females may be more efficient than trapping, particularly in areas where females do not commute to concentrated foraging sites. Shooting females may also be used in conjunction with trapping, as not all females will readily enter traps. Robinson et al. (1993) cautioned that shooting should not occur in the immediate vicinity of a trap because it may decease subsequent captures in the trap. Trapping and subsequent removal of cowbirds in areas of endangered species has met with varying success. However, trapping and removal programs are usually only effective as long as the trapping is carried out.

Cowbird demographics in a targeted area should be understood before a trapping program is initiated. In particular, it is important to know population size, sex ratios, and site fidelity. It is equally important to understand the effects that cowbirds have on the targeted species. Effects of long-term trapping on the mating system and territoriality are not known but are worth investigating. Generally, females are more trap shy, and over many years of removing both males and females, one might expect that the sex ratio may become less male biased and theoretically could become female biased. Unless new males are recruited, a switch in sex ratio could eventually result in a switch from a monogamous mating system to a promiscuous or polyandrous system. In the longest-standing trapping efforts, sex ratios appear to approach unity, and in some years are female biased (DeCapita 1993; Griffith and Griffith, in press; J. Griffith, pers. comm.). Male Brown-headed Cowbirds can be extremely site faithful, whereas females may be more mobile; therefore, new males may come into a new area more slowly than females.

Kennard (1978) suggested that only females need to be disposed of, based upon five assumptions: (1) cowbirds are monogamous; (2) males are territorial; (3) males are site faithful; (4) if both members are destroyed, a new pair will assume the territory; and (5) if only the female is destroyed, the male will remain in the territory without a new mate. Since this suggestion was made, we have learned several aspects of cowbird behavior that shed doubt on these assumptions. Cowbirds are not always monogamous, and reports of male territoriality are rare. Nevertheless, Laymon (1987) pointed out that females could be selectively shot, leaving a more unbalanced sex ratio. While eliminating unnecessary sacrifice of lives may have appeal, we do not adequately understand how changing the sex ratio would affect the cowbird's mating system, territoriality, and fecundity. Understanding these effects should become an integral part of management plans.

Trap Design. Decoy traps are extremely effective because cowbirds are gregarious; they are attracted not only to abundant food and water but to other cowbirds kept as decoys. A funnel or slit (3.5–4 cm) in the top of the trap allows cowbirds to walk into the trap, but makes escape difficult. Traps vary in size and can be constructed as a permanent solid unit or as one with panels that break apart for compact storage and portability. Panels can easily be constructed with 2 × 2 lumber and 1/2″ hardware cloth. Perches and food shelves should also be provided. Cowbirds can be herded into a holding cage located at the trap's side for easy removal. The holding cage can be designed to be removable from the trap, but if it is not removable, it should be no longer than an arm's length and located at a height for easy removal of birds.

Methods. A variety of grains can be offered, including cracked corn, millet, and wild bird seed. It is not known if providing oyster shells or shells from local molluscs would increase female captures, but it may be worth considering. Easy care, inexpensive water containers made for domestic chickens can be purchased at farm supply stores. Decoy birds should be changed periodically, preferably every day or two. Using a single male or a couple of males is not especially effective for attracting females. Beezley and Rieger (1987) and Griffith and Griffith (1993) suggest a leklike situation (2 males and 3 females) may maximize female captures. Robinson et al. (1993) suggested that traps should be placed in semiopen areas with high perches nearby but not directly above the trap.

Placement of traps appears to be critical in maximizing female captures. Traps in foraging areas appear to yield a higher number of cowbirds (Beezley and Rieger 1987; J. Griffith, pers. comm.); however, traps located on the perimeter of breeding areas may result in a higher yield of females. Robinson et al. (1993) noted that Black-capped Vireo success was higher when traps were placed on the perimeter of breeding sites than within breeding sites. Females may become more trap shy after mid-May with the onset of egg laying (Beezley and Rieger 1987; Dufour and Weatherhead 1991); therefore, it is more effective to initiate trapping programs before the breeding season. The danger in initiating trapping too far in advance is in capturing migrants instead of the targeted population. Females may be more easily caught in Potter traps (Darley 1971; Beezley and Rieger 1987; pers. obs.); therefore, Potter traps used in conjunction with decoys may maximize female captures.

Trapping During the Nonbreeding Season. Laymon (1987) suggested that cowbirds be captured on their wintering grounds because the large concentrations could facilitate trapping efforts. At the North American Research Workshop on the Ecology and Management of Cowbirds, held in Austin, Texas, in 1993, Jane and John Griffith also suggested that trapping on wintering grounds should be seriously considered. They were met with instantaneous opposition and what they interpreted as hostility. While I doubt that those in opposition meant to express hostility toward or offend the Griffiths, those opposed were expressing some serious and valid concerns. Shortly after these meetings, Schram (1994) suggested that 30 million Brown-headed Cowbirds in a Louisiana winter roost represented the possibility of reducing the success of 150–180 million broods. The title of his article, "Open solicitation for cowbird recipes," reflects his opinion regarding how cowbirds ought to be managed. He further called upon American Birding Association members with direct connections with wildlife authorities to express their concerns (presumably reiterating his suggestion of winter trapping). However, Schram did not consider other implications.

In a counterpoint reply, Smith (1994) pointed out that Schram's assumptions that songbirds are declining and that cowbirds are the principal cause of the declines may both be faulty. Smith (1994:257) cited numerous papers in Hagan and Johnston's (1993) book and noted, "What Hagan and Johnston's book as a whole reveals is that songbird population trends across the continent are like Wall Street on a typical business day: there are increases in some places and declines in others. There are endangered species and populations, and declining trends, but these are not the rule."

The assumption that cowbirds are the principal cause for the decline of migratory birds, as Smith pointed out, is correct for some hosts and incorrect for others. Numerous songbirds are able to raise their own young along with a cowbird; therefore, the assumption that all parasitized clutches are doomed is unequivocally incorrect. Additionally, although Brown-headed Cowbirds probably increased dramatically prior to 1950, they have remained relatively stable since 1950 and have even declined in some areas. Smith (1994:258) noted that "We are no longer dealing with an aggressively colonizing brood parasite and new hosts with no previous contact with cowbirds."

Smith (1994:258) cited three reasons for not implementing a widespread winter cowbird eradication program: (1) "Exterminating a native brood parasite because of its cunning and successful life-style is a bit like exterminating the Peregrine Falcon (*Falco peregrinus*) because it is a predator on ducks and seabirds"; (2) Brown-headed Cowbirds are not a major problem throughout their range; and (3) a widespread eradication program could be counterproductive because it may distract attention from the more critical issue of habitat conservation. As Schram called upon members of the birding community to join in his campaign for winter eradication programs, Smith called upon members to collect nest records and *not* remove cowbird eggs from host nests, so that information can be collected on reproductive success. It should also be pointed out that removing cowbird eggs from host nests without appropriate permits (both from the U.S. Fish and Wildlife Service and from the state) is an offense that carries stiff penalties. Brown-headed Cowbirds are as protected as any other native species under the Migratory Bird Treaty of 1916.

There are additional concerns that Smith did not bring up. One concern is that cowbirds disperse widely from breeding grounds to wintering grounds and vice versa (chapter 7). Therefore, cowbirds captured on their wintering grounds will not represent the targeted population. In California, where migration may not be as extensive as other areas of North America, and where cowbirds spend their winters, such methods may be worth consideration. However, migration patterns in California are not well understood, and Laymon (1987) pointed out that banding studies would need to be initiated to determine where targeted populations winter. In addition, it should be determined where birds that winter in particular areas disperse to breed. The efforts and financial support needed for such extensive long-term studies may outstrip the perceived benefits of winter trapping, particularly because trapping during the breeding season would need to be continued regardless of winter trapping. While the feasibility of winter trapping in the Least Bell's Vireo breeding range is questionable at best, winter trapping targeting populations that parasitize Black-capped Vireos, Kirtland's Warblers, and Southwestern Willow Flycatchers is even less feasible. Cowbirds that breed in the Kirtland's Warbler range and part of the Southwestern Willow Flycatcher range probably disperse widely to wintering grounds, and winter trapping within Black-capped Vireo breeding range would result in the sacrifice of many nontargeted individuals.

Effects of nontargeted sacrifices also need to be addressed. One effect may be an outcry by animal rights activists. Steve Rothstein pointed out in his keynote address at the North American Research Workshop on the Ecology and Management of Cowbirds that such attention could undermine cowbird control programs in general, including those conducted during the breeding season.

Additionally, a widespread trapping program such as suggested by Schram (1994) could have a serious impact on the cowbird population, which of course is the objective. However, as Smith (1994) noted, the same ecological conditions that allowed cowbirds to spread would still exist and they would quickly recolonize; their recolonization would likely be aggressive and result in major disequilibrium, potentially harming hosts even further than they are harmed today. Another potential effect that warrants further evaluation is that some hosts have evolved low levels of antiparasite defenses, and within some host species, certain populations have high levels of antiparasite defenses (Sealy 1996). If the selection pressure for these defenses is removed, prevalence of these defenses may decrease, and some hosts might lose these defenses altogether. If widescale eradication programs were implemented, when cowbirds eventually recolonize, these hosts may be more susceptible to the negative effects of brood parasitism than they would have been before such a widespread trapping program.

It is understandable how people casually acquainted with cowbirds and their hosts might be lured into the apparent appeal of capturing cowbirds while they are concentrated in high numbers. However, lessons from the past when we have manipulated populations without consideration of effects should be remembered. Widescale winter eradication programs should not be implemented without full evaluation of the potential effects considered above.

Vandalism and Animal Rights Activists. Vandals and animal rights activists can undermine trapping efforts. This may be especially evident in urban areas. All traps are supposed have a sign provided by the U.S. Fish and Wildlife Service (USFWS) indicating that the trap is operated by the USFWS and that it is a federal offense to interfere with operations of the trap. The sign should be in a conspicuous location and kept in good shape, replaced with a new one if it becomes tattered. Although this warning might keep some potential offenders from interfering with the operations, others are undeterred by the warning (Beezley and Rieger 1987). All evidence of vandalism should be removed as quickly as possible as it may encourage additional vandalism. If animal rights activists are identified as the source of interference, a nonconfrontational resolution should be sought immediately. Educational programs emphasizing the effects cowbirds have on targeted hosts may be especially effective.

Determining the Need for a Control Program

In 1995, an individual representing the Black Canyon Audubon Society in Paonia, Colorado, applied for a permit to initiate a Brown-headed Cowbird control program near Delta, Colorado. According to Stephanie Jones of the USFWS, the society had funds from the Bureau of Reclamation and enthusiastic permission from the BLM to conduct the control program on their land. Initially, the society planned to begin the program without a permit, because they did not realize one was needed. Their concern was based upon five or six singing male Willow Flycatchers and a few cowbirds. The justification for the proposed control was that reproductive success of Willow Flycatchers in California is lowered by parasitism. The proposed trapping methods were to set up mist nets baited with dead cowbirds or study skins. The proposal was flatly denied by the USFWS based upon five criteria: (1) no previous studies had been conducted to determine the effects of cowbird parasitism, (2) the protocol would probably be more damaging than beneficial, (3) Brown-headed Cowbirds are not a depredating species (i.e., eggs do not prey upon hosts), (4) the proposal had no scientific merit, and (5) the personnel were untrained. Additionally, it was highly questionable whether the small population of Willow Flycatchers was, indeed, the endangered southwestern form.

Before cowbird programs are implemented, certain baseline data should be collected to ascertain the need for control, to consider changes that will likely occur, and to determine the efficacy of the control program. Robinson et al. (1993) suggested that the ratio of female cowbirds to hosts may be used as a crude index of parasitism intensity on the avian community. It should be emphasized that male cowbirds should not be used in this ratio because many populations are male biased, and use of males could inflate the actual need for management. Ratios of 0.05:0.10 female cowbirds to male hosts correspond with high levels (60–80%) of parasitism. This method is useful only at the community level and cannot be used to estimate intensity of parasitism of one particular species. This method may be most useful in identification of areas that are suspected to be a problem and in need of further study. Robinson et al. (1993) did not state, either implicitly or explicitly, that this method can be used to identify areas in need of active cowbird control.

Once the need for further study has been established, baseline data on intensity and location of parasitism need to be evaluated. High intensity parasitism does not, by itself, establish the need for control because hosts respond differently to parasitism. Before initiating a control program, it should be established that host reproductive success is lower than levels needed to maintain the population. In addition to quantifying how parasites and predators affect the success of individual nests, it is important to determine how parasites and predators affect seasonal fecundity of the host because many hosts renest after failure due to predation or parasitism, and they may ultimately produce as many young as those unparasitized or not preyed upon. Pease and Grzybowski (1995) provided an excellent model to assess both brood parasitism and predation on seasonal fecundity. The following parameters are used in their simple model to calculate seasonal fecundity: brood parasitism rate (per day), nest predation rate (per day), probability host nest abandoned when parasitized, beginning of both windows of susceptibility, end of window of susceptibility to brood parasitism, end of window of susceptibility to nest predation, time when successful females renest after terminating parental care, time in breeding season the last nesting cycle initiated, number of host young fledged from successful unparasitized nests, and number of host young fledged from successful parasitized nests. The following parameters are used in their more complex models to calculate cohort parasitism, seasonal parasitism fraction, and snapshot parasitism fraction: rate unparasitized nests become parasitized (per day), rate parasitized nests become parasitized again (per day), rate unparasitized nests are lost to nest predation (per day), rate parasitized nests are lost to nest predation (per day), probability unparasitized nest abandoned when parasitized (per day), probability parasitized nest abandoned when parasitized again (per day), time when nonrenesting parent terminates parental care, number of host young fledged from successful nest containing i parasite eggs, probability of renesting at day s of breeding season, and rate previously unreproductive females enter breeding pool on day s of breeding season (per day).

One of the most difficult complications in their model is determining renesting attempts, because to determine this for a particular population, the population must be banded with auxiliary markers. However, this information can sometimes be estimated from other studies (J. Grzybowski, pers. comm.). Generally, species with high rates of predation, low rates of abandonment of parasitized nests, long incubation periods, and short breeding seasons can tolerate only low rates of parasitism in order to maintain positive population growth, whereas hosts with low rates of predation, higher rates of abandonment of parasitized nests, shorter incubation periods, and longer breeding seasons can tolerate higher rates of parasitism and still maintain positive population growth (Robinson et al. 1993).

Ideally, radiotelemetry studies should also be conducted to establish patterns of breeding and feeding. Radiotelemetry is labor intensive and expensive. At roughly $140 per transmitter, Thompson (in Robinson et al. 1993) estimated the cost of tracking 35–40 cowbirds to be $25,000–35,000 per year for one site, including salaries for two months for three researchers. While temptation may exist to forgo this expensive part of gathering baseline data, it can provide useful information that could improve efficacy and cost of the program.

Assessing Efficacy of Management Programs

Part of any cowbird management plan should include how efficacy of cow-bird control will be evaluated. This necessitates that funds be allocated for re-search to evaluate success of hosts both before and after implementation of a management program, and funds for research should continue for the dura-tion of the management program. Allocating funds for on-going research should also be considered for any mitigation involving cowbird trapping.

In 1996, the Justice Department ruled that all federal agencies are exempt from the Migratory Bird Treaty of 1916, and therefore, no longer require a sci-entific collection permit from the U.S. Fish and Wildlife Service (W. Howe, USFWS, pers. comm.). This ruling has at least two negative implications. First, federal agencies are no longer required to justify the need for cowbird trapping programs, and some federal agencies may not have the expertise to adequately evaluate the need for cowbird control. Several cowbird trap-ping programs have already been implemented by federal agencies without adequately addressing the actual need for cowbird control. Second, there is no longer a single entity that houses all information for the on-the-ground management programs and management programs in the planning phase. Without a central location for this information, we cannot obtain a large pic-ture of cowbird control, which should be important in making each manage-ment decision. In other words, it is conceivable that small cowbird trapping programs could become so numerous as to annihilate entire populations of cowbirds. With this unfortunate ruling by the Justice Department, fed-eral agencies should cooperate to unite information by informing USFWS of their actions. Furthermore, before federal agencies that do not have the expertise implement a cowbird control program, they should have USFWS or researchers experienced with cowbird and host populations evaluate the need for cowbird control.

Summary

The perceived need to manage cowbirds is often based on assumptions that cowbirds have increased and hosts have declined dramatically throughout this century and that disequilibrium occurs in areas of recent expansion of the parasite because hosts are naive. While Brown-headed Cowbirds may have increased during the first half of the century, BBS data indicate that overall, from 1966 through 1992, cowbirds have declined and are no longer the aggressive colonizing species they once were. Shiny Cowbirds, on the other hand, are still rapidly expanding. Some host populations are declining while others are not; declines, however, may not be as much due to brood parasitism as to increased predation rates and habitat loss on both breeding and wintering grounds. In areas of recent range expansion, disequilibrium conditions may result in high levels of parasitism, high rates of egg puncturing and multiple parasitism, and low nesting success for hosts. However, there is little evidence that supports the assumption that this behavior by cowbirds has much to do with the hosts' historic lack of exposure to brood parasitism.

Since European settlement, we have created conditions favoring cowbirds and predators in North and South America, and we are now faced with the ethical dilemmas of managing parasites who are an integral part of the ecosystem. Cowbird removal programs should not be viewed as a long-term solution but only as a stop-gap measure. Control programs, which have met with varying success, can be extremely useful, however, in buying time and a second chance for some endangered hosts when used in conjunction with managing the conditions that favor cowbirds. Even when cowbird management targets a particular threatened or endangered host species, the effects the management program could have on the other members of the community should unequivocally be considered.

Basic strategies for controlling cowbird parasitism include: (1) preservation of undisturbed forests with large core areas; (2) restoration of large expanses of native grasslands grazed on long-term rotation of perhaps up to 20–25 years; (3) reduction of edge habitat both within forests and tallgrass prairies; (4) reduction of further development that creates islands within forests and tallgrass prairies; (5) follow principles of shape and size of refuges: compact shapes that reduce edges; (6) reduce or control foraging conditions for cowbirds, including grazing and mowing; (7) restructure grazing regimes; (8) eliminate grazing near areas of logging; (9) revegetate logging roads and clearcuts; (10) relocate dairies and stockyards from areas of high risk; (11) increase efficiency of harvesting techniques that would reduce winter food availability; (12) increase the amount of federal land that excludes grazing; (13) work with landowners to help define and fulfill their role in bird conservation within the context of stewardship; and (14) remove cowbirds only in areas where aggressive control measures are needed.

Before cowbird programs are implemented, baseline data should be collected to ascertain the need for control, to consider changes that will likely occur, and to determine the efficacy of the control program. If aggressive control must be initiated, it is important to know the cowbird population size, sex ratio, and site fidelity.

Parasitism Rates of Some Shiny Cowbird Hosts

Host Parasitism	%	(n)	Year	Location	Source
Rufous Hornero	14.6	(48)	1979–84	Argentina	Salvador 1984
	17.6	(17)	1977–79	Argentina	Mason 1986b
	29.5	(61)	<1929	Argentina	Friedmann 1929b
	100.0	(?)	<1964	Argentina	Hoy and Ottow 1964
Crested Hornero	14.3	(7)	<1929	Argentina	Friedmann 1929b
Pale-breasted Spinetail	0.0	(11)	1982–84	Venezuela	Cruz and Andrews 1989
	4.3	(46)	1979–84	Argentina	Salvador 1984
Stripe-crowned Spinetail	0.0	(2)	1979–83	Argentina	Salvador 1983
Rusty-backed Spinetail	0.0	(2)	1982–84	Venezuela	Cruz and Andrews 1989
Yellow-chinned Spinetail	0.0	(33)	1950–84	Trinidad, Tobago	Cruz et al. 1995
	0.0	(32)	1982–84	Venezuela	Cruz and Andrews 1989
Short-billed Canastero	0.0	(3)	1979–83	Argentina	Salvador 1983
	100.0	(?)	<1964	Argentina	Hoy and Ottow 1964
Common Thornbird	0.0	(4)	1982–84	Venezuela	Cruz and Andrews 1989
Little Thornbird	0.0	(15)	1979–83	Argentina	Salvador 1983
Freckle-breasted Thornbird	14.3	(7)	1977–79	Argentina	Mason 1986b
Wren-like Rushbird	0.0	(22)	1977–79	Argentina	Mason 1986b
	0.0	(14)	1979–83	Argentina	Salvador 1983
Leñatero	0.0	(11)	1977–79	Argentina	Mason 1986b
	0.0	(12)	1979–83	Argentina	Salvador 1983
Lark-like Brushrunner	0.0	(16)	1979–83	Argentina	Salvador 1983
	3.7	(27)	1980	Argentina	Contreras in FK
White-throated Cachalote	5.0	(20)	1980	Argentina	Contreras in FK
Caribbean Elaenia	0.0	(14)	1973–83	Puerto Rico	Cruz et al. 1989
	0.0	(53)	1982–88	Puerto Rico	Nakamura and Cruz, in press
	9.1	(22)	1984–85	St. Lucia	Nakamura 1995
	11.1	(18)	1983–85	St. Lucia	Cruz et al. 1989
Yellow-bellied Elaenia	0.0	(41)	1950–84	Trinidad, Tobago	Cruz et al. 1995
White-crested Tyrannulet	0.0	(4)	1977–79	Argentina	Mason 1986b
Bran-colored Flycatcher	0.0	(9)	1950–84	Trinidad, Tobago	Cruz et al. 1995

Host Parasitism	%	(n)	Year	Location	Source
Vermillion Flycatcher	0.0	(22)	1977–79	Argentina	Mason 1986b
	0.0	(3)	1979–83	Argentina	Salvador 1983
	0.0	(5)	1982–84	Venezuela	Cruz and Andrews 1989
White Monjita	0.0	(2)	1979–83	Argentina	Salvador 1983
	100.0	(?)	<1964	Argentina	Hoy and Ottow 1964
Short-tailed Field-Tyrant	2.8	(36)		Ecuador	Marchant in FK
	41.9	(31)	1978, 81	Peru	Williams in FK
Pied Water-Tyrant	0.0	(4)	1950–84	Trinidad, Tobago	Cruz et al. 1995
	10.7	(28)	1980	Venezuela	Ramo and Busto in FK
	52.9	(102)	1982–84	Venezuela	Cruz and Andrews 1989
Black-backed Water-Tyrant	0.0	(20)	<1929	Argentina	Friedmann 1929b
White-headed Marsh-Tyrant	0.0	(5)	1950–84	Trinidad Tobago	Cruz et al. 1995
	80.0	(5)	<1981	Venezuela	Ramo and Busto in FK
	92.3	(13)	1982–84	Venezuela	Cruz and Andrews 1989
Yellow-browed Tyrant	40.0	(10)	1977–79	Argentina	Mason 1986b
Cattle Tyrant	0.0	(14)	1982–84	Venezuela	Cruz and Andrews 1989
	40.0	(10)	<1929	Argentina	Friedmann 1929b
	50.0	(8)	1979–84	Argentina	Salvador 1984
	66.7	(3)	1977–79	Argentina	Mason 1986b
Puerto Rican Flycatcher	16.3	(76)	1982–88	Puerto Rico	Nakamura and Cruz, in press
	34.0	(47)	1973–83	Puerto Rico	Cruz et al. 1989
	52.0	(25)	1973–82	Puerto Rico	Cruz et al. 1985
Snowy-throated Kingbird	8.3	(12)	1978, 81	Peru	Williams in FK
Tropical Kingbird	0.0	(6)	1982–84	Venezuela	Cruz and Andrews 1989
	5.0	(20)	1950–84	Trinidad, Tobago	Cruz et al. 1995
Fork-tailed Flycatcher	0.0	(2)	1982–84	Venezuela	Cruz and Andrews 1989
	1.1	(87)	1982–84	Brazil	Cavalcanti and Pimentel 1988
	>50.0		1970–<78	Argentina	Fraga 1978
	51.9	(27)	1979–84	Argentina	Salvador 1984
	57.1	(7)	1977–79	Argentina	Mason 1986b
	80.0	(10)	<1874	Argentina	Hudson 1874
Gray Kingbird	0.0	(45)	1974–75	Puerto Rico	Post and Wiley 1977b
	0.0	(125)	1982–88	Puerto Rico	Nakamura and Cruz, in press
	0.0	(2)	1982–84	Venezuela	Cruz and Andrews 1989
	0.0	(38)	1975–76	Puerto Rico	Post and Wiley 1977a
	0.9	(109)	1973–83	Puerto Rico	Cruz et al. 1989
	1.1	(92)	1973–82	Puerto Rico	Cruz et al. 1985
	1.6	(61)	1975–81	Puerto Rico	Wiley 1985
White-bearded Flycatcher	0.0	(2)	1982–84	Venezuela	Cruz and Andrews 1989
	41.7	(36)	1977, 79	Venezuela	Thomas in FK
Baird's Flycatcher	3.8	(53)	1978, 81	Peru	Williams in FK
Great Kiskadee	0.0	(7)	1977–79	Argentina	Mason 1986b
	0.0	(14)	1982–84	Venezuela	Cruz and Andrews 1989
	21.7	(23)	1979–84	Argentina	Salvador 1984

Host Parasitism	%	(n)	Year	Location	Source
White-naped Xenopsaris	0.0	(4)	1982–84	Venezuela	Cruz and Andrews 1989
White-winged Becard	0.0	(2)	1982–84	Venezuela	Cruz and Andrews 1989
White-rumped Swallow	40.0	(10)	1977–79	Argentina	Mason 1986b
Stripe-backed Wren	16.3	(56)	1990–91	Venezuela	Piper 1994
Rufous-breasted Wren	9.5	(21)	1950–84	Trinidad, Tobago	Cruz et al. 1995
Superciliated Wren	18.2	(33)	<1983	Peru	Williams in FK
House Wren	0.0	(43)	1979–83	Argentina	Salvador 1983
	31.7	(41)	1950–84	Trinidad, Tobago	Cruz et al. 1995
	58.8	(17)	1977–79	Argentina	Mason 1986b
Masked Gnatcatcher	0.0	(17)	<1929	Argentina	Friedmann 1929b
	20.0	(10)	1979–84	Argentina	Salvador 1984
Red-legged Thrush	0.0	(2)	1974–75	Puerto Rico	Post and Wiley 1977b
	0.0	(5)	1982–88	Puerto Rico	Nakamura and Cruz, in press
	3.7	(27)	1973–82	Puerto Rico	Cruz et al. 1985
	4.0	(25)	1975–81	Puerto Rico	Wiley 1985
Rufous-bellied Thrush	100.0	(4)	1977–79	Argentina	Mason 1986b
Creamy-bellied Thrush	62.5	(8)	1923	Argentina	Friedmann 1929b
	100.0	(1)	1977–79	Argentina	Mason 1986b
	100.0	(2)	1979–84	Argentina	Salvador 1984
Tropical Mockingbird	1.3	(79)	1950–84	Trinidad Tobago	Cruz et al. 1995
Northern Mockingbird	0.0	(5)	1974–75	Puerto Rico	Post and Wiley 1977b
	0.0	(135)	1982–88	Puerto Rico	Nakamura and Cruz, in press
	1.9	(52)	1973–82	Puerto Rico	Cruz et al. 1985
Chalk-browed Mockingbird	26.7	(15)	1982–84	Brazil	Cavalcanti and Pimentel 1988
	72.5	(69)	1972–79	Argentina	Fraga 1985
	73.5	(68)	1977–79	Argentina	Mason 1986b
	88.0	(92)	1979–84	Argentina	Salvador 1984
Long-tailed Mockingbird	53.2	(188)	1978, 81	Peru	Williams in FK
	85.7	(7)	1978?	Peru	Williams 1981
Short-billed Pipit	0.0	(3)	1979–83	Argentina	Salvador 1983
Yellowish Pipit	0.0	(2)	1982–84	Venezuela	Cruz and Andrews 1989
Palmchat	5.3	(243)	1974–77	Dominican Republic	Cruz et al. 1989
	12.8	(305)	1974–78	Dominican Republic	Cruz et al. 1989
	25.8	(62)	1982	Dominican Republic	Cruz et al. 1989
Puerto Rican Vireo	52.9	(17)	1973–83	Puerto Rico	Cruz et al. 1989
	≈80.0	(?)	<1993	Puerto Rico	Woodworth 1993
	87.1	(39)	1982–88	Puerto Rico	Nakamura and Cruz, in press
Red-eyed Vireo	0.0	(5)	1950–84	Trinidad, Tobago	Cruz et al. 1995

Host Parasitism	%	(n)	Year	Location	Source
Black-whiskered Vireo	14.3	(14)	1974–77	Dominican Republic	Cruz et al. 1989
	34.8	(23)	1974–78	Dominican Republic	Cruz et al. 1989
	66.7	(9)	1982	Dominican Republic	Cruz et al. 1989
	73.0	(111)	1982–88	Puerto Rico	Nakamura and Cruz, in press
	81.8	(11)	1975–81	Puerto Rico	Wiley 1985
	86.7	(15)	1973–82	Puerto Rico	Cruz et al. 1985
	87.5	(24)	1983–85	St. Lucia	Cruz et al. 1989
	89.5	(19)	1973–83	Puerto Rico	Cruz et al. 1989
Golden-fronted Greenlet	0.0	(2)	1950–84	Trinidad and Tobago	Cruz et al. 1995
Yellow Warbler	0.0	(25)	1974–75	Puerto Rico	Post and Wiley 1977b
	9.1	(22)	1975–76	Puerto Rico	Post and Wiley 1977a
	10.5	(19)	1974–77	Dominican Republic	Cruz et al. 1989
	44.1	(34)	1984–85	St. Lucia	Nakamura 1995
	46.9	(262)	1982–88	Puerto Rico	Nakamura and Cruz, in press
	63.9	(144)	1973–82	Puerto Rico	Cruz et al. 1985
	74.7	(87)	1975–81	Puerto Rico	Wiley 1985
	83.3	(12)	1982	Dominican Republic	Cruz et al. 1989
Adelaide's Warbler	3.0	(33)	1982–88	Puerto Rico	Nakamura and Cruz, in press
	≈12.0	(?)	<1993	Puerto Rico	Woodworth 1993
	30.8	(13)	1973–83	Puerto Rico	Cruz et al. 1989
Cinereous Conebill	18.2	(11)	1978, 81	Peru	Williams in FK
White-banded Tanager	83.3	(6)	1982–84	Brazil	Cavalcanti and Pimentel 1988
White-rumped Tanager	100.0	(3)	1982–84	Brazil	Cavalcanti and Pimentel 1988
White-lined Tanager	0.0	(21)	1950–84	Trinidad, Tobago	Cruz et al. 1995
Silver-beaked Tanager	0.9	(106)	1950–84	Trinidad, Tobago	Cruz et al. 1995
Blue-gray Tanager	0.0	(120)	1950–84	Trinidad, Tobago	Cruz et al. 1995
Blue Dacnis	0.0	(6)	1950–84	Trinidad and Tobago	Cruz et al. 1995
Grayish Saltator	22.2	(9)	<1929	Argentina	Friedmann 1929b
Rufous-collared Sparrow	15.4	(13)	1970–71	Argentina	Gochfeld 1979a, 1979b
	21.6	(51)	1977–79	Argentina	Mason 1986b
	38.5	(26)	1979–84	Argentina	Salvador 1984
	61.4	(83)	<1958	Brazil	Sick and Ottow in King 1973b
	66.0	(50)	1969–70	Argentina	King 1973b

Host Parasitism	%	(n)	Year	Location	Source
	72.5	(40)	1970–<78	Argentina	Fraga 1978
	72.7	(33)	<1929	Argentina	Friedmann 1929b
	83.3	(6)	1982–84	Brazil	Cavalcanti and Pimentel 1988
Grassland Sparrow	0.0	(9)	1982–84	Venezuela	Cruz and Andrews 1989
	28.6	(7)	1982–84	Brazil	Cavalcanti and Pimentel 1988
Tumbes Sparrow	45.5	(22)	1978, 81	Peru	Williams in FK
Common Diuca-Finch	72	(100)	<1946	Chile	Goodall et al. in Friedmann 1963
Cinereous Finch	86.1	(36)	1978, 81	Peru	Williams in FK
	100.0	(4)[1]	1978	Peru	Williams 1981
Saffron Finch	0.0	(3)	1979–83	Argentina	Salvador 1983
	0.0	(14)	1982–84	Venezuela	Cruz and Andrews 1989
	2.2	(45)	1977–79	Argentina	Mason 1986b
	11.8	(17)	1978, 81	Peru	Williams in FK
Grassland Yellow-Finch	0.0	(3)	1977–79	Argentina	Mason 1986b
	27.8	(18)	1979–84	Argentina	Salvador 1984
Double-collared Seedeater	0.0	(16)	1979–83	Argentina	Salvador 1983
Ruddy-breasted Seedeater	0.0	(12)	1950–84	Trinidad, Tobago	Cruz et al. 1995
Yellow Oriole	0.0	(13)	1950–84	Trinidad, Tobago	Cruz et al. 1995
	0.0	(16)	1982–84	Venezuela	Cruz and Andrews 1989
Troupial	0.0	(2)	1982–84	Venezuela	Cruz and Andrews 1989
	8.3	(12)	1982–88	Puerto Rico	Nakamura and Cruz, in press
	100.0	(5)	1973–82	Puerto Rico	Cruz et al. 1985
Black-cowled Oriole	0.0	(2)	1982–88	Puerto Rico	Nakamura and Cruz, in press
	29.2	(24)	1974–77	Dominican Republic	Cruz et al. 1989
	100.0	(6)	1982	Dominican Republic	Cruz et al. 1989
	100.0	(12)	1975–81	Puerto Rico	Wiley 1985
	100.0	(14)	1973–82	Puerto Rico	Cruz et al. 1985
Yellow-hooded Blackbird	33.3	(15)	1971	Trinidad	Wiley and Wiley 1980
	33.3	(72)	1974–76	Venezuela	Wiley and Wiley 1980
	33.3	(3)	1982–84	Venezuela	Cruz and Andrews 1989
	40.6	(377)	1979,81,84	Trinidad	Cruz et al. 1990
Yellow-shouldered Blackbird	77.8	(72)	1974–75	Puerto Rico	Post and Wiley 1977b
	86.3	(194)	1982–88	Puerto Rico	Nakamura and Cruz, in press
	90.6	(266)	1973–83	Puerto Rico	Cruz et al. 1989
	100.0	(61)	1975–76	Puerto Rico	Post and Wiley 1977a
	100.0	(44)	1982	Puerto Rico	Cruz et al. 1985
Chestnut-capped Blackbird	6.2	(32)		Argentina	Klimaitis 1971
	22.5	(213)	1979–84	Argentina	Salvador 1984

Host Parasitism	%	(n)	Year	Location	Source
Greater Red-breasted Meadowlark	71.4	(14)	<1981	Venezuela	Ramo and Busto in FK
	95.8	(24)	1970–71	Argentina	Gochfeld 1979a, 1979b
	100.0	(2)	1950–84	Trinidad, Tobago	Cruz et al. 1995
Pampas Meadowlark	0.0	(11)	1970–71	Argentina	Gochfeld 1979a, 1979b
White-browed Blackbird	33.3	(6)	1977–79	Argentina	Mason 1986b
Brown-and-Yellow Marshbird	74.3	(55)	1992	Argentina	Mermoz and Reboreda 1994
Greater Antillean Grackle	0.0	(18)	1974–75	Puerto Rico	Post and Wiley 1977b
	2.9	(137)	1982–88	Puerto Rico	Nakamura and Cruz, in press
	6.7	(15)	1975–76	Puerto Rico	Post and Wiley 1977a
	8.9	(281)	1973–83	Puerto Rico	Cruz et al. 1989
Carib Grackle	0.0	(41)	1950–84	Trinidad, Tobago	Cruz et al. 1995
	2.1	(96)	1983–85	St. Lucia	Cruz et al. 1989
	30.0	(20)	<1981	Venezuela	Ramo and Busto in FK
	100.0	(4)	1982–84	Venezuela	Cruz and Andrews 1989
Bay-winged Cowbird	11.1	(9)	1979–84	Argentina	Salvador 1984
	22.4	(85)	1973–80	Argentina	Fraga 1986
	28.6	(14)	1977–79	Argentina	Mason 1986b
Hooded Siskin	0.0	(7)	1977–79	Argentina	Mason 1986b
Village Weaver	1.3	(936)	1974–77	Dominican Republic	Cruz and Wiley 1989a
	3.1	(1074)	1983–85	Dominican Republic	Cruz et al. 1989
	15.7	(134)	1982	Dominican Republic	Cruz and Wiley 1989a
House Sparrow	0.0	(3)	1977–79	Argentina	Mason 1986b
	4.4	(45)	1979–84	Argentina	Salvador 1984
Bronzed Mannikin	0.0	(79)	1982–88	Puerto Rico	Nakamura and Cruz, in press
	16.7	(6)	1975–81	Puerto Rico	Wiley 1985
	20.0	(10)	1973–82	Puerto Rico	Cruz et al. 1985

Abbreviation: FK: Friedman and Kiff 1985.

[1] Friedmann and Kiff (1985) reported a parasitism rate of 66.7%, but they included 2 nests that Williams found empty.

Known Victims of Brown-headed Cowbirds

Order Anseriformes, Family Anatidae
Blue-winged Teal, *Anas discors*
Order Falconiformes, Family Accipitridae
Ferruginous Hawk, *Buteo regalis*
Order Gruiformes, Family Rallidae
Virginia Rail, *Rallus limicola*
Order Charadriiformes, Family Charadriidae
Killdeer, *Charadrius vociferus*
Family Scolopacidae
Spotted Sandpiper, *Actitis macularia*
Upland Sandpiper, *Bartramia longicauda*
Wilson's Phalarope, *Phalaropus tricolor*
Family Laridae
California Gull, *Larus californicus*
Common Tern, *Sterna hirundo*
Order Columbiformes, Family Columbidae
*Mourning Dove, *Zenaida macroura*
Common Ground Dove, *Columbina passerina*
Order Cuculiformes, Family Cuculidae
Black-billed Cuckoo, *Coccyzus
erythropthalmus*
Yellow-billed Cuckoo, *Coccyzus americanus*
Order Apodiformes, Family Trochilidae
Ruby-throated Hummingbird, *Archilochus
colubris*
Order Piciformes, Family Picidae
Red-headed Woodpecker, *Melanerpes
erythrocephalus*
Order Passeriformes, Family Tyrannidae
*Olive-sided Flycatcher, *Contopus cooperi*
[*borealis*]
*Western Wood-Pewee, *Contopus sordidulus*
*Eastern Wood-Pewee, *Contopus virens*
Yellow-bellied Flycatcher, *Empidonax
flaviventris*

*Acadian Flycatcher, *Empidonax virescens*
*Alder Flycatcher, *Empidonax alnorum*
*Willow Flycatcher, *Empidonax traillii*
Least Flycatcher, *Empidonax minimus*
Hammond's Flycatcher, *Empidonax
hammondii*
*Dusky Flycatcher, *Empidonax oberholseri*
*Gray Flycatcher, *Empidonax wrightii*
*Western Flycatcher, *Empidonax difficilis*
Buff-breasted Flycatcher, *Empidonax
fulvifrons*
Black Phoebe, *Sayornis nigricans*
*Eastern Phoebe, *Sayornis phoebe*
Say's Phoebe, *Sayornis saya*
*Vermilion Flycatcher, *Pyrocephalus rubinus*
Great Crested Flycatcher, *Myiarchus crinitus*
Cassin's Kingbird, *Tyrannus vociferans*
*†Western Kingbird, *Tyrannus verticalis*
*†Eastern Kingbird, *Tyrannus tyrannus*
*†Scissor-tailed Flycatcher, *Tyrannus forficatus*
Family Alaudidae
*Horned Lark, *Eremophila alpestris*
Family Hirundinidae
Purple Martin, *Progne subis*
*Tree Swallow, *Tachycineta bicolor*
Bank Swallow, *Riparia riparia*
Cliff Swallow, *Hirundo pyrrhonota*
*Barn Swallow, *Hirundo rustica*
Family Corvidae
† Blue Jay, *Cyanocitta cristata*
American Crow, *Corvus brachyrynchos*
Family Paridae
*Black-capped Chickadee, *Poecile* [*Parus*]
atricapillus

*Carolina Chickadee, *Poecile* [*Parus*]
   carolinensis
Tufted Titmouse, *Parus bicolor*
Family Remizidae
 *Verdin, *Auriparus flaviceps*
Family Aegithalidae
   Bushtit, *Psaltriparus minimus*
Family Sittidae
 *Red-breasted Nuthatch, *Sitta canadensis*
   White-breasted Nuthatch, *Sitta carolinensis*
Family Certhiidae
 *Brown Creeper, *Certhia americana*
Family Troglodytidae
   Boucard's Wren, *Campylorhynchus jocosus*
 *Rock Wren, *Salpinctes obsoletus*
 *Carolina Wren, *Thryothorus ludovicianus*
 *Bewick's Wren, *Thryomanes bewickii*
 *House Wren, *Troglodytes aedon*
   Marsh Wren, *Cistothorus palustris* (Picman
   1986)
Family Muscicapidae, Subfamily Sylviinae
 *Golden-crowned Kinglet, *Regulus satrapa*
 *Ruby-crowned Kinglet, *Regulus calendula*
 *Blue-gray Gnatcatcher, *Polioptila caerulea*
 *Black-tailed Gnatcatcher, *Polioptila melanura*
   White-lored Gnatcatcher, *Polioptila albiloris*
Family Muscicapidae, Subfamily Turdinae
   Eastern Bluebird, *Sialia sialis*
   Western Bluebird, *Sialia mexicana*
   Mountain Bluebird, *Sialia currucoides*
   Townsend's Solitaire, *Myadestes townsendi*
 *Veery, *Catharus fuscescens*
   Swainson's Thrush, *Catharus ustulatus*
 *Hermit Thrush, *Catharus guttatus*
 *Wood Thrush, *Hylocichla mustelina*
 *†American Robin, *Turdus migratorius*
Family Muscicapidae, Subfamily Timaliinae
 *Wrentit, *Chamaea fasciata*
Family Mimidae
 *†Gray Catbird, *Dumetella carolinensis*
 *Northern Mockingbird, *Mimus polyglottos*
 †Sage Thrasher, *Oreoscoptes montanus*
 *†Brown Thrasher, *Toxostoma rufum*
   Long-billed Thrasher, *Toxostoma longirostre*
   Bendire's Thrasher, *Toxostoma bendirei*
 †Curve-billed Thrasher, *Toxostoma curvirostre*
Family Motacillidae
   Sprague's Pipit, *Anthus spragueii*
Family Bombycillidae
   Bohemian Waxwing, *Bombicilla garrulus*
 *†Cedar Waxwing, *Bombicilla cedrorum*

Family Ptilogonatidae
 *Phainopepla, *Phainopepla nitens*
Family Sturnidae
 *European Starling, *Sturnus vulgaris*
Family Vireonidae
   Slaty Vireo, *Vireo brevipennis*
 *White-eyed Vireo, *Vireo griseus*
   Flat-billed Vireo, *Vireo nanus*
 *Bell's Vireo, *Vireo bellii*
 *Black-capped Vireo, *Vireo atricapillus*
 *Gray Vireo, *Vireo vicinior*
 *Blue-headed Vireo, *Vireo solitarius*
 *Cassin's Vireo (*Vireo cassinii*)
 *Plumbeous Vireo (*Vireo plumbeus*)
 *Yellow-throated Vireo, *Vireo flavifrons*
 *Hutton's Vireo, *Vireo huttoni*
 *Warbling Vireo, *Vireo gilvus*
 *Philadelphia Vireo, *Vireo philadelphicus*
 *Red-eyed Vireo, *Vireo olivaceus*
Family Emberizidae, Subfamily Parulinae
 *Blue-winged Warbler, *Vermivora pinus*
 *Golden-winged Warbler, *Vermivora*
   chrysoptera
 *Tennessee Warbler, *Vermivora peregrina*
 *Orange-crowned Warbler, *Vermivora celata*
 *Nashville Warbler, *Vermivora ruficapilla*
 *Virginia's Warbler, *Vermivora virginiae*
 *Lucy's Warbler, *Vermivora luciae*
 *Northern Parula, *Parula americana*
 *Tropical Parula, *Parula pitiayumi*
 *Yellow Warbler, *Dendroica petechia*
 *Chestnut-sided Warbler, *Dendroica*
   pensylvanica
 *Magnolia Warbler, *Dendroica magnolia*
   Cape May Warbler, *Dendroica tigrina*
 *Black-throated Blue Warbler, *Dendroica*
   caerulescens
 *Yellow-rumped Warbler, *Dendroica coronata*
 *Black-throated Gray Warbler, *Dendroica*
   nigrescens
 *Townsend's Warbler, *Dendroica townsendi*
 *Hermit Warbler, *Dendroica occidentalis*
 *Black-throated Green Warbler, *Dendroica*
   virens
 *Golden-cheeked Warbler, *Dendroica*
   chrysoparia
 *Blackburnian Warbler, *Dendroica fusca*
 *Yellow-throated Warbler, *Dendroica dominica*
 *Grace's Warbler, *Dendroica graciae*
 *Pine Warbler, *Dendroica pinus*
 *Kirtland's Warbler, *Dendroica kirtlandii*

*Prairie Warbler, *Dendroica discolor*
*Palm Warbler, *Dendroica palmarum*
*Bay-breasted Warbler, *Dendroica castanea*
*Blackpoll Warbler, *Dendroica striata*
*Cerulean Warbler, *Dendroica cerulea*
*Black-and-white Warbler, *Mniotilta varia*
*American Redstart, *Setophaga ruticilla*
*Prothonotary Warbler, *Protonotaria citrea*
*Worm-eating Warbler, *Helmitheros
    vermivorus*
*Swainson's Warbler, *Limnothlypis swainsonii*
*Ovenbird, *Seiurus aurocapillus*
*Northern Waterthrush, *Seiurus
    noveboracensis*
*Louisiana Waterthrush, *Seiurus motacilla*
*Kentucky Warbler, *Oporornis formosus*
*Mourning Warbler, *Oporornis philadelphia*
*MacGillivray's Warbler, *Oporornis tolmiei*
*Common Yellowthroat, *Geothlypis trichas*
Gray-crowned Yellowthroat, *Geothlypis
    poliocephala*
*Hooded Warbler, *Wilsonia citrina*
*Wilson's Warbler, *Wilsonia pusilla*
*Canada Warbler, *Wilsonia canadensis*
Painted Redstart, *Myioborus pictus*
*Yellow-breasted Chat, *Icteria virens*
Family Emberizidae, Subfamily Thraupinae
*Summer Tanager, *Piranga rubra*
*Scarlet Tanager, *Piranga olivacea*
*Western Tanager, *Piranga ludoviciana*
Family Emberizidae, Subfamily Cardinalinae
*Northern Cardinal, *Cardinalis cardinalis*
Pyrrhuloxia, *Cardinalis sinuatus*
*Rose-breasted Grosbeak, *Pheucticus
    ludovicianus*
Black-headed Grosbeak, *Pheucticus
    melanocephalus*
*Blue Grosbeak, *Guiraca caerulea*
*Lazuli Bunting, *Passerina amoena*
*Indigo Bunting, *Passerina cyanea*
Varied Bunting, *Passerina versicolor*
*Painted Bunting, *Passerina ciris*
*Dickcissel, *Spiza americana*
Family Emberizidae, Subfamily Emberizinae
Rusty-crowned Ground-Sparrow, *Melozone
    kieneri*
Olive Sparrow, *Arremonops rufivirgatus*
Green-tailed Towhee, *Pipilo chlorurus*
*Eastern Towhee, *Pipilo erythrophthalmus*
*Spotted Towhee, *Pipilo maculatus*
*Canyon Towhee, *Pipilo fuscus*

*California Towhee, *Pipilo crissalis*
*Abert's Towhee, *Pipilo aberti*
White-collared Seedeater, *Sporophila
    torqueola*
Bachman's Sparrow, *Aimophila aestivalis*
Cassin's Sparrow, *Aimophila cassinii*
*Rufous-winged Sparrow, *Aimophila carpalis*
Rufous-crowned Sparrow, *Aimophila ruficeps*
*Five-striped Sparrow, *Aimophila
    quinquestriata*
*Chipping Sparrow, *Spizella passerina*
*Clay-colored Sparrow, *Spizella pallida*
*Brewer's Sparrow, *Spizella breweri*
*Field Sparrow, *Spizella pusilla*
Black-chinned Sparrow, *Spizella atrogularis*
*Vesper Sparrow, *Pooecetes gramineus*
*Lark Sparrow, *Chondestes grammacus*
Black-throated Sparrow, *Amphispiza bilineata*
Sage Sparrow, *Amphispiza belli*
Lark Bunting, *Calamospiza melanocorys*
*Savannah Sparrow, *Passerculus sandwichensis*
Baird's Sparrow, *Ammodramus bairdii*
*Grasshopper Sparrow, *Ammodramus
    savannarum*
Henslow's Sparrow, *Ammodramus henslowii*
Le Conte's Sparrow, *Ammodramus leconteii*
*Sharp-tailed Sparrow, *Ammodramus
    caudacutus*
*Seaside Sparrow, *Ammodramus maritimus*
*Fox Sparrow, *Passerella iliaca*
*Song Sparrow, *Melospiza melodia*
*Lincoln's Sparrow, *Melospiza lincolnii*
*Swamp Sparrow, *Melospiza georgiana*
*White-throated Sparrow, *Zonotrichia
    albicollis*
*White-crowned Sparrow, *Zonotrichia
    leucophrys*
*Dark-eyed Junco, *Junco hyemalis*
McCown's Longspur, *Calcarius mccownii*
*Chestnut-collared Longspur, *Calcarius
    ornatus*
Family Emberizidae, Subfamily Icterinae
*Bobolink, *Dolichonyx oryzivorus*
*Red-winged Blackbird, *Agelaius phoeniceus*
Tri-colored Blackbird, *Agelaius tricolor*
*Eastern Meadowlark, *Sturnella magna*
Western Meadowlark, *Sturnella neglecta*
Yellow-headed Blackbird, *Xanthocephalus
    xanthocephalus*[a]
Rusty Blackbird, *Euphagus carolinus*

*Brewer's Blackbird, *Euphagus cyanocephalus*
Common Grackle, *Quisculus quiscula*
*Orchard Oriole, *Icterus spurius*
*Hooded Oriole, *Icterus cucullatus*
Audubon's Oriole, *Icterus graduacauda*
*Baltimore Oriole, *Icterus galbula*
*Bullock's Oriole, *Icterus bullockii*
Scott's Oriole, *Icterus parisorum*
Family Fringillidae
*Purple Finch, *Carpodacus purpureus*

*House Finch, *Carpodacus mexicanus*
Red Crossbill, *Loxia curvirostra*
Common Redpoll, *Carduelis flammea*
*Pine Siskin, *Carduelis pinus*
*Lesser Goldfinch, *Carduelis psaltria*
*Lawrence's Goldfinch, *Carduelis lawrencei*
American Goldfinch, *Carduelis tristis*
*Evening Grosbeak, *Coccothraustes vespertinus*
Family Passeridae
*House Sparrow, *Passer domesticus*

Source: Most were listed by Friedmann and Kiff 1985.
Note: Hosts known to have raised Brown-headed Cowbirds are identified with "*", and victims known to eject Brown-headed Cowbird eggs are identified with "†".

[a] Yellow-headed Blackbirds successfully raised 7 Brown-headed Cowbirds in cross-fostering experiments (Ortega 1991; Ortega and Cruz 1991).

Parasitism Rates of Some Brown-headed Cowbird Hosts

Host Parasitism	%	(n)	Year	Location	Source
Mourning Dove	0.0	(30)	1977–78	CO WY	Hanka 1979
	0.0	(15)	1992–95	CO	Ortega and Ortega, unpubl.
	0.0	(102)	1992–93	IL	Peer and Bolinger 1997
	0.0	(1,023)	1973–74	KS	Hill 1976a
	0.0	(48)	1960	NE	Hergenrader 1962
	0.0	(110)	1966	NE	Holcomb 1968
Olive-sided Flycatcher	0.0	(4)	1928–67	MI	Southern and Southern 1980
Western Wood-Pewee	0.0	(10)	1980–81	CA	Verner and Ritter 1983
	0.0	(8)	1977–78	CO WY	Hanka 1979
	0.0	(16)	1992–95	CO	Ortega and Ortega, unpubl.
	14.3	(7)	1980–81	CA	Verner and Ritter 1983
Eastern Wood-Pewee	0.0	(4)	1963–75	LA	Goertz 1977
	0.0	(4)	1991–92	NY	Hahn and Hatfield 1995
	15.0	(40)	1911–66	MI	Southern and Southern 1980
	66.7	(3)	1980s	IL	Robinson 1992
Yellow-bellied Flycatcher	0.0	(6)	1945	MI	Walkinshaw 1961
	0.0	(3)	1959–74	MI	Southern and Southern 1980
Acadian Flycatcher	3.0	(67)	1991–93	MO[1]	T. Donovan (pers. comm.)
	4.2	(24)	<1975	Ontario	Peck in FKR
	23.9	(67)	1945	MI	Walkinshaw 1961
	50.0	(12)	<1974	IL	Graber et al. in FKR
	50.0	(2)	1980s	IL	Robinson 1992
	50.0	(6)	1991–93	MO[2]	T. Donovan (pers. comm.)
Alder Flycatcher[3]	6.2	(146)	1896–1958	Quebec	Terril 1961
Willow Flycatcher[3]	0.0	(7)	1928–70	MI	Southern and Southern 1980
	0.0	(5)	1991–92	NY	Hahn and Hatfield 1995
	4.5	(44)	1952–53	WA	King 1954
	7.5	(53)	1945	MI	Walkinshaw 1961
	8.1	(37)	1946–49	MI	Berger 1951

Host Parasitism	%	(n)	Year	Location	Source
	20.8	(48)	1951	MI	Berger and Parmelee 1952
	21.3	(108)	1918–34	OH	Hicks 1934
	40.7	(27)	1985–86	CO	Sedgwick and Knopf 1988
	50.0	(8)	1982–87	AZ	Brown 1994
	68.4	(19)	1987	CA	Harris 1991
	100.0	(3)	1982–83	IA	Lowther 1983
Least Flycatcher	0.0	(2)	1991–93	WI[2]	T. Donovan (pers. comm.)
	0.0	(5)	1991–93	WI[1]	T. Donovan (pers. comm.)
	2.8	(262)	1984–86	Manitoba	Briskie et al. 1990
	2.9	(478)	1974–92	Manitoba	Neudorf and Sealy 1994
	9.1	(22)	1945	MI	Walkinshaw 1961
	11.9	(143)	1924–73	MI	Southern and Southern 1980
	13.3	(15)	1896–1958	Quebec	Terril 1961
	60.0	(5)	1991–92	NY	Hahn and Hatfield 1995
Dusky Flycatcher	0.0	(12)	1980–81	CA	Verner and Ritter 1983
	0.0	(3)	1992–95	CO	Ortega and Ortega, unpubl.
	50.0	(2)[4]	1980–81	CA	Verner and Ritter 1983
Gray Flycatcher	25.0	(28)	1970–71	OR	Haislip in FKR
Western Flycatcher	41.7	(12)	<1984	MT	Dolan and Wright 1984
Black Phoebe	0.0	(2)	1982–87	AZ	Brown 1994
Eastern Phoebe	0.0	(97)	1916–78	MI	Southern and Southern 1980
	0.0	(22)	1991–92	NY	Hahn and Hatfield 1995
	6.5	(31)	1899–1968	KS NE MO	Lowther 1977
	9.9	(66)	<1974	KS	Schukman in FKR
	10.3	(68)	1973–74	KS	Hill 1976a
	20.6	(136)	<1975	CT MI	Rothstein 1975b
	24.3	(391)	1962–65	KS	Klaas 1975
	25.9	(108)	1896–1958	Quebec	Terril 1961
	33.3	(15)	1946–49	MI	Berger 1951
	42.1	(19)	1991–92	NY	Hahn and Hatfield 1995
Say's Phoebe	0.0	(4)	1982–87	AZ	Brown 1994
	0.0	(8)	1977–78	CO WY	Hanka 1979
	0.0	(3)	1963–75	LA	Goertz 1977
	2.7	(37)	1973–74	KS	Hill 1976a
	3.3	(30)	<1974	KS	Schukman in FKR
	50.0	(2)	1956	OK	Ely in Wiens 1963
Great Crested Flycatcher	0.0	(8)	1963–75	LA	Goertz 1977
Western Kingbird	0.0	(51)	1973–74	KS	Hill 1976a
Eastern Kingbird	0.0	(16)	1973–74	KS	Hill 1976a
	0.0	(43)	1963–75	LA	Goertz 1977
	0.01	(402)	1974–92	Manitoba	Neudorf and Sealy 1994
	0.7	(143)	1923–73	MI	Southern and Southern 1980
	4.8	(42)	1979	NY	Murphy 1986
	9.5	(220)	1980–83	KS	Murphy 1986
Scissor-tailed Flycatcher	0.0	(7)	1973–74	KS	Hill 1976a
	0.0	(8)	1963–75	LA	Goertz 1977

Host Parasitism	%	(n)	Year	Location	Source
Horned Lark	1.5	(201)	1896–1958	Quebec	Terril 1961
	18.7	(16)	1899–1968	KS NE MO	Lowther 1977
	45.2	(31)	1973–74	KS	Hill 1976a
Purple Martin	0.0	(218)	1963–75	LA	Goertz 1977
Barn Swallow	0.0	(3)	1980–81	CA	Verner and Ritter 1983
	0.0	(284)	1973–74	KS	Hill 1976a
	0.0	(322)	1963–75	LA	Goertz 1977
Blue Jay	0.0	(293)	1963–75	LA	Goertz 1977
	0.0	(9)	1925–68	MI	Southern and Southern 1980
American Crow	0.0	(7)	1963–75	LA	Goertz 1977
Carolina Chickadee	0.6	(181)	1963–75	LA	Goertz 1977
Tufted Titmouse	0.0	(51)	1963–75	LA	Goertz 1977
Bushtit	1.9	(54)	1977	WA	Smith and Atkins 1979
Brown Creeper	0.0	(4)	1980–81	CA	Verner and Ritter 1983
Carolina Wren	0.5	(203)	1963–75	LA	Goertz 1977
	20.0	(5)	1899–1968	KS NE MO	Lowther 1977
	33.3	(3)	1991–92	NY	Hahn and Hatfield 1995
House Wren	0.2	(>900)	<1977	DE OH	Murphy in FK
Golden-crowned Kinglet	6.7	(15)[4]	1980–81	CA	Verner and Ritter 1983
Ruby-crowned Kinglet	8.7	(23)	1896–1958	Quebec	Terril 1961
Blue-gray Gnatcatcher	0.0	(4)	1991–92	NY	Hahn and Hatfield 1995
	6.7	(15)	1963–75	LA	Goertz 1977
	32.1	(28)	1982–87	AZ	Brown 1994
	75.9	(83)	1992–95	NM	Goguen and Mathews 1996
Black-tailed Gnatcatcher	80.0	(5)	1965	AZ	Taylor 1966
	100.0	(3)	1977	NV	Kingery 1977
Eastern Bluebird	0.4	(240)	1963–75	LA	Goertz 1977
	1.9	(54)	1896–1958	Quebec	Terril 1961
	2.6	(268)	1945	IL	Musselman in FKR
	22.2	(27)	<1979	VA	Woodwards in FK
Western Bluebird	0.0	(4)[4]	1980–81	CA	Verner and Ritter 1983
Mountain Bluebird	0.0	(115)	<1977	BC	BCNR in FKR
	0.0	(646)	1971	Manitoba	FKR
Veery	0.0	(8)	1991–93	WI[1]	T. Donovan (pers. comm.)
	13.3	(128)	1896–1958	Quebec	Terril 1961
	16.7	(6)	1952–53	WA	King 1954
	28.2	(39)	1991–92	NY	Hahn and Hatfield 1995
	51.5	(33)	1923–74	MI	Southern and Southern 1980
	63.7	(11)	1944–45	PA	Norris 1947
Swainson's Thrush	0.0	(6)	1914–53	MI	Southern and Southern 1980
Hermit Thrush	0.0	(11)	1991–93	WI[1]	T. Donovan (pers. comm.)
	5.0	(120)	1896–1958	Quebec	Terril 1961
	21.7	(60)	1919–73	MI	Southern and Southern 1980
Wood Thrush	0.0	(20)	1963–75	LA	Goertz 1977
	0.0	(9)	1991–93	WI[1]	T. Donovan (pers. comm.)

Host Parasitism	%	(n)	Year	Location	Source
	1.8	(55)	1991–93	MO[1]	T. Donovan (pers. comm.)
	8.3	(72)	1991–92	NY	Hahn and Hatfield 1995
	9.0	(171)	1990–91	PA	Hoover et al. 1995
	11.1	(18)	1899–1968	KS NE MO	Lowther 1977
	14.3	(7)	1946–49	MI	Berger 1951
	14.3	(7)	1930–67	MI	Southern and Southern 1980
	18.2	(11)	1944–45	PA	Norris 1947
	19.0		1976–77	DE	Roth and Johnson 1993
	35–65		1983–87	DE	Roth and Johnson 1993
	44.4	(18)	1991–93	WI[2]	T. Donovan (pers. comm.)
	47.8	(69)	<1971	IL	Graber et al. in FKR
	78.2	(32)	1991–93	MO[2]	T. Donovan (pers. comm.)
	91.7	(12)	<1975	PA	Harrison in FKR
	100.0	(19)	1980s	IL	Robinson 1992
American Robin	0.0	(17)	1980–81	CA	Verner and Ritter 1983
	0.0	(33)[4]	1980–81	CA	Verner and Ritter 1983
	0.0	(25)	1977–78	CO WY	Hanka 1979
	0.0	(240)	1992–95	CO	Ortega and Ortega, unpubl.
	0.0	(14)	1992–93	IL	Peer and Bolinger 1997
	0.0	(32)	1973–74	KS	Hill 1976a
	0.0	(351)	1916–78	MI	Southern and Southern 1980
	0.5	(216)	1963–75	LA	Goertz 1977
	4.4	(92)	1974–92	Manitoba	Neudorf and Sealy 1994
	40.0	(5)	1974–75	KS	Elliott 1978
Gray Catbird	0.0	(8)	1992–93	IL	Peer and Bolinger 1997
	0.0	(15)	1963–75	LA	Goertz 1977
	0.0	(40)	1916–74	MI	Southern and Southern 1980
	0.3	(≈3,000)	1939–57	MI	Nickell 1958
	0.6	(163)	1896–1958	Quebec	Terril 1961
	1.4	(71)	1946–49	MI	Berger 1951
	2.1	(47)	1944–45	PA	Norris 1947
	5.0	(101)	1974–92	Manitoba	Neudorf and Sealy 1994
	43.7	(16)	1969	Ontario	Scott 1977
Northern Mockingbird	0.0	(12)	1973–74	KS	Hill 1976a
	0.0	(607)	1963–75	LA	Goertz 1977
	0.0	(6)	1958–65	MI	Southern and Southern 1980
	0.0	(48)	<1985	Texas	Mason in FK
Brown Thrasher	0.0	(17)	1960	NE	Hergenrader 1962
	0.6	(525)	1963–75	LA	Goertz 1977
	4.7	(43)	1911–75	MI	Southern and Southern 1980
	5.3	(38)	1899–1968	KS NE MO	Lowther 1977
	5.6	(18)	1896–1958	Quebec	Terril 1961
	6.1	(49)	1973–74	KS	Hill 1976a

Host Parasitism	%	(n)	Year	Location	Source
	7.1	(14)	1944–45	PA	Norris 1947
	37.5	(8)	1974–75	KS	Elliott 1978
Long-billed Thrasher	0.0	(14)	1977–78	TX	Fischer 1980
Curve-billed Thrasher	0.0	(76)	1977–78	TX	Fischer 1980
Cedar Waxwing	0.0	(60)	1939–43	OH	Putnam 1949
	1.2	(329)	1896–1958	Quebec	Terril 1961
	5.6	(467)	1911–78	MI	Southern and Southern 1980
	5.6	(108)	1968–69	MI	Rothstein 1976a
	6.8	(103)	1933–38	MI	Hinds in Rothstein 1976a
	18.2	(11)	1940–41	MI	Lea 1942
Phainopepla	0.0	(3)	1982–87	AZ	Brown 1994
European Starling	0.0	(218)	1963–75	LA	Goertz 1977
White-eyed Vireo	40.0	(15)	1963–75	LA	Goertz 1977
Bell's Vireo	0.7	(939)	1988–95	CA	Griffith and Griffith, in press
	7.0	(57)	1982–87	AZ	Brown 1994
	12.5	(8)	1958–59	IN	Nolan 1960
	13.0	(292)	1983–87	CA	Griffith and Griffith, in press
	27.6	(29)	1899–1968	KS NE MO	Lowther 1977
	29.0	(61)	1960–61	OK	Overmire 1962
	47.2	(108)	1981–82	CA	Griffith and Griffith, in press
	50.0	(14)	1977	CA	Goldwasser et al. 1980
	50.0	(2)	1973–74	KS	Hill 1976a
	53.8	(13)	<1952	IN	Mumford 1952
	68.6	(35)	<1962	KS	Barlow 1962
	70.6	(17)	1960–61	OK	Wiens 1963
	71.4	(17)	1956	OK	Ely in Wiens 1963
	73.9	(23)	1971	TX	Webster in FKR
	80.0	(5)	1939	IL	Pitelka and Koestner 1942
Black-capped Vireo	93.3	(15)	1983–84	TX	Grzybowski et al. 1986
	93.8	(16)	1984–86	OK	Grzybowski et al. 1986
Solitary Vireo	0.0	(2)	1991–92	NY	Hahn and Hatfield 1995
	25.0	(4)	1941–74	MI	Southern and Southern 1980
	48.7	(78)	1984–86	CO	Marvil and Cruz 1989
	80.0	(5)	1992–95	CO	Ortega and Ortega, unpubl.
Yellow-throated Vireo	50.0	(2)	1963–75	LA	Goertz 1977
Warbling Vireo	0.0	(6)	1943–69	MI	Southern and Southern 1980
	16.7	(6)[4]	1980–81	CA	Verner and Ritter 1983
	70.0	(20)	1992–95	CO	Ortega and Ortega, unpubl.
Red-eyed Vireo	0.0	(44)	1941–50	Ontario	Lawrence 1953
	0.0	(11)	1991–93	WI[1]	T. Donovan (pers. comm.)
	10.7	(28)	1991–93	MO[1]	T. Donovan (pers. comm.)
	36.4	(231)	1918–34	OH	Hicks 1934

Host Parasitism	%	(n)	Year	Location	Source
	42.2	(64)	1896–1958	Quebec	Terril 1961
	50.0	(2)	1991–93	MO[2]	T. Donovan (pers. comm.)
	55.0	(20)	1991–92	NY	Hahn and Hatfield 1995
	66.7	(12)	1991–93	WI[2]	T. Donovan (pers. comm.)
	69.3	(257)	1922–78	MI	Southern and Southern 1980
	75.0	(4)	1899–1968	KS NE MI	Lowther 1977
	85.7	(14)	1944–45	PA	Norris 1947
	100.0	(1)	1946–49	MI	Berger 1951
	100.0	(3)	1969	Ontario	Scott 1977
Blue-winged Warbler	22.2	(9)	1991–92	NY	Hahn and Hatfield 1995
	50.0	(2)	1980	IL	Graber et al. 1983
Nashville Warbler	0.0	(3)[4]	1980–81	CA	Verner and Ritter 1983
	7.2	(83)	1896–1958	Quebec	Terril 1961
	22.2	(9)	1922–74	MI	Southern and Southern 1980
Lucy's Warbler	23.1	(13)	1982–87	AZ	Brown 1994
Northern Parula	0.0	(3)	1963–75	LA	Goertz 1977
	0.0	(5)	1949–67	MI	Southern and Southern 1980
Yellow Warbler	0.0	(3)	1980–81	CA	Verner and Ritter 1983
	0.0	(7)	1977–78	CO, WY	Hanka 1979
	0.0	(20)	1990–93	FL	Prather and Cruz 1995
	6.3	(96)	1991–92	NY	Hahn and Hatfield 1995
	7.4	(27)	1899–1968	KS NE MO	Lowther 1977
	13.7	(307)	1896–1958	Quebec	Terril 1961
	14.3	(7)	1952–53	WA	King 1954
	17.8	(682)	1984–86	Manitoba	Briskie et al. 1990
	19.1	(2,163)	1974–92	Manitoba	Neudorf and Sealy 1994
	20.0	(5)[4]	1980–81	CA	Verner and Ritter 1983
	21.0	(1,885)	1974–87	Manitoba	Sealy 1995
	22.7	(22)	1982–87	AZ	Brown 1994
	≈25.0	(315)	1974–76	Manitoba	Goossen and Sealy 1982
	28.6	(21)	1922–69	MI	Southern and Southern 1980
	29.3	(41)	1938–39	IA	Schrantz 1943
	30.3	(310)	1983–84	Manitoba	Weatherhead 1989
	32.7	(257)	1948–54	MI	McGeen 1972
	33.9	(56)	1992–95	CO	Ortega and Ortega, unpubl.
	35.0	(20)	1982–83	IA	Lowther 1983
	40.9	(44)	1946–49	MI	Berger 1951
	41.3	(109)	<1981	Ontario	Clark and Robertson 1981
	42.5	(146)	1918–34	OH	Hicks 1934
	63.3	(49)	<1989	Ontario	Burgham and Picman 1989
	70.0	(10)	1969	Ontario	Scott 1977
Chestnut-sided Warbler	29.1	(55)	1896–1958	Quebec	Terril 1961
	33.3	(12)	1918–34	OH	Hicks 1934
	50.0	(2)	1991–92	NY	Hahn and Hatfield 1995
	50.0	(2)	1944–45	PA	Norris 1947

Host Parasitism	%	(n)	Year	Location	Source
	52.9	(17)	1915–66	MI	Southern and Southern 1980
	71.4	(7)	1958	IL	Johnson in FK
Magnolia Warbler	0.0	(3)	1947–71	MI	Southern and Southern 1980
	4.1	(147)	1896–1958	Quebec	Terril 1961
	12.5	(16)	<1975	Maine	Harrison in FKR
	100.0	(2)	<1975	PA	Harrison in FKR
Black-throated Blue Warbler	0.0	(125)	1986–89	NH	Holmes et al. 1992
	9.5	(21)	<1975	Ontario	Peck in FKR
	25.0	(4)	1924–49	MI	Southern and Southern 1980
	42.9	(7)	1896–1958	Quebec	Terril 1961
Yellow-rumped Warbler	6.1	(33)	1896–1958	Quebec	Terril 1961
	22.6	(31)[4]	1980–81	CA	Verner and Ritter 1983
	27.3	(77)	<1975	Ontario	Peck in FKR
	36.8	(19)	1924–68	MI	Southern and Southern 1980
Hermit Warbler	0.0	(1)	1980–81	CA	Verner and Ritter 1983
	12.5	(8)[4]	1980–81	CA	Verner and Ritter 1983
	46.2	(13)	1991–92	NY	Hahn and Hatfield 1995
Black-throated Green Warbler	12.5	(16)	1896–1958	Quebec	Terril 1961
	50.0	(6)	1934–49	MI	Southern and Southern 1980
Pine Warbler	25.0	(4)	1963–75	LA	Goertz 1977
	50.0	(2)	1911–19	MI	Southern and Southern 1980
Kirtland's Warbler	0.0	(33)	1973	MI	Shake and Mattsson 1975
	6.5	(31)	1972	MI	Shake and Mattsson 1975
	9.5	(63)	1974	MI	Shake and Mattsson 1975
	52.8	(142)		MI	Mayfield 1961a
	86.0	(?)	<1972	MI	Shake and Mattsson 1975
Prairie Warbler	0.0	(42)	1990–93	FL	Prather and Cruz 1995
	0.0	(4)	1991–92	NY	Hahn and Hatfield 1995
	20.0	(10)	1967–73	IL	Graber et al. 1983
	23.5	(17)		MI	Walkinshaw in Young 1963
	30.0	(10)	1963–75	LA	Goertz 1977
Bay-breasted Warbler	13.3	(15)	1896–1958	Quebec	Terril 1961
Black-and-white Warbler	16.7	(6)	1928–74	MI	Southern and Southern 1980
	25.0	(4)	1991–93	MO[1]	T. Donovan (pers. comm.)
	25.0	(4)	1991–92	NY	Hahn and Hatfield 1995
	28.6	(14)	1896–1958	Quebec	Terril 1961
American Redstart	5.6	(18)	1940–42	OH	Sturm 1945
	7.4	(27)	<1900	IL	Graber et al. 1983
	15.9	(145)	1896–1958	Quebec	Terril 1961
	30.6	(49)	1991–92	NY	Hahn and Hatfield 1995
	31.8	(22)	1918–34	OH	Hicks 1934
	36.6	(41)	>1900	IL	Graber et al. 1983

Host Parasitism	%	(n)	Year	Location	Source
	35.7	(42)	1916–74	MI	Southern and Southern 1980
Prothonotary Warbler	6.7	(15)	1899–1968	KS NE MO	Lowther 1977
	12.3	(57)	1963–75	LA	Goertz 1977
	15.6	(154)	< 1900	IL	Graber et al. 1983
	20.9	(172)	1985	TN	Petit 1991
Worm-eating Warbler	0.0	(27)	1991–93	MO[1]	T. Donovan (pers. comm.)
	20.0	(10)	1991–92	NY	Hahn and Hatfield 1995
	75.0	(12)	1991–93	MO[2]	T. Donovan (pers. comm.)
Swainson's Warbler	100.0	(1)	1963–75	LA	Goertz 1977
Ovenbird	2.5	(40)	1991–93	MO[1]	T. Donovan (pers. comm.)
	4.0	(50)	1991–93	WI[1]	T. Donovan (pers. comm.)
	9.8	(61)	1896–1958	Quebec	Terril 1961
	19.4	(31)	1991–93	WI[2]	T. Donovan (pers. comm.)
	29.5	(44)	1916–68	MI	Southern and Southern 1980
	36.6	(112)	1918–34	OH	Hicks 1934
	40.0	(5)	1980s	IL	Robinson 1992
	52.4	(42)	1932–35	MI	Hann 1937
	55.2	(29)	1991–92	NY	Hahn and Hatfield 1995
	60.0	(5)	1991–93	MO[2]	T. Donovan (pers. comm.)
	85.7	(7)	1944–45	PA	Norris 1947
	100.0	(2)	1946–49	MI	Berger 1951
Northern Waterthrush	16.7	(18)	1896–1958	Quebec	Terril 1961
Louisiana Waterthrush	0.0	(5)	1991–93	MO[2]	T. Donovan (pers. comm.)
	22.2	(9)	1991–92	NY	Hahn and Hatfield 1995
	50.0	(2)	1980s	IL	Robinson 1992
	56.3	(16)	1947–49	NY	Eaton 1958
Kentucky Warbler	0.0	(2)	1991–93	MO[1]	T. Donovan (pers. comm.)
	33.3	(6)	1980s	IL	Robinson 1992
	41.7	(12)	1991–93	MO[2]	T. Donovan (pers. comm.)
	45.1	(133)	1887–1937	PA	Jacobs 1938
	50.0	(2)	1963–75	LA	Goertz 1977
Mourning Warbler	0.0	(1)	1949	MI	Southern and Southern 1980
	32.0	(25)	1896–1958	Quebec	Terril 1961
MacGillivray's Warbler	20.0	(5)	1952–53	WA	King 1954
	25.0	(4)[4]	1980–81	CA	Verner and Ritter 1983
Common Yellowthroat	0.0	(6)	1963–75	LA	Goertz 1977
	0.0	(8)	1924–64	MI	Southern and Southern 1980
	7.1	(113)	1896–1958	Quebec	Terril 1961
	14.3	(14)	1950–51	MN	Hofslund 1957
	16.7	(12)	1899–1968	KS NE MO	Lowther 1977
	22.4	(67)	<1983	IL	Graber et al. 1983
	25.0	(4)	1991–92	NY	Hahn and Hatfield 1995
	45.5	(22)	1938	MI	Stewart 1953
	46.3	(41)	1918–34	OH	Hicks 1934
	47.4	(38)	1948–49	MI	Hofslund 1957

Host Parasitism	%	(n)	Year	Location	Source
	55.6	(9)	1982–87	AZ	Brown 1994
	66.7	(3)	<1977	CA	Steele in FKR
	100.0	(1)	1946–49	MI	Berger 1951
	100.0	(2)	1944–45	PA	Norris 1947
Wilson's Warbler	0.0	(4)	1980–81	CA	Verner and Ritter 1983
	0.0	(5)[4]	1980–81	CA	Verner and Ritter 1983
	54.5	(11)	<1977	CA	FK
Canada Warbler	0.0	(4)	1935–74	MI	Southern and Southern 1980
Yellow-breasted Chat	6.7	(15)	1963–75	LA	Goertz 1977
	10.8	(37)	1982–87	AZ	Brown 1994
	21.4	(14)	1899–1968	KS NE MO	Lowther 1977
	33.3	(3)	1946–49	MI	Berger 1951
	90.9	(11)	1937–53	MI	Nickell 1955
Summer Tanager	41.2	(17)	<1965	KY	Mengel in FK
	100.0	(6)	1980s	IL	Robinson 1992
	100.0	(2)	1899–1968	KS NE MO	Lowther 1977
Scarlet Tanager	63.6	(11)	1920–74	MI	Southern and Southern 1980
	66.7	(6)	1991–92	NY	Hahn and Hatfield 1995
	75.0	(4)	1980s	IL	Robinson 1992
	78.6	(14)	<1965	MI	Prescott in FKR
	100.0	(1)	1899–1968	KS NE MO	Lowther 1977
	100.0	(4)	1944–45	PA	Norris 1947
Western Tanager	0.0	(8)	1980–81	CA	Verner and Ritter 1983
	0.0	(5)[4]	1980–81	CA	Verner and Ritter 1983
Northern Cardinal	2.7	(259)	1963–75	LA	Goertz 1977
	8.3	(12)	1956	OK	Ely in Wiens 1963
	14.3	(7)	1991–92	NY	Hahn and Hatfield 1995
	25.6	(39)	1899–1968	KS NE MO	Lowther 1977
	33.3	(15)	1991–93	MO[2]	T. Donovan (pers. comm.)
	37.5	(8)	1944–45	PA	Norris 1947
	45.5	(22)	1946–49	MI	Berger 1951
	48.4	(64)	1993	OH	Eckerle and Breitwisch 1997
	55.0	(20)	1980s	IL	Robinson 1992
	59.9	(187)	1955–61	Ontario	Scott 1963
	80.8	(52)	<1983	TX	Mason in FK
	85.7	(14)	1992–93	IL	Peer and Bolinger 1997
	100.0	(3)	1973–74	KS	Hill 1976a
	100.0	(4)	1960–61	OK	Wiens 1963
Rose-breasted Grosbeak	0.0	(2)	1980s	IL	Robinson 1992
	0.0	(16)	1941–74	MI	Southern and Southern 1980
	6.7	(15)	1991–92	NY	Hahn and Hatfield 1995
	7.1	(42)	1896–1958	Quebec	Terril 1961
	50.0	(2)	1899–1968	KS NE MO	Lowther 1977
	50.0	(2)	1946–49	MI	Berger 1951
Black-headed Grosbeak	0.0	(1)	1982–87	AZ	Brown 1994

Host Parasitism	%	(n)	Year	Location	Source
	0.0	(4)	1980–81	CA	Verner and Ritter 1983
	0.0	(3)	1980–81	CA	Verner and Ritter 1983
	0.0	(14)	1992–95	CO	Ortega and Ortega, unpubl.
Blue Grosbeak	0.0	(6)	1963–75	LA	Goertz 1977
	40.0	(5)	1960–61	OK	Wiens 1963
	60.0	(5)	1982–87	AZ	Brown 1994
	75.0	(8)	1956	OK	Ely in Wiens 1963
	100.0	(7)	1949	CA	Bleitz 1956
Lazuli Bunting	0.0	(5)[4]	1980–81	CA	Verner and Ritter 1983
	66.7	(3)	1952–53	WA	King 1954
Indigo Bunting	0.0	(2)	1982–87	AZ	Brown 1994
	16.7	(6)	1963–75	LA	Goertz 1977
	20.0	(30)	1896–1958	Quebec	Terril 1961
	21.6	(88)	<1975	Ontario	Peck in FKR
	39.5	(43)	1918–34	OH	Hicks 1934
	42.9	(14)	1948–50	OH	Phillips 1951
	43.2	(37)	1911–73	MI	Southern and Southern 1980
	50.0	(4)	1991–93	MO[2]	T. Donovan (pers. comm.)
	55.6	(9)	1946–49	MI	Berger 1951
	71.4	(14)	1899–1968	KS NE MO	Lowther 1977
Painted Bunting	28.9	(45)	<1959	OK	Parmelee in Wiens 1963
	50.0	(2)	1960–61	OK	Wiens 1963
	80.0	(5)	1956	OK	Ely in Wiens 1963
Dickcissel	7.1	(14)	1956	OK	Ely in Wiens 1963
	8.3	(24)	1899–1968	KS NE MO	Lowther 1977
	10.0	(10)	1963–75	LA	Goertz 1977
	31.1	(61)	<1962	OK	Overmire in Wiens 1963
	33.3	(15)	1960–61	OK	Wiens 1963
	50.0	(28)	1973–74	KS	Hill 1976a
	52.9	(17)	1960	NE	Hergenrader 1962
	65.2	(23)	1981–82	KS	Fleischer 1986
	69.4	(620)	1965–79	KS	Zimmerman 1983
	90.8	(65)	1974	KS	Hatch 1983
	94.7	(19)	1974–75	KS	Elliott 1978
	100.0	(2)	1982–83	IA	Lowther 1983
Green-tailed Towhee	0.0	(4)	1980–81	CA	Verner and Ritter 1983
	0.0	(12)[4]	1980–81	CA	Verner and Ritter 1983
Eastern Towhee	0.0	(3)	1991–93	MO[2]	T. Donovan (pers. comm.)
	18.2	(11)	1899–1968	KS NE MO	Lowther 1977
	21.4	(103)	1918–34	OH	Hicks 1934
	22.2	(9)	1991–92	NY	Hahn and Hatfield 1995
	24.1	(29)	1911–68	MI	Southern and Southern 1980
	33.3	(6)	1946–49	MI	Berger 1951
	50.0	(4)	1980s	IL	Robinson 1992
	54.2	(24)	1944–45	PA	Norris 1947

Host Parasitism	%	(n)	Year	Location	Source
Abert's Towhee	31.9	(69)	1979–80	AZ	Finch 1983
Chipping Sparrow	0.0	(4)	1980–81	CA	Verner and Ritter 1983
	0.0	(40)[4]	1980–81	CA	Verner and Ritter 1983
	0.0	(26)	1963–75	LA	Goertz 1977
	6.3	(32)	1991–92	NY	Hahn and Hatfield 1995
	6.9	(29)	<1975	PA	Harrison in FKR
	11.1	(9)	1944–45	PA	Norris 1947
	11.6	(138)	1896–1958	Quebec	Terril 1961
	14.3	(7)	1899–1968	KS NE MO	Lowther 1977
	14.3	(14)	1952–53	WA	King 1954
	24.5	(216)	1920–76	MI	Southern and Southern 1980
	25.0	(12)	1976	MN	Buech 1982
	36.1	(83)	1968–69	MI	Rothstein 1976a
	40.0	(5)	1992–95	CO	Ortega and Ortega, unpubl.
	52.2	(115)	1918–34	OH	Hicks 1934
	62.5	(8)	1946–49	MI	Berger 1951
Clay-colored Sparrow	10.4	(135)	1983–84	MN	Johnson and Temple 1990
	10.8	(204)	1987–91	Manitoba	Hill and Sealy 1994
	32.5	(40)	1976	MN	Buech 1982
	39.4	(33)	<1977	ND	Stewart in FKR
	88.9	(9)	1959	Saskatche-wan	Fox 1961
	89.4	(94)	<1979	Manitoba	Knapton in FK
Brewer's Sparrow	12.5	(16)	1976	ID	Rich and Rothstein 1985
Field Sparrow	0.0	(15)	1991–92	NY	Hahn and Hatfield 1995
	10.9	(147)	1971–72	IL	Best 1978
	15.8	(57)	1944–45	PA	Norris 1947
	18.2	(33)	1946–49	MI	Berger 1951
	30.0	(10)	1976	MN	Buech 1982
	32.1	(159)	1918–34	OH	Hicks 1934
	35.7	(14)	1899–1968	KS NE MO	Lowther 1977
	50.0	(2)	1947–67	MI	Southern and Southern 1980
	75.0	(4)	1956	OK	Ely in Wiens 1963
	100.0	(1)	1960–61	OK	Wiens 1963
Vesper Sparrow	4.1	(74)	1896–1958	Quebec	Terril 1961
	8.0	(112)	1918–34	OH	Hicks 1934
	10.1	(89)	1916–72	MI	Southern and Southern 1980
	33.3	(6)	1946–49	MI	Berger 1951
Lark Sparrow	0.0	(10)	1963–75	LA	Goertz 1977
	5.9	(17)	1956	OK	Ely in Wiens 1963
	20.0	(5)	1899–1968	KS NE MO	Lowther 1977
	45.5	(33)	1968	OK	Newman 1970
	75.0	(4)	1960–61	OK	Wiens 1963
	81.8	(11)	1973–74	KS	Hill 1976a
Sage Sparrow	50.0	(4)	1976	ID	Rich and Rothstein 1985

Host Parasitism	%	(n)	Year	Location	Source
Five-striped Sparrow	22.8	(22)	<1975	AZ	Mills in FK
Lark Bunting	15.5	(142)	1973–74	KS	Hill 1976a
Savannah Sparrow	0.0	(26)	1924–72	MI	Southern and Southern 1980
	1.3	(398)	<1978	New Brunswick	Dixon in FK
	1.9	(54)	<1974	MI	Potter in FKR
	3.6	(140)	1896–1958	Quebec	Terril 1961
	37.0	(46)	1983–84	MN	Johnson and Temple 1990
Grasshopper Sparrow	0.0	(1)	1951	MI	Southern and Southern 1980
	6.5	(46)	1983–84	MN	Johnson and Temple 1990
	11.8	(51)	<1977	Ontario	Peck in FKR
	22.2	(18)	1973–74	KS	Hill 1976a
	50.0	(18)	1974–75	KS	Elliott 1978
Fox Sparrow	0.0	(2)	1980–81	CA	Verner and Ritter 1983
	0.0	(19)[4]	1980–81	CA	Verner and Ritter 1983
Song Sparrow	0.0	(2)	1980–81	CA	Verner and Ritter 1983
	0.0	(11)[4]	1980–81	CA	Verner and Ritter 1983
	4.0	(50)	1991–92	NY	Hahn and Hatfield 1995
	6.6	(61)	1911–75	MI	Southern and Southern 1980
	12.8	(486)	1896–1958	Quebec	Terril 1961
	25.0	(8)	1960–65	CA	Payne 1973b
	33.3	(10)	1952–53	WA	King 1954
	33.9	(398)	1918–34	OH	Hicks 1934
	35.7	(277)	1975–79	Vancouver	Smith 1981
	40.7	(27)	1944–45	PA	Norris 1947
	62.7	(59)	1946–49	MI	Berger 1951
	73.3	(30)	1948–54	MI	McGeen 1972
	90.9	(11)	1969	Ontario	Scott 1977
Lincoln's Sparrow	0.0	(6)	1980–81	CA	Verner and Ritter 1983
	0.0	(4)[4]	1980–81	CA	Verner and Ritter 1983
Swamp Sparrow	0.0	(4)	1925–65	MI	Southern and Southern 1980
	10.6	(322)	1896–1958	Quebec	Terril 1961
	80.0	(5)	1946–49	MI	Berger 1951
White-throated Sparrow	0.0	(11)	1934–75	MI	Southern and Southern 1980
	3.9	(507)	1896–1958	Quebec	Terril 1961
White-crowned Sparrow	0.0	(42)	1934–38	CA	Trail and Baptista 1993
	2.2	(648)	1975–85	CA	Trail and Baptista 1993
	32.6	(242)	1990–91	CA	Trail and Baptista 1993
Dark-eyed Junco	0.0	(78)[4]	1980–81	CA	Verner and Ritter 1983
	1.3	(75)	1896–1958	Quebec	Terril 1961
	2.1	(48)	1980–81	CA	Verner and Ritter 1983
	33.3	(9)	1915–40	MI	Southern and Southern 1980
	38.9	(175)	1983–84	VA	Wolf 1987

Host Parasitism	%	(n)	Year	Location	Source
Chestnut-collared Longspur	22.6	(62)	<1975	ND	Stewart in FK
Bobolink	0.0	(19)	1930–69	MI	Southern and Southern 1980
	5.4	(184)	1918–34	OH	Hicks 1934
	10.3	(58)	<1975	Ontario	Peck in FKR
	33.3	(3)	1982–83	IA	Lowther 1983
	34.0	(47)	1983–84	MN	Johnson and Temple 1990
Red-winged Blackbird	0.0	(20)	<1977	CA	Steele in FKR
	0.0	(4)	1980–81	CA	Verner and Ritter 1983
	0.0	(6)	1980–81	CA	Verner and Ritter 1983
	0.0	(44)	1922–68	MI	Southern and Southern 1980
	0.0	(35)	1991–92	NY	Hahn and Hatfield 1995
	0.0	(33)	1960–61	OK	Wiens 1963
	0.2	(≈1,400)	1976–82	BC	Picman 1986
	1.5	(344)	1992–93	IL	Peer and Bolinger 1997
	1.6	(754)	1963–75	LA	Goertz 1977
	2.1	(653)	1941–48	IL, AK	Smith 1949
	2.7	(113)	< 1995	BC	Ward et al. 1996
	2.7	(37)	1992–95	CO	Ortega and Ortega, unpubl.
	2.7	(73)	1956	OK	Ely in Wiens 1963
	5.1	(99)	1946–49	MI	Berger 1951
	5.9	(51)	1899–1968	KS, NE, MO	Lowther 1977
	7.1	(85)	1960–65	CA	Payne 1973b
	7.7	(1,325)	1978–88	WA	Freeman et al. 1990
	9.7	(802)	1984–86	CO	Ortega 1991
	9.7	(2,039)	1977–83	WA	Orians et al. 1989
	12.5	(24)	1952–53	WA	King 1954
	17.5	(154)	1977–78	CO WY	Hanka 1979
	19.7	(213)	1974–92	Manitoba	Neudorf and Sealy 1994
	21.9	(228)	1973–74	KS	Hill 1976a
	30.1	(73)	1981–82	KS	Fleischer 1986
	31.0	(29)	1978	KS	Facemire 1980
	32.7	(98)	1982–83	IA	Lowther 1983
	35.0	(382)	1983–84	Manitoba	Weatherhead 1989
	42.2	(258)	1962–65	ND	Linz and Bolin 1982
	52.8	(53)	1975–76	SD	Blankespoor et al. 1982
	54.2	(59)	1960	NE	Hergenrader 1962
	76.5	(17)	1973	ND	Houston in FKR
Eastern Meadowlark	0.0	(5)	1931–69	MI	Southern and Southern 1980
	1.1	(91)	1963–75	LA	Goertz 1977
	1.9	(52)	1896–1958	Quebec	Terril 1961
	10.1	(69)	1899–1968	KS NE MO	Lowther 1977
	50.0	(10)	1981–82	KS	Fleischer 1986
	70.0	(40)	1974–75	KS	Elliott 1978
Western Meadowlark	6.9	(29)	1973–74	KS	Hill 1976a
	18.4	(76)	1983–84	MN	Johnson and Temple 1990

Host Parasitism	%	(n)	Year	Location	Source
Yellow-headed Blackbird	0.0	(356)	1985–86	CO	Ortega 1991
	0.5	(381)	1977–78	CO, WY	Hanka 1979
	0.8	(120)	1959–60	WI	Young 1963
	1.5	(67)	1974–92	Manitoba	Neudorf and Sealy 1994
Brewer's Blackbird	0.0	(21)	1980–81	CA	Verner and Ritter 1983
	0.0	(7)[+]	1980–81	CA	Verner and Ritter 1983
	0.0	(1)	1966	MI	Southern and Southern 1980
	3.6	(83)	1974–75	CA	FKR
	6.7	(15)	1952–53	WA	King 1954
	19.4	(217)	1977–78	CO WY	Hanka 1979
	30.8	(13)	1992–95	CO	Ortega and Ortega, unpubl.
	32.0	(837)	1969–70	WA	Furrer in FKR
	37.0	(27)	1974–92	Manitoba	Neudorf and Sealy 1994
Common Grackle	0.0	(18)	1977–78	CO WY	Hanka 1979
	0.0	(401)	1992–93	IL	Peer and Bollinger 1997
	0.0	(371)	1963–75	LA	Goertz 1977
	0.0	(79)	1973–74	KS	Hill 1976a
	0.0	(5)	1927–32	MI	Southern and Southern 1980
	0.0	(62)	1947–49	WI	Petersen and Young 1950
	2.3	(44)	1974–92	Manitoba	Neudorf and Sealy 1994
	6.2	(16)	1981–82	KS	Fleischer 1986
Orchard Oriole	6.7	(15)	1899–1968	KS NE MO	Lowther 1977
	28.2	(71)	1963–75	LA	Goertz 1977
	46.2	(13)	1956	OK	Ely in Wiens 1963
	53.3	(15)	1973–74	KS	Hill 1976a
	100.0	(3)	1960–61	OK	Wiens 1963
Hooded Oriole	71.4	(21)	1975	CA	FKR
Baltimore Oriole	2.2	(136)	1975–85	Manitoba	Hobson and Sealy 1987
	2.5	(318)	1956–1974	Ontario	Peck 1974
	3.3	(153)	1974–92	Manitoba	Neudorf and Sealy 1994
	12.5	(8)	1934–78	MI	Southern and Southern 1980
Bullock's Oriole	0.0	(1)	1982–87	AZ	Brown 1994
	8.8	(34)	1974	CA	FKR
Purple Finch	0.0	(6)[+]	1980–81	CA	Verner and Ritter 1983
	6.2	(16)	1896–1958	Quebec	Terril 1961
	42.1	(19)	1926–74	MI	Southern and Southern 1980
	57.1	(14)	<1977	MI	UMBS in FKR
House Finch	0.0	(7)	1982–87	AZ	Brown 1994
	0.0	(11)	1991–92	NY	Hahn and Hatfield 1995
	58.3	(12)	1992–93	IL	Peer and Bollinger 1997
Pine Siskin	54.9	(51)	1973–74	KS	Hill 1976a
Lesser Goldfinch	0.0	(8)	1982–87	AZ	Brown 1994
	0.0	(3)[+]	1980–81	CA	Verner and Ritter 1983
American Goldfinch	0.0	(25)	1991–92	NY	Hahn and Hatfield 1995
	2.2	(318)	1896–1958	Quebec	Terril 1961

Host Parasitism	%	(n)	Year	Location	Source
	2.9	(70)	1946–49	MI	Berger 1951
	5.4	(74)	1916–69	MI	Southern and Southern 1980
	5.9	(802)	1968–89	Ontario	Middleton 1991
	8.3	(12)	1968–69	MI	FKR
	9.4	(234)	1968–75	Ontario	Middleton 1977
House Sparrow	0.0	(26)	1973–74	KS	Hill 1976a
	0.0	(229)	1963–75	LA	Goertz 1977

Abbreviations: BCNR: British Columbia Nest Records, University of British Columbia, Vancouver, British Columbia; FKR: Friedmann et al. 1977; FK: Friedmann and Kiff 1985; UMBS: University of Michigan Biological Station, Pellston, Michigan.

[1] Unfragmented

[2] Fragmented

[3] Alder Flycatchers and Trail's Flycatchers were considered one species prior to 1973; in earlier studies, where the two species overlap (particularly in Michigan), species identification is unclear.

[4] From fledged family group data.

Brown-headed Cowbird Egg Distribution in Nests of Some Hosts

Host Location	Number of Brown-headed Cowbird Eggs							Source
	0	1	2	3	4	5	≥6	
Alder Flycatcher								
Michigan	38	6	0	0	0	0	0	Berger and Parmelee 1952
Willow Flycatcher								
California	6	11	2	0	0	0	0	Harris 1991
Washington	42	2	0	0	0	0	0	King 1954
Least Flycatcher								
Manitoba	255	5	2	0	0	0	0	Briskie et al. 1990
Eastern Phoebe								
Kansas	296	55	19	6	1	3	0	Klaas 1975
Kansas	61	5	1	1	0	0	0	Hill 1976a
Say's Phoebe								
Michigan	10	3	1	1	0	0	0	Berger 1951
Eastern Kingbird								
New York and Kansas	262	15	3	1	0	0	0	Murphy 1986
Horned Lark								
Kansas	16	6	6	2	0	0	0	Hill 1976a
Blue-gray Gnatcatcher								
New Mexico	20	48	11	3	1	0	0	Goguen and Mathews 1996
Veery								
Pennsylvania	4	6	1	0	0	0	0	Norris 1947
Wood Thrush								
Iowa	?	2	1	2	3	3	0	Stoner 1919
Michigan	6	0		1	0	0	0	Berger 1951
Pennsylvania	9	2	0	0	0	0	0	Norris 1947
Catbird								
Michigan	70	0	1	0	0	0	0	Berger 1951
Michigan	≈3,000	6	2	0	0	0	0	Nickell 1958
Ontario	9	6	0	1	0	0	0	Scott 1977
Pennsylvania	46	1	0	0	0	0	0	Norris 1947

Host	Number of Brown-headed Cowbird Eggs							Source
Location	0	1	2	3	4	5	≥6	
Brown Thrasher								
Kansas	19	0	0	1	0	0	0	Hill 1976a
Pennsylvania	13	0	1	0	0	0	0	Norris 1947
Bell's Vireo								
California		3	0	0	0	0	0	Hanna 1918
Red-eyed Vireo								
Michigan	0	0	0	0	1	0	0	Berger 1951
Ontario	0	0	0	1	1	1	0	Scott 1977
Pennsylvania	2	7	4	1	0	0	0	Norris 1947
Yellow Warbler								
Manitoba	83	9	2	0	0	0	0	Weatherhead 1989
Manitoba	563	108	11	0	0	0	0	Briskie et al. 1990
Manitoba	1,489	354	38	4	0	0	0	Sealy 1992
Michigan	26	9	3	6	0	0	0	Berger 1951
Ontario	3	0	1	2	3	0	1	Scott 1977
Prothonotary Warbler								
Tennessee	30	5	1	0	0	0	0	Petit 1991
Ovenbird								
Michigan	20	10	8	2	2	0	0	Hann 1937
Michigan	0	0	1	1	0	0	0	Berger 1951
Pennsylvania	1	2	2	2	0	0	0	Norris 1947
Louisiana Waterthrush								
New York	7	4	2	2	1	0	0	Eaton 1958
Kentucky Warbler								
Pennsylvania	73	43	14	3	0	0	0	Jacobs 1938
Common Yellowthroat								
Michigan	0	1	0	0	0	0	0	Berger 1951
Yellow-breasted Chat								
Michigan	2	1	0	0	0	0	0	Berger 1951
Michigan	1	5	2	1	1	1	0	Nickell 1955
Scarlet Tanager								
Pennsylvania	0	3	1	0	0	0	0	Norris 1947
Northern Cardinal								
Illinois	2	6	4	2	0	0	0	Peer and Bollinger 1997
Ohio	60	41	11	3	0	0	0	Eckerle and Breitwisch 1997
Michigan	12	3	5	2	0	0	0	Berger 1951
Pennsylvania	5	2	1	0	0	0	0	Norris 1947
Rose-breasted Grosbeak								
Michigan	1	1	0	0	0	0	0	Berger 1951
Indigo Bunting								
Michigan	4	3	1	0	1	0	0	Berger 1951
Lazuli Bunting								
Washington	1	1	1	0	0	0	0	King 1954
Dickcissel								
Kansas	13	12	1	0	0	0	0	Hill 1976a

| Host | Number of Brown-headed Cowbird Eggs | | | | | | | |
Location	0	1	2	3	4	5	≥6	Source
Kansas								
Prairie	240	17	29	19	17	7	6	Zimmerman 1983
Old field	184	72	56	38	20	8	7	Zimmerman 1983
Eastern Towhee								
Michigan	4	1	1	0	0	0	0	Berger 1951
Pennsylvania	11	6	5	2	0	0	0	Norris 1947
Abert's Towhee								
Arizona	47	14	5	3	0	0	0	Finch 1983
Chipping Sparrow								
Michigan	3	3	2	0	0	0	0	Berger 1951
Pennsylvania	8	1	0	0	0	0	0	Norris 1947
Washington	12	2	0	0	0	0	0	King 1954
Clay-colored Sparrow								
Manitoba	182	18	3	1	0	0	0	Hill and Sealy 1994
Saskatchewan	1	1	4	1	2	0	0	Fox 1961
Field Sparrow								
Michigan	27	5	1	0	0	0	0	Berger 1951
Pennsylvania	48	5	4	0	0	0	0	Norris 1947
Vesper Sparrow								
Michigan	4	2	0	0	0	0	0	Berger 1951
Lark Sparrow								
Kansas	0	4	3	2	0	0	0	Hill 1976a
Oklahoma	18	11	4	0	0	0	0	Newman 1970
Lark Bunting								
Kansas	113	19	3	0	0	0	0	Hill 1976a
Grasshopper Sparrow								
Kansas	11	3	1	0	0	0	0	Hill 1976a
Song Sparrow								
Illinois	2	0	0	1	2(≥4)			Peer and Bollinger 1997
Michigan	22	10	17	4	4	2	0	Berger 1951
Ontario	1	2	4	0	3	1	0	Scott 1977
Pennsylvania	16	9	2	0	0	0	0	Norris 1947
Vancouver	1,568	327	15	0	0	0	0	Smith and Arcese 1994
Washington	7	2	1	0	0	0	0	King 1954
Swamp Sparrow								
Michigan	1	3	1	0	0	0	0	Berger 1951
Dark-eyed Junco								
Virginia	90	37	20	5	1	2	1	Wolf 1987
Red-winged Blackbird								
Colorado	557	62	14	1	0	0	0	Ortega 1991
Illinois and Arkansas	639	13	1	0	0	0	0	Smith 1949
Illinois	339	5	0	0	0	0	0	Peer and Bollinger 1997
Kansas	72	17	1	0	0	0	0	Hill 1976a
Manitoba	93	28	7	1	1	1	1	Weatherhead 1989
Michigan	94	5	0	0	0	0	0	Berger 1951
North Dakota	149	66	28	10	2	0	3	Linz and Bolin 1982

| Host | Number of Brown-headed Cowbird Eggs | | | | | | | Source |
Location	0	1	2	3	4	5	≥6	Source
Washington	21	3	0	0	0	0	0	King 1954
Washington	2,039	156	27	10	4	5	0	Orians et al. 1989
Orchard Oriole								
Kansas	3	1	1	0	0	0	0	Hill 1976a
House Finch								
Illinois	5	5	2	0	0	0	0	Peer and Bollinger 1997
Pine Siskin								
Kansas	7	12	4	0	0	0	0	Hill 1976a
American Goldfinch								
Michigan	68	1	0	1	0	0	0	Berger 1951
3 spp. ground nesters[1]								
Kansas	19	14	13	14	5	7	0	Elliott 1977

[1] Grasshopper Sparrows, Dickcissels, and Eastern Meadowlarks.

Appendix E

Known Victims of Shiny Cowbirds

Order Columbiformes, Family Columbidae
Eared Dove, *Zenaida auriculata*
Picui Ground-Dove, *Columbina picui*
Black-winged Ground-Dove, *Metriopelia melanoptera*
Order Piciformes, Family Picidae
Green-barred Woodpecker, *Colaptes [Chrysoptilus] melanochloros*
Order Passeriformes, Family Dendrocolaptidae
Straight-billed Woodcreeper, *Xiphorhynchus picus*
Family Furnariidae
*†Rufous Hornero, *Furnarius rufus*
Crested Hornero, *Furnarius cristatus*
Plain-mantled Tit-Spinetail, *Leptasthenura aegithaloides*
Pale-breasted Spinetail, *Synallaxis albescens*
Chicli Spinetail, *Synallaxis spixi*
Stripe-crowned Spinetail, *Cranioleuca pyrrhophia*
*Olive Spinetail, *Cranioleuca obsoleta*
Rusty-backed Spinetail, *Cranioleuca vulpina*
Yellow-chinned Spinetail, *Certhiaxis cinnamomea*
*Short-billed Canastero, *Asthenes baeri*
Creamy-breasted Canastero, *Asthenes dorbignyi*
Patagonian Canastero, *Asthenes patagonica*
Córdoba Canastero, *Asthenes sclateri*
Hudson's Canastero, *Asthenes hudsoni*
Common Thornbird, *Phacellodomus rufifrons*
Little Thornbird, *Phacellodomus sibilatrix*
Streak-fronted Thornbird, *Phacellodomus striaticeps*
Freckle-breasted Thornbird, *Phacellodomus striaticollis*
Greater Thornbird, *Phacellodomus ruber*

Wren-like Rushbird, *Phleocryptes melanops*
Curve-billed Reedhaunter, *Limnornis curvirostris*
*Leñatero, *Anumbius annumbi*
Lark-like Brushrunner, *Coryphistera alaudina*
Rufous Cacholote, *Pseudoseisura cristata*
White-throated Cacholote, *Pseudoseisura gutturalis*
Family Thamnophilidae
Great Antshrike, *Taraba major*
*Collared Antshrike, *Sakesphorus bernardi*
Rufous-capped Antshrike, *Thamnophilus ruficapillus*
Family Rhinocryptidae
Crested Gallito, *Rhinocrypta lanceolata*
Family Cotingidae
White-tipped Plantcutter, *Phytotoma rutila*
Family Tyrannidae
Suiriri Flycatcher, *Suiriri suiriri*
*[1]Caribbean Elaenia, *Elaenia martinica* (Cruz et al. 1989; Post et al. 1990)
Yellow-bellied Elaenia, *Elaenia flavogaster*
White-crested Elaenia, *Elaenia albiceps*
Sooty Tyrannulet, *Serpophaga nigricans*
White-crested Tyrannulet, *Serpophaga subcristata*
*Warbling Doradito, *Pseudocolopteryx flaviventris*
Bran-colored Flycatcher, *Myiophobus fasciatus*
Vermillion Flycatcher, *Pyrocephalus rubinus* (Mason 1986a)
Fire-eyed Diucon, *Xolmis pyrope*
Gray Monjita, *Xolmis cinerea*
White Monjita, *Xolmis irupero*
*Short-tailed Field-Tyrant, *Muscigralla brevicauda*

Blue-billed Black-Tyrant, *Knipolegus cyanirostris*
Cabanis' Black Tyrant, *Knipolegus cabanisi*
Spectacled Tyrant, *Hymenops perspicillatus*
Pied Water-Tyrant, *Fluvicola pica*
Black-backed Water-Tyrant, *Fluvicola albiventer*[2]
*White-headed Marsh-Tyrant, *Arundinicola leucocephala*
Cock-tailed Tyrant, *Alectrurus tricolor*
*Yellow-browed Tyrant, *Satrapa icterophrys*
*Cattle Tyrant, *Machetornis rixosus*
Short-crested Flycatcher, *Myiarchus ferox*
Brown-crested Flycatcher, *Myiarchus tyrannulus*
*[1]Puerto Rican Flycatcher, *Myiarchus antillarum* (Cruz et al. 1985; Post et al. 1990)
Snowy-throated Kingbird, *Tyrannus niveigularis*
*Tropical Kingbird, *Tyrannus melancholicus*
*[†]Fork-tailed Flycatcher, *Tyrannus savana*
[†]Gray Kingbird, *Tyrannus dominicensis*
Variegated Flycatcher, *Empidonomus varius*
*Crowned Slaty Flycatcher, *Empidonomus aurantioatrocristatus*
*White-bearded Flycatcher, *Conopias inornatus*
Baird's Flycatcher, *Myiodynastes bairdii*
Streaked Flycatcher, *Myiodynastes maculatus*
*[†]Great Kiskadee, *Pitangus sulphuratus*
White-naped Xenopsaris, *Xenopsaris albinucha*
White-winged Becard, *Pachyramphus polychopterus*
*Black-tailed Tityra, *Tityra cayana*
Family Hirundinidae
*White-rumped Swallow, *Tachycineta leucorrhoa*
Brown-chested Martin, *Phaeoprogne tapera*
Family Troglodytidae
*Bicolored Wren, *Campylorhynchus griseus*
Thrush-like Wren, *Campylorhynchus turdinus*
*Stripe-backed Wren, *Campylorhynchus nuchalis*
Fasciated Wren, *Campylorhynchus fasciatus*
*Rufous-breasted Wren, *Thryothorus rutilus*
Buff-breasted Wren, *Thryothorus leucotis*
*Superciliated Wren, *Thryothorus superciliaris*
*House Wren, *Troglodytes aedon*

Family Muscicapidae, Subfamily Sylviinae
*Masked Gnatcatcher, *Polioptila dumicola*
Family Muscicapidae, Subfamily Turdinae
Orange-billed Nightingale-Thrush, *Catharus aurantiirostris*
[†]Red-legged Thrush, *Turdus plumbeus* (Cruz et al. 1985)
Chiguanco Thrush, *Turdus chiguanco*
Glossy-black Thrush, *Turdus serranus*
Andean Slaty-Thrush, *Turdus nigriceps*
*Rufous-bellied Thrush, *Turdus rufiventris*
Austral Thrush, *Turdus falcklandii*
Pale-breasted Thrush, *Turdus leucomelas*
*Creamy-bellied Thrush, *Turdus amaurochalinus*
Bare-eyed Thrush, *Turdus nudigenis*
Family Mimidae
[†]Northern Mockingbird, *Mimus polyglottos* (Cruz et al. 1985)
Tropical Mockingbird, *Mimus gilvus*
*[†]Chalk-browed Mockingbird, *Mimus saturninus*
*Patagonian Mockingbird, *Mimus patagonicus*
*White-banded Mockingbird, *Mimus triurus*
*Long-tailed Mockingbird, *Mimus longicaudatus*
Chilean Mockingbird, *Mimus thenca*
Family Motacillidae
Correndera Pipit, *Anthus correndera*
Short-billed Pipit, *Anthus furcatus*
Yellowish Pipit, *Anthus lutescens*
Family Dulidae
Palmchat, *Dulus dulus [dominicus]* (Cruz, pers. comm. in Arendt and Vargas Mora 1984)
Family Vireonidae
*Rufous-browed Peppershrike, *Cyclarhis gujanensis*
*Puerto Rican Vireo, *Vireo latimeri* (Cruz et al. 1989)
Red-eyed Vireo, *Vireo olivaceus*
*[1]Black-whiskered Vireo, *Vireo altiloquus* (Wiley 1985; Cruz and Wiley 1989; Post et al. 1990)
Rufous-crowned Greenlet, *Hylophilus poicilotis*
Golden-fronted Greenlet, *Hylophilus aurantiifrons*
Family Emberizidae, Subfamily Parulinae
*Yellow Warbler, *Dendroica petechia*

*[1]Adelaide's Warbler, *Dendroica adelaidae*
    (Wiley 1985; Cruz et al. 1989)
Masked Yellowthroat, *Geothlypis*
    aequinoctialis
White-striped Warbler, *Basileuterus*
    leucophrys
Flavescent Warbler, *Basileuterus flaveolus*
Family Emberizidae, Subfamily Coerubinae
    *Bicolored Conebill, *Conirostrum bicolor*
    Cinereous Conebill, *Conirostrum cinereum*
Family Emberizidae, Subfamily Thraupinae
    Black-faced Tanager, *Schistochlamys*
        melanopis
    *[1]White-banded Tanager, *Neothraupis fasciata*
        (Cavalcanti and Pimentel 1988)
    White-rumped Tanager, *Cypsnagra*
        hirundinacea
    Orange-headed Tanager, *Thlypopsis sordida*
    *[1]Guira Tanager, *Hemithraupis guira*(Cavalcanti
        and Pimentel 1988)
    Rosy Thrush-Tanager, *Rhodinocichla rosea*
    White-lined Tanager, *Tachyphonus rufus*
    Hepatic Tanager, *Piranga flava*
    *Silver-beaked Tanager, *Ramphocelus carbo*
    Brazilian Tanager, *Ramphocelus bresilius*
    Blue-gray Tanager, *Thraupis episcopus*
    *Sayaca Tanager, *Thraupis sayaca*
    Golden-chevroned Tanager, *Thraupis ornata*
    *[1]Palm Tanager, *Thraupis palmarum*
        (Cavalcanti and Pimentel 1988)
    *Blue-and-yellow Tanager, *Thraupis*
        bonariensis
Family Emberizidae, Subfamily Dacninae
    Blue Dacnis, *Dacnis cayana*
Family Emberizidae, Subfamily Cardinalinae
    Yellow Grosbeak, *Pheucticus chrysopeplus*
    Black-backed Grosbeak, *Pheucticus*
        aureoventris
    Black-throated Grosbeak, *Pitylus fuliginosus*
    *Grayish Saltator, *Saltator coerulescens*
    Green-winged Saltator, *Saltator similis*
    *Golden-billed Saltator, *Saltator aurantiirostris*
    *Streaked Saltator, *Saltator albicollis*
    Ultramarine Grosbeak, *Cyanocompsa*
        [brissonii] cyanea
Family Emberizidae, Subfamily Emberizinae
    *Rufous-collared Sparrow, *Zonotrichia*
        capensis
    Grassland Sparrow, *Ammodramus*
        [Myiospiza] humeralis

Stripe-capped Sparrow, *Aimophila strigiceps*
*Tumbes Sparrow, *Aimophila [Rhynchospiza]*
    stolzmanni
Saffron-billed Sparrow, *Arremon flavirostris*
Black-striped Sparrow, *Arremonops*
    conirostris
*Ochre-breasted Brush-Finch, *Atlapetes*
    semirufus
Yellow-striped Brush-Finch, *Atlapetes*
    citrinellus
Yellow Cardinal, *Gubernatrix cristata*
Red-crested Cardinal, *Paroaria coronata*
Red-capped Cardinal, *Paroaria gularis*
Yellow-billed Cardinal, *Paroaria capitata*
Many-colored Chaco-Finch, *Saltatricula*
    multicolor
Pileated Finch, *Coryphospingus pileatus*
Red-crested Finch, *Coryphospingus cucullatus*
Crimson Finch, *Rhodospingus cruentus*
Patagonian Sierra-Finch, *Phrygilus*
    patagonicus
Plumbeous Sierra-Finch, *Phrygilus unicolor*
*Long-tailed Reed-Finch, *Donacospiza*
    albifrons
*Common Diuca-Finch, *Diuca diuca*
*Cinereous Finch, *Piezorhina cinerea*
*Black-and-Rufous Warbling-Finch, *Poospiza*
    nigrorufa
Red-rumped Warbling-Finch, *Poospiza*
    lateralis
Collared Warbling-Finch, *Poospiza*
    hispaniolensis
Ringed Warbling-Finch, *Poospiza torquata*
Black-capped Warbling-Finch, *Poospiza*
    melanoleuca[3]
Cinereous Warbling-Finch, *Poospiza cinerea*
Greater Yellow-Finch, *Sicalis auriventris*
Saffron Finch, *Sicalis flaveola*
Grassland Yellow-Finch, *Sicalis luteola*
Wedge-tailed Grass-Finch, *Emberizoides*
    herbicola
Great Pampa-Finch, *Embernagra platensis*
Lined Seedeater, *Sporophila lineola*
*Double-collared Seedeater, *Sporophila*
    caerulescens
Ruddy-breasted Seedeater, *Sporophila minuta*
Large-billed Seed-Finch, *Oryzoborus*
    crassirostris
Lesser Seed-Finch, *Oryzoborus angolensis*
Lesser Antillean Bullfinch, *Loxigilla noctis*

Family Emberizidae, Subfamily Icterinae
  Crested Oropendola, *Psarocolius decumanus*
  Yellow-rumped Cacique, *Cacicus cela*
  Golden-winged Cacique, *Cacicus chrysopterus*
  Moriche Oriole, *Icterus chrysocephalus*
  Epaulet Oriole, *Icterus cayanensis*
  Yellow-backed Oriole, *Icterus chrysater*
  Yellow Oriole, *Icterus nigrogularis*
  *White-edged Oriole, *Icterus graceannae*
  *[1]Troupial, *Icterus icterus* (Wiley 1985)
  *[1]Black-cowled Oriole, *Icterus dominicensis* (Cruz, pers. comm. in Arendt and Vargas Mora 1984; Wiley 1985; Baltz 1995)
  *[4]St. Lucia Oriole, *Icterus laudabilis* (Post et al. 1990)
  Martinique Oriole, *Icterus bonana*
  Oriole Blackbird, *Gymnomystax mexicanus*
  Saffron-cowled Blackbird, *Xanthospar flavus*
  *[1]Yellow-winged Blackbird, *Agelaius thilius* (see Mason 1986a)
  Unicolored Blackbird, *Agelaius cyanopus*
  *Yellow-hooded Blackbird, *Agelaius icterocephalus* (Cruz et al. 1990)
  *Yellow-shouldered Blackbird, *Agelaius xanthomus*
  *Chestnut-capped Blackbird, *Agelaius ruficapillus*
  Red-breasted Blackbird, *Leistes militaris*
  Pampas Meadowlark, *Sturnella defillippi* (Withington and Orians in Gochfeld 1979b)

  *Greater Red-breasted Meadowlark, *Sturnella loyca*
  White-browed Blackbird, *Sturnella superciliaris* (Mason 1986b)
  Yellow-rumped Marshbird, *Pseudoleistes guirahuro*
  †Brown-and-Yellow Marshbird, *Pseudoleistes virescens*
  Scarlet-headed Blackbird, *Amblyramphus holosericeus*
  Chopi Blackbird, *Gnorimopsar chopi*
  Golden-tufted Grackle, *Macroagelaius imthurni*
  *Scrub Blackbird, *Dives warszewiczi*
  *[1]†Greater Antillean Grackle, *Quisculus niger* (Wiley 1985)
  *†Carib Grackle, *Quisculus lugubris*
  *[1]†Bay-winged Cowbird, *Molothrus badius* (Fraga 1986; Mason 1986a)
Family Fringillidae
  *Hooded Siskin, *Carduelis magellanica*
  Black-chinned Siskin, *Carduelis barbata*
Family Ploceidae
  Village Weaver, *Ploceus cucullatus* (Cruz and Wiley 1989; Cruz, pers. comm. in Arendt and Vargas Mora 1984)
Family Passeridae
  House Sparrow, *Passer domesticus*
Family Estrildidae
  Bronzed Mannikin, *Lonchura cucullata*
  Spice Finch, *Lonchura punctulata* (Wiley by T. Nakamura, pers. comm. 1996)

Note: Most victims were listed by Friedmann and Kiff 1985. Hosts known to have raised Shiny Cowbirds are identified with an asterisk, and victims known to eject Shiny Cowbird eggs are identified with a dagger.

*[1] Hosts known to have successfully raised Shiny Cowbirds since Friedmann and Kiff (1985)

[2] Treated as a subspecies of *Fluvicola pica* by Friedmann and Kiff 1985. Clements (1991) and Friedmann et al. (1977) list *Fluvicola albiventer* and *Fluvicola pica* as separate species.

[3] Clements (1991) considered the Black-capped Warbling-Finch (*Poospiza melanoleuca*) as a separate species, whereas Friedmann and Kiff (1985) listed *melanoleuca* as a subspecies of *Poospiza cinerea;* both are parasitized by Shiny Cowbirds.

*[4] Hosts presumed to have been parasitized and to have successfully raised Shiny Cowbirds since Friedmann and Kiff (1985) but based on observations of hosts feeding Shiny Cowbirds out of the nest

Notes

Chapter 1. Introduction

1. Also called Honey-badgers.

2. Formerly family Icteridae, now subfamily Icterinae.

3. The family Estrildidae has also been classified as Estrildinae, a subfamily of Ploceidae. According to Clements (1991), all parasitic finches except *Anomalospiza* are in Estrildidae, and *Anomalospiza* is in the family Ploceidae.

4. Also called Steel Combassou.

5. Compared to Rosybill eggs, Black-headed Duck eggs are rounder (Daguerre 1923), and Weller (1968) reported the eggshells to be more granular.

6. The Greater Red-breasted Meadowlark is also called Loyca Meadowlark (Gochfeld 1979a) and Long-tailed Meadowlark (Clements 1991). Friedmann and Kiff (1985) refer to the scientific name, *Pezites militari*.

7. To eliminate the unusual or "accidental" cases of parasitism from the terminology, Jensen and Jensen (1969) used the term "biological hosts" to refer to successful hosts.

Chapter 2. Defense Mechanisms

1. Black-billed Magpies are parasitized by several cuckoos in Europe, but they are not parasitized by Brown-headed Cowbirds in North America.

Chapter 3. Evolution of Brood Parasitism

1. This is sometimes referred to as the evolutionary-equilibrium hypothesis (Ward et al. 1996); I recommend use of more specific terms (bill-size constraint or ejection inability) because evolutionary equilibrium has also been used in other explanations for parasitic egg acceptance in populations that conflict with Rohwer and Spaw's hypothesis (Lotem et al. 1992).

Chapter 6. Bronzed Cowbirds

1. The type locality listed in Blake (1968) is Laguna, Vera Cruz. This was an erroneous statement made by Blake, not a discrepancy in the determination of type locality (K.C. Parkes, pers. comm.).

2. Breast feathers of Shiny Cowbirds also have a hairlike appearance (Parkes and Blake 1965). The hairlike quality arises from barbules on the distal barbs laying flat against the barbs, giving them the appearance of very thick barbs or hairs. The plumage of Bay-winged Cowbirds appears even more hairlike than Bronzed or Shiny Cowbirds, but for a different reason. The barbs in Bay-winged Cowbirds are ivory (almost white), while the barbules are a mousey gray, and the contrast in color makes the plumage look hairlike. In Screaming, Giant, and Brown-headed Cowbirds, barbules on the distal barbs are reduced, but they are not pressed against the barbs.

3. The operational sex ratio includes only after-second-year males but all adult females (SY plus ASY), based on the assumption that SY males do not breed and SY females do breed.

4. Carter (1984) was unable to distinguish between SY and ASY females.

5. Five ha can be considered large for a nesting passerine but small for even a monogamous brood parasite.

Chapter 7. Brown-headed Cowbirds

1. The counting methods for the west-central area differed from the southeastern area in that some sites were sampled up to ten times. In the figure (17.7%) only one count per site (the middle) was used. Using all sites, cowbirds were recorded at 42.7% of their counting sites.

2. Long-term trapping and removal programs may eventually produce a more balanced or even a female-biased sex ratio (chapter 9).

3. This did not include foraging sites, which may be up to 6.7 km from breeding range.

4. Captive males responded to artificially increased photoperiod from early January to mid-February with increased testes weight, but males exposed to artificially increased photoperiod from late September to mid-November, when they were in a refractory period, did not respond (Middleton and Scott 1965). The mean testes volume of males housed in outdoor aviaries with 17 hours of artificial light for a month, between 8 December and 3 January, increased from 1.14 mm^3 to 12.5 mm^3 ($n = 7$), whereas mean testes volume of males housed under natural light increased from 1.05 mm^3 to 1.61 mm^3 ($n = 4$, Payne 1967a).

5. We have caught individual males up to 135 times and individual females up to 76 times during the breeding season.

6. Six of 20 (30.0%) second-year males produced the local flight whistle dia-

lect, whereas 16 of 18 (88.9%) after-second-year males did (O'Loghlen and Rothstein 1995).

7. Yearling males in O'Loghlen and Rothstein's (1993) study were kept in captivity for only 3–4 days before recordings were made, and they were in audio contact with other (older) males.

8. Common Grackle eggs (28.8 ± 1.4 × 21.4 ± 0.8 mm, n = 1,614) and Yellow-headed Blackbird eggs (26.33 ± 1.16 × 18.1 ± 0.57 m, n = 842, Ortega and Cruz 1991).

9. Brown-headed Cowbirds do very occasionally parasitize other cavity nesters, including Black-capped Chickadees (Packard 1936).

Chapter 8. Shiny Cowbirds

1. Post and Wiley (1977b) suspected that Shiny Cowbirds actually arrived on Puerto Rico at an earlier date than 1955 because of the large numbers that were seen at widely scattered sites and the rapidity with which they spread.

2. Rufous-collared Sparrows are accepter species and were, therefore, probably not responsible for Shiny Cowbird eggs being punctured.

3. After collecting data a couple more years, Fraga (1985) reported that 37 of 76 (48.7%) Shiny Cowbird eggs in the nests of Rufous-collared Sparrows were immaculate.

Chapter 9. The Management Challenge

1. The Breeding Bird Survey also has biases in estimates of relative abundance, and these biases may change over time as a consequence of changes in habitat (chapter 7; Bart et al. 1995).

2. Wolf (1987:339) defined closed canopy as "along a rarely traveled dirt road that followed a canyon stream through relatively dense secondary riparian deciduous forest." Open canopy "included all remaining sites in the study area, none of which included dense forest."

References

Journal Abbreviations Used

African J. Ecol.	African Journal of Ecology
Am. Birds	American Birds
Am. Midl. Nat.	American Midland Naturalists
Am. Nat.	American Naturalist
Am. Sci.	American Scientist
Am. Zool.	American Zoology
Anim. Behav.	Animal Behavior
Ann. Carnegie Mus.	Annals of the Carnegie Museum
Ann. Rep. Smiths. Inst.	Annual Report of the Smithsonian Institution
Ann. Rev. Ecol. Syst.	Annual Review of Ecology and Systematics
A.O.U. Monogr.	American Ornithologists' Union Monographs
Audubon Bull.	Audubon Bulletin
Australian J. Biol. Sci.	Australian Journal of Biological Sciences
Aust. J. Zool.	Australian Journal of Zoology
Aust. Zool. Rev.	Australian Zoological Reviews
Avicultural Mag.	Avicultural Magazine
Behav. Ecol.	Behavioral Ecology
Behav. Ecol. Sociobiol.	Behavioral Ecology and Sociobiology
Behav. Res. Meth. Inst.	Behavior Research Methods and Instrumentation
Biog. of the West Indies	Biogeography of the West Indies
Biol. Conserv.	Biological Conservation
Biol. Notes Ill. Nat. Hist. Survey	Biological Notes. Illinois Natural History Survey
Biol. Rev.	Biological Reviews
Bird Behav.	Bird Behavior
Br. Birds	British Birds
Bull. Am. Mus. of Nat. Hist.	Bulletin. American Museum of Natural History
Bull. Br. Ornithol. Club	Bulletin. British Ornithologists' Club
Bull. Northeastern Bird-Banding Assoc.	Bulletin. Northeastern Bird-Banding Association
Bull. Nuttall Ornithological Club	Bulletin. Nuttall Ornithological Club
Can. Field-Nat.	Canadian Field-Naturalist

Can. J. Zool.	Canadian Journal of Zoology
Caribbean J. of Sci.	Caribbean Journal of Science
Conserv. Biol.	Conservation Biology
Curr. Ornithol.	Current Ornithology
Ecol. Monogr.	Ecological Monographs
Evol. Ecol.	Evolutionary Ecology
Horm. Behav.	Hormones and Behavior
J. Anim. Ecol.	Journal of Animal Ecology
J. Avian Biology	Journal of Avian Biology
J. Colorado-Wyoming Acad. Sci.	Journal of the Colorado-Wyoming Academy of Science
J. Comp. Physiol. Psychol.	Journal of Comparative and Physiological Psychology
J. Comp. Psych.	Journal of Comparative Psychology
J. Endocrinology	Journal of Endocrinology
J. Field Ornithol.	Journal of Field Ornithology
J. Forestry	Journal of Forestry
J. Ornithol.	Journal fuer Ornithologie
J. Range Manage.	Journal of Range Management
J. South African Ornithological Union	Journal of the South African Ornithological Union
J. Theor. Biol.	Journal of Theoretical Biology
J. Wildl. Manage.	Journal of Wildlife Management
J. Zool.	Journal of Zoology
Jpn. J. Zool.	Japanese Journal of Zoology
Kans. Ornithol. Soc. Bull.	Kansas Ornithological Society Bulletin
Mus. Comp. Zool.	Museum of Comparative Zoology
Nat. Wildl.	National Wildlife
N. Am. Bird Band.	North American Bird Banding
North Amer. Bird Bander	North American Bird Bander
Occas. Pap. Mus. Nat. Hist. Univ. Kansas	Occasional Papers, Museum of Natural History, University of Kansas
Ohio J. Sci.	Ohio Journal of Science
Ornith. Monat.	Ornithologische Monatsberichte
Ornith. Monogr.	Ornithological Monographs
Philos. Trans. R. Soc. London	Philosophical Transactions, Royal Society of London
Proc. West. Found. Vert. Zool.	Proceedings. Western Foundation of Vertebrate Zoology
Proc. Zool. Soc. Lond.	Proceedings. Zoological Society of London
Rep. U.S. Natl. Mus.	Reports. United States National Museum
Rev. Bras. Biol.	Revista Brasileria de Biología

Smithsonian Contrib. Zool.	Smithsonian Contributions to Zoology
Syst. Biol.	Systematic Biology
Trans. Linn. Soc. of London	Transactions. Linnean Society of London
Trans. Missouri Acad. of Sci.	Transactions. Missouri Academy of Science
Trans. of North American Wildlife Conference	Transactions. North American Wildlife Conference
Univ. of Kansas Mus. of Nat. Hist. Misc. Publ.	University of Kansas. Museum of Natural History. Miscellaneous Publications
Utah Mus. Nat. Hist. Occas. Publ.	Utah Museum of Natural History. Occasional Publications
West. Birds	Western Birds
Wildl. Monogr.	Wildlife Monographs
Wilson Bull.	Wilson Bulletin
Zool. Anz.	Zoologischer Anzeiger
Z. Tierpsychol.	Zeitschrift fuer Tierpsychologie

Airola, D. A. 1986. Brown-headed Cowbird parasitism and habitat disturbance in the Sierra Nevada. *J. Wildl. Manage.* 50:571–575.

Alvarez, F., L. Arias de Reyna, and M. Segura. 1976. Experimental brood parasitism of the magpie *(Pica pica)*. *Anim. Behav.* 24:907–916.

Amat, J. A. 1985. Nest parasitism of Pochard *Aythya ferina* by Red-crested Pochard *Netta rufina*. *Ibis* 127:255–262.

———. 1987. Is nest parasitism among ducks advantageous to the host? *Am. Nat.* 130:454–457.

Ambuel, B., and S. A. Temple. 1982. Songbird populations in southern Wisconsin forests: 1954 and 1979. *J. Field Ornithol.* 53:149–158.

———. 1983. Area-dependent changes in the bird communities and vegetation of southern Wisconsin forests. *Ecology* 64:1057–1068.

American Ornithologists' Union. 1957. *Check-list of North American Birds*, 5th ed. Baltimore: Lord Baltimore Press.

Anderson, R. C. 1982. An evolutionary model summarizing the roles of fire, climate, and grazing animals in the origin and maintenance of grasslands: An end paper. In *Grasses and grasslands: Systematics and ecology*, ed. J. R. Estes, R. J. Tyrl, and J. N. Brunken. Norman: University of Oklahoma Press.

Anderson, W. L., and R. W. Storer. 1976. Factors influencing Kirtland's Warbler nesting success. *Jack-Pine Warbler* 54:105–115.

Andersson, M., and M. O. G. Eriksson. 1982. Nest parasitism in goldeneyes *Bucephala clangula*: Some evolutionary aspects. *Am. Nat.* 120:1–16.

Andrén, H., and P. Angelstam. 1988. Elevated predation rates as an edge effect in habitat islands: Experimental evidence. *Ecology* 69:544–547.

Ankney, C. D., and S. L. Johnson. 1985. Variation in weight and composition of Brown-headed Cowbird eggs. *Condor* 87:296–299.

Ankney, C. D., and D. M. Scott. 1980. Changes in nutrient reserves and diet of breeding Brown-headed Cowbirds. *Auk* 97:684–696.

———. 1982. On the mating system of Brown-headed Cowbirds. *Wilson Bull.* 94:260–268.

Appleby, M. C., and H. R. McRae. 1983. A method for identifying eggs laid by individual birds. *Behav. Res. Meth. Inst.* 15:399–400.

Arcese, P., J. N. M. Smith, and M. I. Hatch. 1996. Nest predation by cowbirds and its consequences for passerine demography. *Proc. Natl. Acad. Sci.* 93:4608–4611.

Arendt, W. J., and T. A. Vargas Mora. 1984. Range expansion of the Shiny Cowbird in the Dominican Republic. *J. Field Ornithol.* 55:104–107.

Arias-de-Reyna, L., and S. J. Hidalgo. 1982. An investigation into egg-acceptance by Azure-winged Magpies and host-recognition by Great Spotted Cuckoo chicks. *Anim. Behav.* 30:819–823.

Arnold, K. A., and D. M. Johnson. 1983. Annual adult survival rates for Brown-headed Cowbirds wintering in southeast Texas. *Wilson Bull.* 95:150–153.

Askins, R. A., M. J. Philbrick, and D. S. Sugeno. 1987. Relationship between the regional abundance of forest and the composition of forest bird communities. *Biol. Conserv.* 39:129–152.

Attiwill, A. R., J. M. Bourne, and S. A. Parker. 1981. Possible nest-parasitism in the Australian Stiff-tailed Ducks (Anatidae: Oxyurini). *Emu* 81:41–42.

Baba, R. 1994. Timing of spawning and host-nest choice for brood parasitism by the Japanese minnow, *Punglungia herzi*, on the Japanese aucha perch, *Siniperca kawamebari*. *Ethology* 98:50.

Baba, R., Y. Nagata, and S. Yamagishi. 1990. Brood parasitism and egg robbing among three freshwater fish. *Anim. Behav.* 40:776–778.

Baerends, G., and A. Hogan-Warburg. 1982. The Herring Gull and its eggs, section 1: The external morphology of the egg and its variability. *Behaviour* 82:1–32.

Bailey, R. E. 1952. The incubation patch of passerine birds. *Condor* 54:121–136.

Baird, J. 1958. The postjuvenile molt of the male Brown-headed Cowbird (*Molothrus ater*). *Bird Banding* 29:224–228.

Balda, R. P., and S. Carothers. 1968. Nest protection by the Brown-headed Cowbird (*Molothrus ater*). *Auk* 85:324–325.

Baltz, M. E. 1995. First records of the Shiny Cowbird (*Molothrus bonariensis*) in the Bahama Archipelago. *Auk* 112:1039–1041.

Barlow, J. C. 1962. Natural history of the Bell Vireo, *Vireo bellii* Audubon. Univ. of Kansas Mus. of Nat. Hist. Publ. 12:241–296.

Bart, J., M. Hofschen, and B. G. Peterjohn. 1995. Reliability of the Breeding Bird Survey: Effects of restricting surveys to roads. *Auk* 112:758–761.

Beane, J. C., and S. L. Alford. 1990. Destruction of a Pine Warbler brood by an adult cowbird. *Chat* 54:85–87.

Beecher, W. J. 1951. Adaptations for food-getting in the American blackbirds. *Auk* 63:411–440.

Beezley, J. A., and J. P. Rieger. 1987. Least Bell's Vireo management by cowbird trapping. *West. Birds* 18:55–61.

Behle, W. H. 1985. Utah birds: Geographic distribution and systematics. *Utah Mus. Nat. Hist. Occas. Publ. No. 5.*

Belcher, C., and G. D. Smooker. 1937. Birds of the colony of Trinidad and Tobago: Part VI. *Ibis* 14:504–550.

Bellrose, F. C. 1980. *Ducks, geese, and swans of North America.* Harrisburg, Penn.: Stackpole Books.

Bendire, C. 1895. The cowbirds. *Rep. U.S. Natl. Mus. for 1893–1894.* Washington, D.C.: Government Printing Office, 589–624.

Bent, A. C. 1958. Life histories of North American blackbirds, orioles, tanagers, and allies. *Bull. U. S. Natl. Mus.* 211.

Berger, A. J. 1948. Early nesting and cowbird parasitism of the Goldfinch in Michigan. *Wilson Bull.* 60:52–53.

———. 1951. The cowbird and certain host species in Michigan. *Wilson Bull.* 63: 26–34.

Berger, A. J., and D. F. Parmalee. 1952. The Alder Flycatcher in Washtenaw County, Michigan: Breeding distribution and cowbird parasitism. *Wilson Bull.* 64:33–38.

Best, L. B. 1978. Field Sparrow reproductive success and nesting ecology. *Auk* 95:9–22.

Best, L. B., and D. F. Stauffer. 1980. Factors affecting nesting success in riparian bird communities. *Condor* 82:149–158.

Bianchi, V. C. 1971. Un ejemplar de color canela del Tordo Bayo. *El Hornero* 11:128.

Biebach, H. 1981. Energetic costs of incubation on different clutch sizes in starlings (*Sturnus vulgaris*). *Ardea* 69:141–142.

Birkenstein, L. R., and R. E. Tomlinson. 1981. Native names of Mexican birds. *Resource Publication 139.* Washington, D.C.: United States Department of Interior, Fish and Wildlife Service.

Bischoff, C. M., and M. T. Murphy. 1993. The detection of and responses to experimental intraspecific brood parasitism in Eastern Kingbirds. *Anim. Behav.* 45: 631–638.

Blake, E. R. 1968. Family Icteridae. In *Checklist of birds of the world,* vol. 14, ed. R. A. Paynter. Cambridge, Mass.: Museum of Comparative Zoology, 138–202.

Blake, J. G, and J. R. Karr. 1984. Species composition of bird communities and the conservation benefit of large versus small forests. *Biol. Conserv.* 30:173–187.

Blankespoor, G. W., J. Oolman, and C. Uthe. 1982. Eggshell strength and cowbird parasitism of Red-winged Blackbirds. *Auk* 99:363–365.

Bleitz, D. 1956. Heavy parasitism of Blue Grosbeaks by cowbirds in California. *Condor* 58:236–238.

Blincoe, B. J. 1935. A cowbird removes a robin's egg. *Wilson Bull.* 47:158.

Bock, C. E., V. A. Saab, T. D. Rich, and D. S. Dobkin. 1993. Effects of livestock grazing on neotropical migratory landbirds in western North America. In *Status and management of neotropical migratory birds*, ed. D. M. Finch and P. W. Stangel. U. S. Department of Agriculture, Forest Service, Gen. Tech. Rep. RM-229, 296–309.

Bolen, E. G., and B. W. Cain. 1968. Mixed Wood Duck–Tree Duck clutch in Texas. *Condor* 70:389–390.

Bond, J. 1936. *Birds of the West Indies*. Philadelphia: Academy of Natural Sciences of Philadelphia.

———. 1956. *Birds of the West Indies*. Philadelphia: Academy of Natural Sciences of Philadelphia.

Bonnicksen, T. M. 1991. Managing biosocial systems: A framework to organize society-environment relationships. *J. Forestry* 89(10):10–15.

Bonwell, J. R. 1895. A strange freak of the cowbird. *Nidiologist* 2:53.

Bouffard, S. H. 1983. Redhead egg parasitism of Canvasback nests. *J. Wildl. Manage.* 47:213–216.

Braa, A. T., A. Moksnes, and E. Røskaft. 1992. Adaptations of Bramblings and Chaffinches towards parasitism by the Common Cuckoo. *Anim. Behav.* 43:67–78.

Brackbill, H. 1976. A Brown-headed Cowbird in postjuvenile molt at age of about 38 days. *Bird-Banding* 47:275–276.

Brandt, A. E. 1947. The rearing of a cowbird by Acadian Flycatchers. *Wilson Bull.* 59:79–82.

Brewer, A. D. 1995. Cowbird warning. *British Birds* 88:157.

Briskie, J. V., and S. G. Sealy. 1987. Responses of Least Flycatchers to experimental inter- and intraspecific brood parasitism. *Condor* 89:899–901.

———. 1989. Changes in nest defense against a brood parasite over the breeding cycle. *Ethology* 82:61–67.

Briskie, J. V., S. G. Sealy, and K. A Hobson. 1990. Differential parasitism of Least Flycatchers and Yellow Warblers by the Brown-headed Cowbird. *Behav. Ecol. Sociobiol.* 27:403–410.

———. 1992. Behavioral defenses against avian brood parasitism in sympatric and allopatric host populations. *Evolution* 46:334–340.

Brittingham, M. C., and S. A. Temple. 1983. Have cowbirds caused forest songbirds to decline? *BioScience* 33:31–35.

Brooke, M. de L., and N. B. Davies. 1988. Egg mimicry by cuckoos *Cuculus canorus* in relation to discrimination by hosts. *Nature* 335:630–632.

———. 1991. A failure to demonstrate host imprinting in the cuckoo (*Cuculus canorus*) and alternative hypotheses for the maintenance of egg mimicry. *Ethology* 89:154–166.

Brooker, L. C., and M. G. Brooker. 1990. Why are cuckoos host specific? *Oikos* 57:301–309.

Brooker, L. C., M. G. Brooker, and A. M. H. Brooker. 1990. An alternative population/genetics model for the evolution of egg mimesis and egg crypsis in cuckoos. *J. Theor. Biol.* 146:123–143.

Brooker, M. G., and L. C. Brooker. 1989a. Cuckoo hosts in Australia. *Aust. Zool. Rev.* 2:1–70.

————. 1989b. The comparative breeding behavior of two sympatric cuckoos, Horsfield's Bronze-Cuckoo *Chrycococcyx basilis* and the Shiny Bronze-Cuckoo *C. lucidus*, in western Australia: A new model for the evolution of egg morphology and host specificity in avian brood parasites. *Ibis* 131:528–547.

————. 1991. Eggshell strength in cuckoos and cowbirds. *Ibis* 133:406–413.

Brooker, M. G., L. C. Brooker, and I. Rowley. 1988. Egg deposition by the Bronze-Cuckoos *Chrysococcyx basalis* and *C. lucidus. Emu* 88:107–109.

Broughton, K. E., A. L. A. Middleton, and E. D. Bailey. 1987. Early vocalizations of the Brown-headed Cowbird and three host species. *Bird Behav.* 7:27–30.

Brown, B. T. 1994. Rates of brood parasitism by Brown-headed Cowbirds on riparian passerines in Arizona. *J. Field Ornithol.* 65:160–168.

Brown, C. R., and M. B. Brown. 1988. A new form of reproductive parasitism in Cliff Swallows. *Nature* 331:66–68.

————. 1989. Behavioural dynamics of intraspecific brood parasitism in colonial Cliff Swallows. *Anim. Behav.* 37:777–796.

Brown, C. R., and L. C. Sherman. 1989. Variation in the appearance of swallow eggs and the detection of intraspecific brood parasitism. *Condor* 91:620–627.

Brown, R. J., M. N. Brown, M. de L. Brooke, and N. B. Davies. 1990. Reactions of parasitized and unparasitized populations of *Acrocephalus* warblers to model cuckoo eggs. *Ibis* 132:109–111.

Buech, R. R. 1982. Nesting ecology and cowbird parasitism of Clay-colored, Chipping and Field Sparrows in a Christmas tree plantation. *J. Field Ornithol.* 53:363–369.

Bureau of Land Management, U.S. Department of the Interior. 1994. *Public land statistics 1993.* U.S. Department of the Interior.

Burgham, M. C. J., and J. Picman. 1989. Effect of Brown-headed Cowbirds on the evolution of Yellow Warbler anti-parasite strategies. *Anim. Behav.* 38:298–308.

Burnell, K., and S. I. Rothstein. 1994. Variation in the structure of female Brown-headed Cowbird vocalizations and its relation to vocal function and development. *Condor* 96:703–715.

Burtt, H. E., and M. L. Giltz. 1970. A study of blackbird repeats at a decoy trap. *Ohio J. Sci.* 70:162–170.

————. 1976. Sex differences in the tendency for Brown-headed Cowbirds and Red-winged Blackbirds to re-enter a decoy trap. *Ohio J. Sci.* 76:264–267.

Byers, G. W. 1950. A Black and White Warbler's nest with eight cowbird eggs. *Wilson Bull.* 62:136–138.

Callenbach, E. 1996. *Bring back the buffalo!* Washington, D.C.: Island Press.

Cannell, P. F., and B. A. Harrington. 1984. Interspecific egg-dumping by a Great Egret and Black-crowned Night Herons. *Auk* 101:889–891.

Carello, C. A. 1995. The Brown-headed Cowbird and its host the Red-winged Blackbird: The effects of nesting demographics on parasitism and comparative incubation between species. Master's thesis, University of Colorado.

Carey, C. 1986. Possible manipulation of eggshell conductance of host eggs by Brown-headed Cowbirds. *Condor* 88:388–390.

Carter, M. D. 1984. The social organization and parasitic behavior of the Bronzed Cowbird in south Texas. Ph.D. diss., University of Minnesota.

———. 1986. The parasitic behavior of the Bronzed Cowbird in south Texas. *Condor* 88:11–25.

———. 1987. An incident of brood parasitism by the Verdin. *Wilson Bull.* 99:136.

Cavalcanti, R. B., and T. M. Pimentel. 1988. Shiny Cowbird parasitism in central Brazil. *Condor* 90:40–43.

Chance, E. 1922. *The cuckoo's secret.* London: Sedgwick and Jackson.

Chance, E. P. 1940. *The truth about the cuckoo.* London: Country Life Ltd.

Chapman, F. M. 1928. The nesting habits of Wagler's Oropendola (*Zarhynchus wagleri*) on Barro Colorado Island. *Bull. Am. Mus. of Nat. Hist.* 58:123–166.

———. 1930. The nesting habits of Wagler's Oropendola on Barro Colorado Island. *Smithsonian Report* 73:347–386.

Chasko, G. G., and J. E. Gates. 1982. Avian habitat suitability along a transmission-line corridor in an oak-hickory forest region. *Wildl. Monogr.* 82:1–41.

Clark, K. L., and R. J. Robertson. 1981. Cowbird parasitism and evolution of anti-parasite strategies in the Yellow Warbler. *Wilson Bull.* 93:249–258.

Clements, J. F. 1991. *Birds of the world: A checklist.* Vista, Calif: Ibis Publishing Company.

Clotfelter, E. D. 1995. Courtship displaying and intrasexual competition in the Bronzed Cowbird. *Condor* 97:816–818.

Clotfelter, E. D., and T. Brush. 1995. Unusual parasitism by the Bronzed Cowbird. *Condor* 97:814–815.

Coker, D. R., and D. E. Capen. 1995. Landscape-level habitat use by Brown-headed Cowbirds in Vermont. *J. Wildl. Manage.* 59:631–637.

Collias, N. E. 1984. Egg measurements and coloration throughout life in the Village Weaver, *Ploceus cucullatus.* In *Proceedings of the 5th Pan-African Ornithological Congress, Held in Malawi in 1980*, ed. J. Ledger. Johannesburg: South African Ornithological Society, 461–473.

Collias, N. E., and E. C. Collias. 1984. *Nest building and bird behavior.* Princeton: Princeton University Press.

Cone, T. 1993. The sounds of silence: Midwestern interloper takes over nests, threatens to kill off California songbirds. *Mercury News*, Oct. 1, 1A.

Conover, M. R. 1985. Foreign objects in bird nests. *Auk* 102:696–700.

Cooke, F., and P. J. Mirsky. 1972. A genetic analysis of Lesser Snow Goose families. *Auk* 89:863–871.

Coon, D. W., and K. A. Arnold. 1977. Origins of Brown-headed Cowbird populations wintering in central Texas. *N. Am. Bird Band.* 2:7–11.

Cooper, C. L., E. L. Troutman, and J. L. Crites. 1973. Helminth parasites of the Brown-headed Cowbird, *Molothrus ater ater*, from Ohio. *Ohio J. Sci.* 73:376–380.

Coulter, M. C. 1980. Stones: An important incubation stimulus for gulls and terns. *Auk* 97:898–899.

Craib, J. 1994. Why do Common Cuckoos resemble raptors? *British Birds* 87:78–79.

Craighead, J. J., and F. C. Craighead. 1956. *Hawks, owls and wildlife.* Harrisburg, Penn.: Stackpole.

Crandall, L. S. 1914. Notes on Costa Rican birds. *Zoologica* 1:325–343.

Crase, F. T., R. W. DeHaven, and P. P. Woronecki. 1972. Movements of Brown-headed Cowbirds banded in the Sacramento Valley, California. *Bird-Banding* 43:197–204.

Cruz, A., and R. W. Andrews. 1989. Observations on the breeding biology of passerines in a seasonally flooded savanna in Venezuela. *Wilson Bull.* 101:62–76.

Cruz, A., T. D. Manolis, and R. W. Andrews. 1990. Reproductive interactions of the Shiny Cowbird *Molothrus bonariensis* and the Yellow-hooded Blackbird *Agelaius icterocephalus* in Trinidad. *Ibis* 132:436–444.

———. 1995. History of Shiny Cowbird *Molothrus bonariensis* brood parasitism in Trinidad and Tobago. *Ibis* 137:317–321.

Cruz, A., T. D. Manolis, and J. W. Wiley. 1985. The Shiny Cowbird: A brood parasite expanding its range in the Caribbean region. *Ornithological Monographs.* 36:607–620.

Cruz, A., and J. W. Wiley. 1989. The decline of an adaptation in the absence of a presumed selection pressure. *Evolution* 43:55–62.

Cruz, A., J. W. Wiley, T. K. Nakamura, and W. Post. 1989. The Shiny Cowbird *Molothrus bonariensis* in the West Indian region: Biogeographical and ecological implications. In *Biogeography of the West Indies: Past, present and future*, ed. C. A. Woods. Gainesville, Fla.: Sandhill Crane Press, 519–540.

Dabbene, R. 1921. Huevos del Pato Pardo (*Heteronetta atricapilla*), en el nido del Carao (*Aramus scolopaceus*). *El Hornero* 2:227.

Daguerre, J. B. 1920. Observaciones sobre los patos *Metopiana peposaca* y *Heteronetta atricapilla. El Hornero* 2:61–62.

———. 1923. Parasitismo del pato, *Heteronetta atricapilla. El Hornero* 3:194–195.

———. 1924. Observaciones sobre la nidificacion de los tordos *Molotrus brevirostris* y *M. badius. El Hornero* 3:285.

Darley, J. A. 1968. The social organization of breeding Brown-headed Cowbirds. Ph.D. diss., University of Western Ontario.

———. 1971. Sex ratio and mortality in the Brown-headed Cowbird. *Auk* 88:560–566.

———. 1978. Pairing in captive Brown-headed Cowbirds (*Molothrus ater*). *Can. J. Zool.* 11:2249–2252.

———. 1982. Territoriality and mating behavior of the male Brown-headed Cowbird. *Condor* 84:15–21.

———. 1983. Territorial behavior of the female Brown-headed Cowbird (*Molothrus ater*). *Can. J. Zool.* 61:65–69.

Davies, N. B., and M. De L. Brooke. 1988. Cuckoos versus Reed Warblers: Adaptations and counteradaptations. *Anim. Behav.* 36:262–284.

———. 1989a. An experimental study of co-evolution between the cuckoo, *Cuculus canorus*, and its hosts: I. Host egg discrimination. *J. Anim. Ecol.* 58:207–224.

———. 1989b. An experimental study of co-evolution between the cuckoo, *Cuculus canorus*, and its hosts: II. Host egg markings, chick discrimination and general discussion. *J. Anim. Ecol.* 58:225–236.

Davis, D. E. 1942. The number of eggs laid by cowbirds. *Condor* 44:10–12.

Davis, L. I. 1972. *A field guide to the birds of Mexico and Central America*. Austin: University of Texas Press.

Davis, R. 1993. Southern Atlantic coast region. *Am. Birds* 47:403–405.

Dawkins, R., and J. R. Krebs. 1979. Arms races between and within species. *Proc. R. Soc. Lond.* B 205:489–511.

Dearborn, D. C. 1996. Video documentation of a Brown-headed Cowbird nestling ejecting an Indigo Bunting nestling from the nest. *Condor* 98:645–649.

Debrot, A. O., and T. G. Prins. 1992. First record and establishment of the Shiny Cowbird in Curaçao. *Caribbean J. of Sci.* 28:104–105.

De Capita, M. E. 1993. Brown-headed Cowbird control on Kirtland's Warbler nesting areas in Michigan, 1972–1993. Abstract 14. Austin, Tex.: North American Workshop on the Ecology and Management of Cowbirds.

de la Peña, M. R. 1983. Habitos parasitarios de algunas especies do aves. *Hornero No. Extraord.* 165–169.

De Schauensee, R. M. 1964. *The birds of Colombia and adjacent areas of South and Central America*. Narberth, Penn.: Livingston Publishing Company.

Dhindsa, M. S. 1983a. Intraspecific nest parasitism in the White-throated Munia. *Notornis* 30:87–92.

———. 1983b. Intraspecific nest parasitism in two species of Indian weaverbirds *Ploceus benghalensis* and *P. manyar*. *Ibis* 125:243–245.

———. 1990. Intraspecific brood parasitism in the Baya Weaverbirds *Ploceus philippinus*. *Bird Behav.* 8:111–113.

Dhindsa, M. S., and P. S. Sandhu. 1988. Response of the Baya Weaverbird (*Ploceus philippinus*) to eggs of the White-throated Munia (*Lonchura malabarica*): Relation to possible incipient brood parasitism. *Zool. Anz.* 220:216–222.

Dhondt, A., R. Eykerman, and J. Huble. 1983. Laying interruptions in tits *Parus* spp. *Ibis.* 125:370–376.

Diamond, J. M. 1975. The island dilemma: Lessons of modern biogeographic studies for the design of nature preserves. *Biol. Conserv.* 7:129–145.

Dickman, C. R., D. H. King, D. C. D. Happold, and M. J. Howell. 1983. Identification of filial relationships of free-living small mammals by ^{35}sulfur. *Aust. J. Zool.* 31:467–474.

Din, N. A. 1992. Hatching synchronization and polymorphism in the eggs of *Ploceus cucullatus* and *P. nigerrimus* with data on nest parasitism. *African J. Ecol.* 30:252–260.

Dixon, J. B. 1934. Records of the nesting of certain birds in eastern California. *Condor* 36:35–36.

Dolan, P. M., and P. L. Wright. 1984. Damaged Western Flycatcher eggs in nest containing Brown-headed Cowbird chicks. *Condor* 86:483–485.

Dolbeer, R. A. 1982. Migration patterns for age and sex classes of blackbirds and starlings. *J. Field Ornithol.* 53:28–46.

Donovan, T., P. Porneluzi, R. L. Clawson, and J. Faaborg. 1993. Variation in Brown-headed Cowbird abundance and parasitism rates in two Missouri landscapes. Abstract 7. Austin, Tex.: North American Workshop on the Ecology and Management of Cowbirds.

Donovan, T., F. Thompson III, and J. Faaborg. In press. Ecological trade-offs and the influence of scale on Brown-headed Cowbird distribution in fragmented and unfragmented Missouri landscapes. In *North American research workshop on the ecology and management of cowbirds.* Austin: University of Texas Press.

Dow, D. D. 1968. Allopreening invitation display of a Brown-headed Cowbird to Cardinals under natural conditions. *Wilson Bull.* 80:494–495.

Drickamer, L. C., and S. H. Vessey. 1982. *Animal Behavior: Concepts, processes, and methods.* Boston: Willard Grant Press.

Du Bois, A. D. 1956. A cowbird incident. *Auk* 73:286.

Duckworth, J. W. 1991. Responses of breeding Reed Warblers *Acrocephalus scirpaceus* to mounts of sparrowhawk *Accipiter nisus*, cuckoo *Cuculus canorus*, and jay *Garrulus glandarius. Ibis* 133:68–74.

Dufour, K. W., and P. J. Weatherhead. 1991. A test of the condition-bias hypothesis using Brown-headed Cowbirds trapped during the breeding season. *Can. J. Zool.* 69:2686–2692.

Dufty, A. M., Jr. 1981. Social organization of the Brown-headed Cowbird, *Molothrus ater*, in New York State. Ph.D. diss., State University of New York.

———. 1982a. Movements and activities of radio-tracked Brown-headed Cowbirds. *Auk* 99:316–327.

———. 1982b. Responses of Brown-headed Cowbirds to simulated conspecific intruders. *Anim. Behav.* 30:1043–1052.

———. 1983. Variation in the egg markings of the Brown-headed Cowbird. *Condor* 85:109–111.

———. 1986. Singing and the establishment and maintenance of dominance hierarchies in captive Brown-headed Cowbirds. *Behav. Ecol. Sociobiol.* 19:49–55.

———. 1988. Flight whistle incorporated in Brown-headed Cowbird song. *Condor* 90:508–510.

———. 1989. Testosterone and survival: A cost of aggressiveness? *Horm. Behav.* 23:185–193.

———. 1993. Testosterone concentrations in males of an atypical species: The brood parasitic Brown-headed Cowbird. *J. Endocrinology* 107:61–72.

———. 1994. Rejection of foreign eggs by Yellow-headed Blackbirds. *Condor* 96:799–801.

Dufty, A. M., Jr., and R. McChrystal. 1992. Vocalizations and copulatory attempts in free-living Brown-headed Cowbirds. *J. Field Ornithol.* 63:16–25.

Dufty, A. M., Jr., and J. C. Wingfield. 1986a. Temporal patterns of circulating LH and steroid hormones in a brood parasite, the Brown-headed Cowbird, *Molothrus ater*: I Males. *J. Zool.* (Lond.). 208:191–203.

———. 1986b. The influences of social cues on the reproductive endocrinology of male Brown-headed Cowbirds: Field and laboratory studies. *Horm. Behav.* 20: 222–234.

———. 1990. Endocrine response of captive male Brown-headed Cowbirds to intrasexual social cues. *Condor* 92:613–620.

Dunning, J. S. 1982. *South American land birds.* Newton Square, Penn.: Harrowood Books.

Eadie, J. M., K. M. Cheng, and C. R. Nichols. 1987. Limitations of tetracycline in tracing multiple maternity. *Auk* 104:330–333.

Eadie, J. M., and H. G. Lumsden. 1985. Is nest parasitism always deleterious to Goldeneyes? *Am. Nat.* 126:859–866.

Earley, C. G. 1991. Brown-headed Cowbird, *Molothrus ater*, seen removing a Chipping Sparrow, *Spizella passerina*, egg. *Can. Field-Nat.* 105:281–282.

Eastzer, D., P. R. Chu, and A. P. King. 1980. The young cowbird: Average or optimal nestling? *Condor* 82:417–425.

Eastzer, D. H., A. P. King, and M. J. West. 1985. Patterns of courtship between cowbird subspecies: Evidence for positive assortment. *Anim. Behav.* 33:30–39.

Eaton, S. W. 1958. A life history study of the Louisiana Water-thrush. *Wilson Bull.* 70:211–236.

Eckerle, K. P., and R. Breitwisch. 1997. Reproductive success of the Northern Cardinal, a large host of Brown-headed Cowbirds. *Condor* 99:169–178.

Edwards, G., E. Hosking, and S. Smith. 1949. Reactions of some passerine birds to a stuffed cuckoo. *Br. Birds* 42:13–19.

Elliott, P. F. 1977. Adaptive significance of cowbird egg distribution. *Auk* 94:590–593.

———. 1978. Cowbird parasitism in the Kansas tallgrass prairie. *Auk* 95:161–167.

———. 1980. Evolution of promiscuity in the Brown-headed Cowbird. *Condor* 82: 138–141.

Erard, C., and G. Armani. 1986. Thoughts on a case of nest parasitism and aid with feeding of *Turdus merula*, *T. philomelos*, and *Erithacus rabecula. Alauda* 54:138–144.

Eriksson, M. O. G., and M. Andersson. 1982. Nest parasitism and hatching success in a population of Goldeneyes *Bucephala clangula. Bird Study* 29:49–54.

Evans, P. G. H. 1988. Intraspecific nest parasitism in the European Starling *Sturnus vulgaris. Anim. Behav.* 36:1282–1294.

Faaborg, J., and W. J. Arendt. 1992. Long-term declines of winter resident warblers in a Puerto Rican dry forest: Which species are in trouble? In *Ecology and conservation of neotropical migrant landbirds,* ed. J. M. Hagan III and D. W. Johnston. Washington, D.C.: Smithsonian Institution Press, 57–63.

Facemire, C. F. 1980. Cowbird parasitism of marsh-nesting Red-winged Blackbirds. *Condor* 82:347–348.

Fankhauser, D. P. 1971. Annual adult survival rates of blackbirds and starlings. *Bird-banding* 42:36–42.

Feekes, F. 1981. Biology and colonial organization of two sympatric caciques, *Cacicus c. cela* and *Cacicus h. haemorrhous* (Icteridae, Aves) in Suriname. *Ardea* 69:83–107.

Ficken, M. S. 1961. Redstarts and cowbirds. *Kingbird* 11:83–85.

———. 1967. Interactions of a crow and fledgling cowbird. *Auk* 84:601–602.

Finch, D. M. 1982. Rejection of cowbird eggs by Crissal Thrashers. *Auk* 99:719–724.

———. 1983. Brood parasitism of the Abert's Towhee: Timing, frequency, and effects. *Condor* 85:355–359.

Finch, D. M., and P. W. Stangel, eds. 1993. Status and management of neotropical migratory birds; 1992 September 21–25; Estes Park, CO, Gen. Tech. Rep. RM-229. Fort Collins, Colo.: U. S. Department of Agriculture, Forest Service, Rocky Mountain Forest and Range Experiment Station.

Fischer, D. H. 1980. Breeding biology of Curve-billed Thrashers and Long-billed Thrashers in southern Texas. *Condor* 82:392–397.

Fleischer, R. C. 1985. A new technique to identify and assess the dispersion of eggs of individual brood parasites. *Behav. Ecol. Sociobiol.* 17:91–99.

———. 1986. Brood parasitism by Brown-headed Cowbirds in a simple host community in eastern Kansas. *Kans. Ornithol. Soc. Bull.* 37:21–29.

Fleischer, R. C., M. T. Murphy, and L. E. Hunt. 1985. Clutch size increase and intraspecific brood parasitism in the Yellow-billed Cuckoo. *Wilson Bull.* 97:125–127.

Fleischer, R. C., and S. I. Rothstein. 1988. Known secondary contact and rapid gene flow among subspecies and dialects in the Brown-headed Cowbird. *Evolution* 42:1146–1158.

Fleischer, R. C., S. I. Rothstein, and L. S. Miller. 1991. Mitochondrial DNA variation indicates gene flow across a zone of known secondary contact between two subspecies of the Brown-headed Cowbird. *Condor* 93:185–189.

Fleischer, R. C., and N. G. Smith. 1992. Giant Cowbird eggs in the nests of two Icterid hosts: The use of morphology and electrophoretic variants to identify individuals and species. *Condor* 94:572–578.

Fleischer, R. C., A. P. Smyth, and S. I. Rothstein. 1987. Temporal and age-related variation in the laying rate of the parasitic Brown-headed Cowbird in the eastern Sierra Nevada, California. *Can. J. Zool.* 65:2724–2730.

Fleming, W. J., and D. R. Petit. 1986. Modified milk carton nest box for studies of Prothonotary Warblers. *J. Field Ornithol.* 57:313–315.

Fletcher, L. B. 1925. A cowbird's maternal instinct. *Bull. Northeastern Bird-Banding Assoc.* 1:22–24.

Folkers, K. L. 1982. Host behavioral defenses to cowbird parasitism. *Kansas Ornithol. Soc. Bull.* 33:32–34.

Folkers, K. L., and P. E. Lowther. 1985. Responses of nesting Red-winged Blackbirds and Yellow Warblers to Brown-headed Cowbirds. *J. Field Ornithol.* 56:175–177.

Fox, A. C. 1940. Observations on the "homing instinct of cowbirds" (*Molothrus ater*). *Bird-Banding* 11:23.

Fox, G. A. 1961. A contribution to the life history of the Clay-colored Sparrow. *Auk* 78:220–224.

Fraga, R. M. 1972. Cooperative breeding and a case of successive polyandry in Bay-winged Cowbirds. *Auk* 89:447–449.

———. 1978. The Rufous-collared Sparrow as host of the Shiny Cowbird. *Wilson Bull.* 90:271–284.

———. 1979. Differences between nestlings and fledglings of Screaming and Bay-winged Cowbirds. *Wilson Bull.* 91:151–154.

———. 1983. The eggs of the parasitic Screaming Cowbird (*Molothrus rufoaxillaris*) and its hosts, the Bay-winged Cowbird (*Molothrus badius*): Is there evidence for mimicry? *J. Ornithol.* 124:187–193.

———. 1984. Bay-winged Cowbirds remove ectoparasites from their brood parasites, the Screaming Cowbirds (*M. rufoaxillaris*). *Biotropica* 16:223–226.

———. 1985. Host-parasite interactions between Chalk-browed Mockingbirds and Shiny Cowbirds. *Ornithol. Monogr.* 36:829–844.

———. 1986. The Bay-winged Cowbird (*Molothrus badius*) and its brood parasites: Interactions, coevolution and comparative efficiency. Ph.D. diss., University of California, Santa Barbara.

———. 1988. Nest sites and breeding success of Baywinged Cowbirds (*Molothrus badius*). *J. Ornithol.* 129:175–183.

———. 1989. Colony sizes and nest trees of Montezuma Oropendolas in Costa Rica. *J. Field Ornithol.* 60:289–295.

———. 1991. The social system of a communal breeder, the Bay-winged Cowbird *Molothrus badius. Ethology* 89:195–210.

———. 1992. Biparental care in Bay-winged Cowbirds *Molothrus badius. Ardea* 80:389–393.

———. 1996. Further evidence of parasitism of Chopi Blackbirds (*Gnorimopsar chopi*) by the specialized Screaming Cowbird (*Molothrus rufoaxillaris*). *Condor* 98:866–867.

Franzreb, K. E. 1988. *Draft Least Bell's Vireo recovery plan.* Portland, Ore.: U.S. Fish and Wildlife Service.

———. 1990. An analysis of options for reintroducing a migratory, native passerine, the endangered Least Bell's Vireo *Vireo bellii pusillus* in the Central Valley, California. *Biol Conserv.* 53:105–123.

Freeberg, T. M. 1996. Assortative mating in captive cowbirds is predicted by social experience. *Anim. Behav.* 52:1129–1142.

Freeman, S., D. F. Gori, and S. Rohwer. 1990. Red-winged Blackbirds and Brown-headed Cowbirds: Some aspects of a host-parasite relationship. *Condor* 92:336–340.

Freeman, S., and R. M. Zink. 1995. A phylogenetic study of the blackbirds based on variation in mitochondrial DNA restriction sites. *Syst. Biol.* 44:409–420.

Fretwell, S. 1973. Why do robins lay blue eggs? *Bird Watch* 1:1–4.

————. 1977. Is the Dickcissel a threatened species? *Am. Birds* 31:923–932.

Friedmann, H. 1925. Notes on the birds observed in the lower Rio Grande Valley of Texas during May, 1924. *Auk* 42:537–554.

————. 1927. A revision of the classification of the cowbirds. *Auk* 44:495–508.

————. 1929a. Social parasitism in birds. *Ann. Rep. Smiths. Inst.* 363–382.

————. 1929b. *The cowbirds: A study in the biology of social parasitism.* Springfield, Ill.: C.C. Thomas Co.

————. 1955. The honey-guides. *Bull. U.S. Natl. Mus.* 208.

————. 1957. The rediscovery of *Tangavius armenti* (Cabanis). *Auk* 74:497–498.

————. 1960. The parasitic weaverbirds. *Bull. U.S. Natl. Mus.* 223.

————. 1963. Host relations of the parasitic cowbirds. *Bull. U.S. Natl. Mus.* 233.

————. 1964. The history of our knowledge of avian brood parasitism. *Centaurus* 10:282–304.

Friedmann, H., and L. F. Kiff. 1985. The parasitic cowbirds and their hosts. *Proc. West. Found. Vert. Zool.* 2:225–302.

Friedmann, H., L. F. Kiff, and S. I. Rothstein. 1977. A further contribution to knowledge of the host relations of the parasitic cowbirds. *Smithsonian Contrib. Zool.* 235: 1–75.

Gaston, A. J. 1976. Brood parasitism by the Pied-crested Cuckoo *(Clamator jacobinus). J. Anim. Ecol.* 45:331–348.

Gates, J. E., and L. W. Gysel. 1978. Avian nest dispersion and fledging success in field-forest ecotones. *Ecology* 59:871–883.

Gibbons, D. W. 1986. Brood parasitism and cooperative nesting in the Moorhen, *Gallenula chloropus. Behav. Ecol. Sociobiol.* 19:221–232.

Gillespie, J. A. 1930. Homing instinct in cowbirds. *Bird-Banding* 1:42.

Giroux, J. F. 1981. Interspecific nest parasitism by Redheads on islands in southeastern Alberta. *Can. J. Zool.* 59:2053–2057.

Given, D. R., and D. A. Norton. 1993. A multivariate approach to assessing threat and for priority setting in threatened species conservation. *Biol. Conserv.* 64:57–66.

Gochfeld, M. 1979a. Begging by nestling Shiny Cowbirds: Adaptive or maladaptive. *Living Bird* 17:41–50.

————. 1979b. Brood parasite and host coevolution: Interactions between Shiny Cowbirds and two species of meadowlarks. *Am. Nat.* 113:855–870.

Goddard, S. V. 1971. Size, migration pattern, structure of fall and early winter blackbird and starling populations in western Oklahoma. *Wilson Bull.* 83:371–382.

Goertz, J. W. 1977. Additional records of Brown-headed Cowbird nest parasitism in Louisiana. *Auk* 94:386–389.

Goguen, C. B., and N. E. Mathews. 1996. Nest desertion by Blue-gray Gnatcatchers in association with Brown-headed Cowbird parasitism. *Anim. Behav.* 52:613–619.

Goldwasser, S., D. Gaines, and S. R. Wilbur. 1980. The Least Bell's Vireo in California: A de facto endangered race. *Am. Birds* 34:742–745.

Goossen, J. P., and S. G. Sealy. 1982. Production of young in a dense nesting population of Yellow Warblers, *Dendroica petechia*, in Manitoba. *Can. Field-Nat.* 96: 189–199.

Gosnell, H. T. 1932. Two cuckoos laying in the same nest without rivalry. *Br. Birds* 26:226.

Gould, S. J., and R. C. Lewontin. 1979. The spandrels of San Marco and the Panglossian paradigm: A critique of the adaptationist programme. *Proc. R. Soc. Lond.* 205: 147–164.

Gowaty, P. A. 1984. Multiple maternity and paternity in single broods of monogamous eastern bluebirds (*Sialia sialis*). *Behav. Ecol. Sociobiol.* 15:91–95.

Graber, J. W., R. R. Graber, and E. L. Kirk. 1983. Illinois birds: Wood warblers. *Biol. Notes Ill. Nat. Hist. Surv.* 118:1–144.

Graham, D. S. 1988. Responses of five host species to cowbird parasitism. *Condor* 90: 588–591.

Graham, D. S., and A. L. A. Middleton. 1989. Conspecific recognition by juvenile Brown-headed Cowbirds. *Bird Behav.* 8:14–22.

Grant, C. H. B. 1911. List of birds collected in Argentina, Paraguay, Bolivia, and southern Brazil, with field notes. *Ibis* 5:80–137.

———. 1912. Notes on some South-American birds. *Ibis* 1:273–280.

Greenwood, P. J. 1980. Mating systems, philopatry and dispersal in birds and mammals. *Anim. Behav.* 28:1140–1162.

Griffith, J. T., and J. C. Griffith. In Press. Cowbird control and the endangered Least Bell's Vireo: A management success story. In *North American Research Workshop on the Ecology and Management of Cowbirds.* Austin: University of Texas Press.

Grinnell, J., and A. H. Miller. 1944. The distribution of the birds of California. *Pacific Coast Avifauna* 27.

Grzybowski, J. A., R. B. Clapp, and J. T. Marshall, Jr. 1986. History and current population status of the Black-capped Vireo in Oklahoma. *Am. Birds* 40:1151–1161.

Gustafson, E. J., and T. R. Crow. 1994. Modeling the effects of forest harvesting on landscape structure and the spatial distribution of cowbird brood parasitism. *Landscape Ecology* 9:237–248.

Hagan, J. M, III, and D. W. Johnston, eds. 1992. *Ecology and conservation of neotropical migrant landbirds.* Washington, D.C.: Smithsonian Institution Press.

Hamas, M. J. 1980. Eastern Kingbird rejection of Brown-headed Cowbird eggs. *Jack-Pine Warbler* 58:33.

Hamilton, W. J., and G. H. Orians. 1965. Evolution of brood parasitism in altricial birds. *Condor* 67:361–382.

Hanka, L. R. 1979. Choice of host nest by the Brown-headed Cowbird in Colorado and Wyoming. *Condor* 81:436–437.

Hann, D. C., and R. C. Fleischer. 1995. DNA fingerprint similarity between female and juvenile Brown-headed Cowbirds trapped together. *Anim. Behav.* 49:1577–1580.

Hann, D. C., and J. S. Hatfield. 1995. Parasitism at the landscape scale: Cowbirds prefer forests. *Conserv. Biol.* 9:1415–1424.

Hann, H. W. 1937. Life history of the Oven-bird in southern Michigan. *Wilson Bull.* 49:145–237.

———. 1941. The cowbird at the nest. *Wilson Bull.* 59:173–174.

———. 1947. An Oven-bird incubates a record number of eggs. *Wilson Bull.* 59:173–174.

Hanna, W. C. 1918. First occurrence of the Dwarf Cowbird in the San Bernardino Valley, California. *Condor* 20:211–212.

Haramis, G. M., W. G. Alliston, and M. E. Richmond. 1983. Dump nesting in the Wood Duck traced by tetracycline. *Auk* 100:729–730.

Hardy, J. W. 1970. Duplex nest construction by Hooded Oriole circumvents cowbird parasitism. *Condor* 72:491.

Harms, K. E., L. D. Beletsky, and G. H. Orians. 1991. Conspecific nest parasitism in three species of New World Blackbirds. *Condor* 93:967–974.

Harris, J. H. 1991. Effects of brood parasitism by Brown-headed Cowbirds on Willow Flycatcher nesting success along the Kern River, California. *West. Birds* 22:13–26.

Harrison, C. 1978. *A field guide to the nests, eggs and nestlings of North American birds.* Glasgow: William Collins Sons & Co. Ltd.

Harrison, C. J. O. 1963. Interspecific preening display by the Rice Grackle, *Psomocolax oryzivorus. Auk* 80:373–374.

———. 1968. Egg mimicry in British cuckoos. *Bird Study* 15:22–28.

Harrison, H. H. 1952. Notes on a cowbird removing an egg at dawn. *Redstart* 19:46–48.

———. 1973. The cowbird strikes at dawn. *Nat. Wildl.* 11:33–37.

Hatch, S. A. 1983. Nestling growth relationships of Brown-headed Cowbirds and Dickcissels. *Wilson Bull.* 95:669–671.

Haverschmidt, F. 1965. *Molothrus bonariensis* parasitizing *Fluvicola pica* and *Arundinicola leucocephala* in Surinam. *Auk* 82:508–509.

———. 1966. The eggs of the Giant Cowbird. *Bull. Br. Ornithol. Club.* 86:144–147.

Henke, M., and C. P. Stone. 1978. Value of riparian vegetation to avian populations along the Sacramento River system. In *Strategies for protection and management of floodplain wetlands and their riparian ecosystems.* USDA Forest Serv. Gen. Tech. Report WO-12, 228–235.

Hergenrader, G. L. 1962. The incidence of nest parasitism by the Brown-headed Cowbird (*Molothrus ater*) on roadside nesting birds in Nebraska. *Auk* 79:85–88.

Heusmann, H. W., R. Bellville, and R. G. Burrell. 1980. Further observation on dump nesting by Wood Ducks. *J. Wildl. Manage.* 44:908–915.

Hickey, J. J. 1940. Territorial aspects of the American Redstart. *Auk* 57:255–256.

Hicks, L. E. 1934. A summary of cowbird host species in Ohio. *Auk* 51:385–386.

Higuchi, H. 1989. Responses of the Bush Warbler *Cettia diphone* to artificial egg of *Cuculus* cuckoos in Japan. *Ibis* 131:94–98.

Hill, D. P., and S. G. Sealy. 1994. Desertion of nests parasitized by cowbirds: Have Clay-coloured sparrows evolved an anti-parasite defence? *Anim. Behav.* 48:1063–1070.

Hill, R. A. 1976a. Host-parasite relationships of the Brown-headed Cowbird in a prairie habitat of west-central Kansas. *Wilson Bull.* 88:555–565.

———. 1976b. Sex ratio and sex determination of immature Brown-headed Cowbirds. *Bird Banding* 47:112–114.

Hobson, K. A., M. L. Bouchart, and S. G. Sealy. 1988. Responses of naive Yellow Warblers to a novel nest predator. *Anim. Behav.* 36:1823–1830.

Hobson, K. A., and S. G. Sealy. 1987. Cowbird egg buried by a Northern Oriole. *J. Field Ornithol.* 58:222–224.

———. 1989. Responses of Yellow Warblers to the threat of cowbird parasitism. *Anim. Behav.* 38:510–519.

Hoffman, W., and G. E. Woolfenden. 1986. A fledgling Brown-headed Cowbird specimen from Pinellas County. *Florida Field Naturalist* 14:18–20.

Hofslund, P. B. 1957. Cowbird parasitism of the Northern Yellow-throat. *Auk* 74:42–48.

Höhn, E. O. 1962. A possible endocrine basis of brood parasitism. *Ibis* 104:418–421.

———. 1972. Nest parasitism at McConnell River, N.W.T. *Can. Field-Nat.* 86:369–372.

———. 1975. Notes on Black-headed Ducks, Painted Snipe, and Spotted Tinamous. *Auk* 92:566–575.

Holcomb, L. C. 1968. Reaction of Mourning Doves to cowbird eggs. *Wilson Bull.* 80:105.

———. 1972. Nest success and age-specific mortality in Traill's Flycatchers. *Auk* 89:837–841.

Holford, K. C., and D. D. Roby. 1993. Factors limiting fecundity of captive Brown-headed Cowbirds. *Condor* 95:536–545.

Holmes, B. 1993. An avian arch-villain gets off easy. *Science* 262:1514–1515.

Holmes, R. T., T. W. Sherry, P. P. Marra, and K. E. Petit. 1992. Multiple brooding and productivity of a neotropical migrant, the Black-throated Blue Warbler (*Dendroica caerulescens*), in an unfragmented temperate forest. *Auk* 109:321–333.

Hoover, J. P., and M. C. Brittingham. 1993. Regional variation in cowbird parasitism of Wood Thrushes. *Wilson Bull.* 105:228–238.

Hoover, J. P., M. C. Brittingham, and L. J. Goodrich. 1995. Effects of forest patch size on nesting success of Wood Thrushes. *Auk* 112:146–155.

Howell, S. N. G., and S. Webb. 1995. *The birds of Mexico and northern Central America.* New York: Oxford University Press.

Hoy, G., and J. Ottow. 1964. Biological and oological studies of the Molothrine cowbirds (Icteridae) of Argentina. *Auk* 81:186–203.

Hudson, W. H. 1874. Notes on the procreant instincts of the three species of *Molothrus* found in Buenos Ayres. *Proc. Zool. Soc. Lond.* 1874:153–174.

———. 1920. *Birds of La Plata*, vol. 1. London: J. M. Dent & Sons Ltd.

Hunter, J. E. 1994. Further observations of head-down displays by Brown-headed Cowbirds. *West. Birds* 25:63–65.

Jackson, G. D. 1992. Central southern region. *Am. Birds* 46:102–107.

———. 1993. Central southern region. *Am. Birds* 47:104–108.

Jackson, N. H., and D. D. Roby. 1992. Fecundity and egg-laying patterns of captive yearling Brown-headed Cowbirds. *Condor* 94:585–589.

Jacobs, J. W. 1938. The Eastern Cowbird vs. the Kentucky Warbler. *Auk* 55:260–262.

James, F. C., D. A. Wiedenfeld, and C. E. McCulloch. 1992. Trends in breeding populations of warblers: Declines in the southern highlands and increases in the lowlands. In *Ecology and conservation of neotropical migrant landbirds*, ed. J. M. Hagan III and D. W. Johnston. Washington, D.C.: Smithsonian Institution Press, 43–56.

Jenner, E. 1788. Observations on the natural history of the cuckoo. *Philos. Trans. R. Soc. London* 78:219–235.

Jensen, R. A. C., and M. K. Jensen. 1969. On breeding biology of southern African cuckoos. *Ostrich* 40:163–181.

Jobin, B., and J. Picman. 1994. Artificial nest parasitized by a Brown-headed Cowbird, *Molothrus ater. Can. Field-Nat.* 108:482–484.

Johnson, A. W. 1967. *The birds of Chile and adjacent regions of Argentina, Bolivia and Peru*, vol. II. Buenos Aires: Platt Establecimientos Gráficos.

Johnson, D. M., G. L. Stewart, M. Corley, R. Ghrist, J. Hagner, A. Ketterer, B. McDonnell, W. Newsom, E. Owen, and P. Samuels. 1980. Brown-headed Cowbird (*Molothrus ater*) mortality in an urban winter roost. *Auk* 97:299–320.

Johnson, L. G. 1950. Yellow Warbler nesting notes. *Audubon Bull.* 74:1–2.

Johnson, R. G., and S. A. Temple. 1986. Assessing habitat quality for birds nesting in fragmented tallgrass prairies. In *Wildlife 2000: Modeling habitat relationships of terrestrial vertebrates*, ed. J. Verner, M. L. Morrison, and C. J. Ralph. Madison: University of Wisconsin Press, 245–249.

———. 1990. Nest predation and brood parasitism of tallgrass prairie birds. *J. Wildl. Manage.* 54:106–111.

Johnson, R. R., and B. Roer. 1968. Changing status of the Bronzed Cowbird in Arizona. *Condor* 70:183.

Johnson, R. R., L. T. Haight, and J. M. Simpson. 1977. Endangered species vs. endangered habitats: A concept. In *Importance, preservation and management of riparian habitat*. U.S.D.A. Forest Serv. Gen. Tech. Rep. RM-43, 68–79.

Johnson, T. W. 1993. Involving the private sector in Georgia's conservation initiatives for neotropical birds. In *Status and management of neotropical migratory birds*, ed. D. M. Finch and P. W. Stangel. U.S.D.A. Forest Serv. Gen. Tech. Rep. RM-229, 45–46.

Johnsrude, I. S., D. M. Weary, L. M. Ratcliffe, and R. G. Weisman. 1994. Effect of motivational context on conspecific song discrimination by Brown-headed Cowbirds (*Molothrus ater*). *J. Comp. Psych.* 108:172–178.

Jones, R. E., and A. S. Leopold. 1967. Nesting interference in a dense population of Wood Ducks. *J. Wildl. Manage.* 31:221–228.

Joyner, D. E. 1975. Nest parasitism and brood-related behavior of the Ruddy Duck (*Oxyura jamaicensis rubida*). Ph.D. diss., University of Nebraska.

———. 1976. Effects of interspecific nest parasitism by Redheads and Ruddy Ducks. *J. Wildl. Manage.* 40:33–38.

———. 1983. Parasitic egg laying in Redheads and Ruddy Ducks in Utah: Incidence and success. *Auk* 100:717–725.

Kattan, G. H. 1993. Extraordinary annual fecundity of Shiny Cowbirds at a tropical locality and its energetic trade-off. Abstract 9. Austin, Tex.: North American Workshop on the Ecology and Management of Cowbirds.

———. 1995. Mechanisms of short incubation period in brood-parasitic cowbirds. *Auk* 112:335–342.

———. 1996. Growth and provisioning of Shiny Cowbird and House Wren host nestlings. *J. Field Ornithol.* 67:434–441.

Kauffman, J. B., and W. C. Krueger. 1984. Livestock impacts on riparian ecosystems and streamside management implications: A review. *J. Range Manage.* 37:430–438.

Kelly, C. 1987. A model to explore the rate of spread of mimicry and rejection in hypothetical populations of cuckoos and their hosts. *J. Theor. Biol.* 125:283–299.

Kemal, R. E., and S. I. Rothstein. 1988. Mechanisms of avian egg recognition: Adaptive responses to eggs with broken shells. *Anim. Behav.* 36:175–183.

Kempenaers, B., R. Pinxten, and M. Eens. 1995. Intraspecific brood parasitism in two tit *Parus* species: Occurrence and responses to experimental parasitism. *J. Avian Biology* 26:114–120.

Kennard, J. H. 1978. A method of local limitation of brood parasitism by the Brown-headed Cowbird. *North Amer. Bird Bander* 3:100–101.

Kennedy, E. D. 1991. Determinate and indeterminate egg-laying patterns: A review. *Condor* 93:106–124.

King, A. P. 1979. Variables affecting parasitism in the North American cowbird (*Molothrus ater*). Ph.D. diss., Cornell University.

King, A. P., and M. J. West. 1977. Species identification in the North American cowbird: Appropriate responses to abnormal song. *Science* 195:1002–1004.

———. 1983. Dissecting cowbird song potency: Assessing a song's geographic identity and relative appeal. *Z. Tierpsychol.* 63:37–50.

———. 1988. Searching for the functional origins of song in eastern Brown-headed Cowbirds, *Molothrus ater ater*. *Anim. Behav.* 36:1575–1588.

King, A. P., M. J. West, and D. H. Eastzer. 1980. Song structure and song development as potential contributors to reproductive isolation in cowbirds (*Molothrus ater*). *J. Comp. Physiol. Psychol.* 94:1028–1036.

King, J. R. 1954. Victims of the Brown-headed Cowbird in Whitman County, Washington. *Condor* 56:150–154.

———. 1973a. Energetics of reproduction in birds. In *Breeding biology of birds*, ed. D. S. Farner. Washington, D.C.: National Academy of Sciences, 78–107.

———. 1973b. Reproductive relationships of the Rufous-collared Sparrow and the Shiny Cowbird. *Auk* 90:19–34.

Kingery, H. E. 1977. Mountain west. *Am. Birds* 36:1166–1170.

Klaas, E. E. 1975. Cowbird parasitism and nesting success in the Eastern Phoebe. *Occas. Pap. Mus. Nat. Hist. Univ. Kansas* 41:1–18.

Klein, N. K., and K. V. Rosenberg. 1986. Feeding of Brown-headed Cowbird *(Molothrus ater)* fledglings by more than one "host" species. *Auk* 103:213–214.

Klimaitis, J. F. 1971. Estudio descriptivo de una colonia de Tordos Varilleros *(Agelaius ruficapillus)*. *El Hornero* 11:193–202.

Knight, R. L., and A. W. Erickson. 1977. Objects incorporated within clutches of the Canada Goose. *West. Birds* 8:108.

Kuroda, N. 1966. A note on the problem of hawk-mimicry in cuckoos. *Jpn. J. Zool.* 1966:173–181.

Kuschel. 1896. Letter to the editor. *Ibis*, 7th ser., 585.

Lack, D. 1968. *Ecological adaptation for breeding in birds*. London: Methuen & Co. Ltd.

Langridge, H. P. 1992. Florida region. *Am. Birds* 46:412–415.

———. 1993. Florida region. *Am. Birds* 47:406–408.

Lank, D. B., P. Mineau, and R. F Rockwell. 1989. Intraspecific nest parasitism and extra-pair copulation in Lesser Snow Geese. *Anim. Behav.* 37:74–89.

Lanyon, S. M. 1992. Interspecific brood parasitism in blackbirds (Icterinae): A phylogenetic perspective. *Science* 255:77–79.

Larsson, K., H. Tegelström, and L. Forslund. 1995. Intraspecific nest parasitism and adoption of young in the Barnacle Goose: Effects on survival and reproductive performance. *Anim. Behav.* 50:1349–1360.

Laskey, A. R. 1944. Cowbird data from a banding station. *Migrant* 15:17–18.

———. 1950. Cowbird behavior. *Wilson Bull.* 62:157–174.

Law, J. E. 1910. Cowbird again noted in Los Angeles County. *Condor* 12:174.

Lawrence, L. deK. 1953. Nesting life and behaviour of the Red-eyed Vireo. *Can. Field-Nat.* 67:47–77.

Laymon, S. A. 1987. Brown-headed Cowbirds in California: Historical perspectives and management opportunities in riparian habitats. *West. Birds* 18:63–70.

Lea, R. B. 1942. A study of the nesting habits of the Cedar Waxwing. *Wilson Bull.* 54:225–237.

Leach, S. W. 1994. Mallard parasitizes Sharp-tailed Grouse nest. *Blue Jay* 52:144.

Leathers, C. L. 1956. Incubating American Robin repels female Brown-headed Cowbird. *Wilson Bull.* 68:68.

Lehrman, D. S. 1958. Effect of female sex hormones on incubation behavior in the Ring Dove *(Streptopelia risoria)*. *J. Comp. Physiol. Psychol.* 51:142–145.

Leopold, F. 1951. A study of nesting Wood Ducks in Iowa. *Condor* 53:209–220.

Lerkelund, H. E., A. Moksnes, E. Røskaft, and T. H. Ringsby. 1993. An experimental test of optimal clutch size of the Fieldfare; with a discussion on why brood parasites remove eggs when they parasitize a host species. *Ornis Scandinavica* 24:95–102.

Linz, G. M., and S. B. Bolin. 1982. Incidence of Brown-headed Cowbird parasitism on Red-winged Blackbirds. *Wilson Bull.* 94:93–95.

Lombardo, M. P., H. W. Power, P. C. Stouffer, L. C. Romagnano, and A. S. Hoffenberg. 1989. Egg removal and intraspecific brood parasitism in the European Starling (*Sturnus vulgaris*). *Behav. Ecol. Sociobiol.* 24:217–223.

Lotem, A., H. Nakamura, and A. Zahavi. 1992. Rejection of cuckoo eggs in relation to host age: A possible evolutionary equilibrium. *Behav. Ecol.* 3:128–132.

Low, J. B. 1941. Nesting of the Ruddy Duck in Iowa. *Auk* 58:506–517.

Lown, B. A. 1980. Reproductive success of the Brown-headed Cowbird: A prognosis based on breeding bird census data. *Am. Birds* 34(1):15–17.

Lowther, P. E. 1975. Geographic and ecological variation in the family Icteridae. *Wilson Bull.* 87:481–495.

———. 1977. Old cowbird breeding records from the Great Plains region. *Bird-Banding* 48:358–369.

———. 1979. Nest selection by Brown-headed Cowbirds. *Wilson Bull.* 91:118–122.

———. 1983. Chickadee, thrasher, and other cowbird hosts from northwest Iowa. *J. Field Ornithol.* 54:414–417.

———. 1986. Breeding biology of House Sparrows: Patterns of intra-clutch variation in egg size. In *Proc. of Gen. Meetings of Granivorous Birds*, ed. J. Pinowski and J. D. Summers. INTECOL, Ottawa, Canada, 137–149.

———. 1988. Spotting pattern of the last laid egg of the House Sparrow. *J. Field Ornithol.* 59:51–54.

———. 1993. Brown-headed Cowbird (*Molothrus ater*). In *The birds of North America*, no. 47, ed. A. Poole and F. Gill. The Academy of Natural Sciences, Philadelphia, and the American Ornithologists' Union, Washington, D.C., 1–23.

———. 1995. Bronzed Cowbird (*Molothrus aeneus*). In *The birds of North America*, no. 144, ed. A. Poole and F. Gill. The Academy of Natural Sciences, Philadelphia, and the American Ornithologists' Union, Washington, D. C., 1–14.

Lyon, B. E., L. D. Hamilton, and M. Magrath. 1992. The frequency of conspecific brood parasitism and the pattern of laying determinancy in Yellow-headed Blackbirds. *Condor* 94:590–597.

MacWhirter, R. B. 1989. On the rarity of intraspecific brood parasitism. *Condor* 91:485–492.

Mallory, M. L., and P. J. Weatherhead. 1993. Responses of nesting mergansers to parasitic Common Goldeneye eggs. *Anim. Behav.* 46:1226–1228.

Manwell, C. L., and C. M. A. Baker. 1975. Molecular genetics of avian proteins: XIII. Protein polymorphism in three species of Australian passerines. *Australian J. Biol. Sci.* 28:545–557.

Mariani, C. L., C. G. Earley, and C. McKinnon. 1993. Early nesting by the American Goldfinch, *Carduelis tristis*, and subsequent parasitism by the Brown-headed Cowbird, *Molothrus ater*, in Ontario. *Can. Field-Nat.* 107:349–350.

Mark, D., and B. J. Stutchbury. 1994. Response of a forest-interior songbird to the threat of cowbird parasitism. *Anim. Behav.* 47:275–280.

Martin, K. 1984. Intraspecific nest parasitism in Willow Ptarmigan. *J. Field Ornithol.* 55:250–251.

Marvil, R. E., and A. Cruz. 1989. Impact of Brown-headed Cowbird parasitism on the reproductive success of the Solitary Vireo. *Auk* 106:476–480.

Mason, P. 1980. Ecological and evolutionary aspects of host selection in cowbirds. Ph.D. diss., University of Texas.

———. 1985. The nesting biology of some passerines of Buenos Aires, Argentina. *Orn. Monogr.* 36:954–972.

———. 1986a. Brood parasitism in a host generalist, the Shiny Cowbird: I. The quality of different species as hosts. *Auk* 103:52–60.

———. 1986b. Brood parasitism in a host generalist, the Shiny Cowbird: II. Host selection. *Auk* 103:61–69.

———. 1987. Pair formation in cowbirds: Evidence found for Screaming but not Shiny Cowbirds. *Condor* 89:349–356.

Mason, P., and S. I. Rothstein. 1986. Coevolution and avian brood parasitism: Cowbird eggs show evolutionary response to host discrimination. *Evolution* 40:1207–1214.

———. 1987. Crypsis versus mimicry and the color of Shiny Cowbird eggs. *Am. Nat.* 130:161–167.

Matteson, R. E. 1970. Bronzed Cowbird taken in Florida. *Auk* 87:588.

May, R. M., and S. K. Robinson. 1985. Population dynamics of avian brood parasitism. *Am. Nat.* 126:475–494.

Mayfield, H. F. 1960. *The Kirtland's Warbler.* Bloomfield Hill, Mich.: Cranbrook Institute of Science.

———. 1961a. Cowbird parasitism and the population of the Kirtland's Warbler. *Evolution* 15:174–179.

———. 1961b. Vestiges of a proprietary interest in nests by the Brown-headed Cowbird parasitizing the Kirtland's Warbler. *Auk* 78:162–166.

———. 1965a. Chance distribution of cowbird eggs. *Condor* 67:257–263.

———. 1965b. The Brown-headed Cowbird with old and new hosts. *Living Bird* 4:13–28.

———. 1973. Census of Kirtland's Warbler in 1972. *Auk* 90:684–685.

———. 1977. Brown-headed Cowbird: Agent of extermination? *Am. Birds* 31:107–113.

Mayr, E. 1974. Behavior programs and evolutionary strategies. *Am. Sci.* 62:650–659.

McGeen, D. S. 1972. Cowbird-host relationships. *Auk* 89:360–380.

McKay, C. R. 1994. Brown-headed Cowbird in Strathclyde: New to Britain and Ireland. *Br. Birds* 87:284–288.

McLaughlin, C. L., and D. Grice. 1952. The effectiveness of large-scale erection of Wood Duck boxes as a management procedure. *Trans. of North American Wildlife Conference* 17:242–259.

McLean, I. G. 1987. Response to a dangerous enemy: Should a brood parasite be mobbed? *Ethology* 75:235–245.

McLean, I. G., and J. R. Waas. 1987. Do cuckoo chicks mimic the begging calls of their hosts? *Anim. Behav.* 35:1896–1898.

McNicholl, M. 1968. Cowbird egg in Mourning Dove nest. *Blue Jay* 26:22–23.

Mengel, R. M., and M. A. Jenkinson. 1970. Parasitism by the Brown-headed Cowbird on a Brown Thrasher and a catbird. *Wilson Bull.* 82:74–78.

Mermoz, M. E., and J. C. Reboreda. 1994. Brood parasitism of the Shiny Cowbird, *Molothrus bonariensis*, on the Brown-and-Yellow Marshbird, *Pseudoleistes virescens*. *Condor* 96:716–721.

———. 1996. New host for a specialized brood parasite, the Screaming Cowbird. *Condor* 98:630–632.

Merrill, J. C. 1877. Notes on *Molothrus aeneus. Wagl. Bull. Nuttall Ornithological Club* 2:85–87.

Middleton, A. L. A. 1977. Effect of cowbird parasitism on American Goldfinch nesting. *Auk* 94:304–307.

———. 1991. Failure of Brown-headed Cowbird parasitism in nests of the American Goldfinch. *J. Field Ornithol.* 62:200–203.

Middleton, A. L. A., and D. M. Scott. 1965. Testicular response to an increased photoperiod in the Brown-headed Cowbird. *Auk* 82:504–505.

Miller, L. E. 1917. Field notes on *Molothrus bonariensis* and *M. badius. Bull. Am. Mus. Nat. Hist.* 37:579–592.

Mills, A. M. 1987. Size of host egg and egg size in Brown-headed Cowbirds. *Wilson Bull.* 99:490–491.

———. 1988. Unsuitability of Tree Swallows as hosts to Brown-headed Cowbirds. *J. Field Ornithol.* 59:331–333.

Mills, W. J. 1957. Summer range of cowbirds in the Atlantic provinces. *Nova Scotia Museum of Science Newsletter* 2:25–27.

Mock, D. W. 1983. On the study of avian mating systems. In *Perspectives in ornithology*, ed. A. H. Brush and G. A. Clark, Jr. Cambridge: Cambridge University Press, 55–91.

Moksnes, A. 1992. Egg recognition in chaffinches and bramblings. *Anim. Behav.* 44:993–995.

Moksnes, A., and E. Røskaft. 1987. Cuckoo host interactions in Norwegian mountain areas. *Ornis Scandinavica* 18:168–172.

———. 1988. Responses of Fieldfares *Turdus pilaris* and Bramblings *Fringilla montifringilla* to experimental parasitism by the cuckoo *Cuculus canorus. Ibis* 130:535.

Moksnes, A., E. Røskaft, A. T. Braa, L. Korsnes, H. M. Lampe, and H. C. Pedersen. 1990. Behavioural responses of potential hosts towards artificial cuckoo eggs and dummies. *Behaviour* 116:64–89.

Moksnes, A., E. Røskaft, and L. Korsnes. 1993. Rejection of cuckoo (*Cuculus canorus*) eggs by Meadow Pipits (*Anthus pratensis*). *Behav. Ecol.* 4:120–127.

Møller, A. P. 1987. Intraspecific nest parasitism and anti-parasite behaviour in swallows, *Hirundo rustica. Anim. Behav.* 35:247–254.

———. 1989. Nest site selection across field-woodland ecotones: The effect of nest predation. *Oikos* 56:240–246.

Monroe, B. L., Jr. 1957. A breeding record of the cowbird for Florida. *Florida Naturalist* 30:124.

Morgan, K. A., and J. E. Gates. 1982. Bird population patterns in forest edge and strip vegetation at Remington Farms, Maryland. *J. Wildl. Manage.* 46:933–944.

Mumford, R. E. 1952. Bell's Vireo in Indiana. *Wilson Bull.* 64:224–233.

Mundy, P. J. 1973. Vocal mimicry of their hosts by nestlings of the Great Spotted Cuckoo and Striped Crested Cuckoo. *Ibis* 115:602–604.

Murphy, M. T. 1986. Brood parasitism of Eastern Kingbirds by Brown-headed Cowbirds. *Auk* 103:626–628.

Nakamura, T. K. 1995. A test of the nest defense paradigm: Are aggressive birds less parasitized? Master's thesis, University of Colorado.

Nakamura, T. K., and A. Cruz. 1993. The ecology of egg puncture by the Shiny Cowbird (*Molothrus bonariensis*) in southwestern Puerto Rico. Austin, Tex.: North American Workshop on the Ecology and Management of Cowbirds.

———. In press. The ecology of egg puncture by the Shiny Cowbird (*Molothrus bonariensis*) in southwestern Puerto Rico. In *North American research workshop on the ecology and management of cowbirds*. Austin: University of Texas Press.

Navas, J. R. 1970. Nuevos registros de aves para la Patagonia. *Neotropica* 16: 11–16.

Navas, J. R., T. Narosky, N. A. Bó, and J. C. Chébez. 1991. *Lista patrón de los nombres comunes de las aves Argentinas*. Argentina: Asociación Ornitológica del Plata.

Neff, J. A. 1926. Misplaced foster devotion. *Bird Lore* 28:334–335.

———. 1943. Homing instinct in the Dwarf Cowbird in Arizona. *Bird-Banding* 14: 1–6.

Neudorf, D. L., and S. G. Sealy. 1992. Reactions of four passerine species to threats of predation and cowbird parasitism: Enemy recognition or generalized responses? *Behaviour* 123:84–105.

———. 1994. Sunrise nest attentiveness in cowbird hosts. *Condor* 96:162–169.

Newman, G. A. 1970. Cowbird parasitism and nesting success of Lark Sparrows in southern Oklahoma. *Wilson Bull.* 82:304–309.

Nice, M. M. 1939. Observations on the behavior of a young cowbird. *Wilson Bull.* 51: 233–239.

———. 1943. Studies in the life history of the Song Sparrow: II. *Trans. Linn. Soc., New York* 6:1–328.

———. 1949. The laying rhythm of cowbirds. *Wilson Bull.* 61:231–234.

———. 1953. The earliest mention of territory. *Condor* 55:316–317.

Nickell, W. P. 1955. Notes on cowbird parasitism on four species. *Auk* 72:88–92.

———. 1958. Brown-headed Cowbird fledged in nest of catbird. *Wilson Bull.* 70: 286–287.

Nicolai, J. 1974. Mimicry in parasitic birds. *Sci. Am.* 231:92–98.

Nielsen, K. 1956. A new Danish breeding result: Breeding of the combasou (*Hypochera chalybeata chalybeata*). *Avicultural Mag.* 62:11–13.

Niles, D. M. 1970. A record of clutch size and breeding in New Mexico for the Bronzed Cowbird. *Condor* 72:500–501.

Nolan, V. A. L., Jr. 1978. The ecology and behavior of the Prairie Warbler (*Dendroica discolor*). *Ornithol. Monogr.* 26.

Nolan, V. A. L., Jr., and C. F. Thompson. 1975. The occurrence and significance of anomalous reproductive activities in two North American non-parasitic cuckoos. *Ibis* 117:496–503.

Norman, R. F., and R. J. Robertson. 1975. Nest-searching behavior in the Brown-headed Cowbird. *Auk* 92:610–611.

Norris, R. T. 1947. The cowbirds of Preston Frith. *Wilson Bull.* 59:83–103.

Nudds, T. D. 1980. Canvasback tolerance of Redhead parasitism: An observation and an hypothesis. *Wilson Bull.* 92:414.

Oberholser, H. C. 1974. *The bird life of Texas*. Austin: University of Texas Press.

O'Conner, R. J., and J. Faaborg. 1992. The relative abundance of the Brown-headed Cowbird (*Molothrus ater*) in relation to exterior and interior edges in forests of Missouri. *Trans. Missouri Acad. of Sci.* 26:1–9.

O'Connor, R. J. 1992. Population variation in relation to migrancy status in some North American birds. In *Ecology and conservation of neotropical migrant landbirds*, ed. J. M. Hagan III and D. W. Johnston. Washington, D.C.: Smithsonian Institution Press, 64–74.

Ogden, J. C. 1992. Florida region. *Am. Birds* 46:255–257.

Øien, I. J., M. Honza, A. Moksnes, and E. Røskaft. 1996. The risk of parasitism in relation to the distance from Reed Warbler nests to cuckoo perches. *J. Anim. Ecol.* 65:147–153.

O'Loghlen, A. L., and S. I. Rothstein. 1993. An extreme example of delayed vocal development: Song learning in a population of wild Brown-headed Cowbirds. *Anim. Behav.* 46:293–304.

————. 1995. Culturally correct song dialects are correlated with male age and female song preferences in wild populations of Brown-headed Cowbirds. *Behav. Ecol. Sociobiol.* 36:251–259.

Olrog, C. C. 1963. *Lista y distribucion de las aves Argentinas*. Tucuman: Universidad Nacional de Tucuman, Instituto Miguel Lillo.

Olson, A. L. 1943. Cowbird carrying away and eating a bird's egg in the evening. *Wilson Bull.* 55:195.

Oniki, Y. 1979. Is nesting success of birds low in the tropics? *Biotropica* 11:60–69.

Orians, G. H. 1985. *Blackbirds of the Americas*. Seattle: University of Washington Press.

Orians, G. H., C. E. Orians, and K. J. Orians. 1977. Helpers at the nest in some Argentine blackbirds. In *Evolutionary Ecology*, ed. B. Stonehouse and C. Perrins. Baltimore: University Park Press, 137–151.

Orians, G. H., E. Røskaft, and L. D. Beletsky. 1989. Do Brown-headed Cowbirds lay their eggs at random in the nests of Red-winged Blackbirds? *Wilson Bull.* 101:599–605.

Ortega, C. P. 1991. The ecology of blackbird/cowbird interactions in Boulder County, Colorado. Ph.D. diss., University of Colorado.

Ortega, C. P., and A. Cruz. 1988. Mechanisms of egg acceptance in marsh dwelling blackbirds. *Condor* 90:349–358.

———. 1991. A comparative study of cowbird parasitism in Yellow-headed Blackbirds and Red-winged Blackbirds. *Auk* 108:16–24.

———. 1992a. Differential growth patterns of nestling Brown-headed Cowbirds and Yellow-headed Blackbirds. *Auk* 109:368–376.

———. 1992b. Gene flow of the *obscurus* race into the Brown-headed Cowbird population of north central Colorado. *J. Field Ornithol.* 63:311–317.

Ortega, C. P., J. C. Ortega, S. A. Backensto, and C. A. Rapp. 1996. Improved methods for aging second-year and after-second-year male Brown-headed Cowbirds (*Molothrus ater*). *J. Field Ornithol.* 67:542–548.

Ortega, C. P., J. C. Ortega, and A. Cruz. 1994. Use of artificial Brown-headed Cowbird eggs as a potential management tool in deterring parasitism. *J. Wildl. Manage.* 58:488–492.

Ortega, J. C, C. P. Ortega, and A. Cruz. 1993. Does Brown-headed Cowbird egg coloration influence Red-winged Blackbird responses towards nest contents? *Condor* 95:217–219.

Ortego, B. 1993. Brown-headed Cowbird population trends at a large winter roost in southwest Louisiana from 1974–1992. Austin, Tex.: North American Workshop on the Ecology and Management of Cowbirds.

Overmire, T. G. 1962. Nesting of the Bell Vireo in Oklahoma. *Condor* 64:75.

Packard, F. M. 1936. A Black-capped Chickadee victimized by the Eastern Cowbird. *Bird-Banding* 7:129–130.

Parkes, K. C., and E. R. Blake. 1965. Taxonomy and nomenclature of the Bronzed Cowbird. *Fieldiana-Zoology* 44:207–216.

Parsons, J., and L. F. Baptista. 1980. Crown color and dominance in the White-crowned Sparrow. *Auk* 97:807–815.

Paton, P. W. C. 1994. The effect of edge on avian nest success: How strong is the evidence? *Conserv. Biol.* 8:17–26.

Paul, R. T., and A. F. Schnapf. 1993. Florida region. *Am. Birds* 47:1101–1103.

Payne, R. B. 1965. Clutch size and number of eggs laid by Brown-headed Cowbirds. *Condor* 67:44–60.

———. 1967a. Gonadal responses of Brown-headed Cowbirds to long day lengths. *Condor* 69:289–297.

———. 1967b. Interspecific communication signals in parasitic birds. *Am. Nat.* 101: 363–375.

———. 1969. Giant Cowbird solicits preening from man. *Auk* 86:751–752.

———. 1973a. Behavior, mimetic songs and song dialects and relationships of the parasitic indigo birds (*Vidua*) of Africa. *A.O.U. Monogr.* 11:1–333.

———. 1973b. The breeding season of a parasitic bird, the Brown-headed Cowbird, in central California. *Condor* 75:80–99.

———. 1977. The ecology of brood parasitism in birds. *Ann. Rev. Ecol. Syst.* 8:1–28.

Payne, R. B., and L. L. Payne. 1990. Evicting and vocal begging behaviors in Australian Glossy Cuckoos. Abstract 221. American Ornithologists' Union 108th and Cooper Ornithological Society 60th Joint Annual Meeting, Los Angeles, Calif.

———. 1995. Song mimicry and association of brood parasitic indigobirds (*Vidua*) with Dybowski's Twinspot (*Eustichospiza dybowskii*). *Auk* 112:649–658.

Pearson, T. G., C. S. Brimley, and H. H. Brimley (revised by D. L. Wray and H. T. Davis). 1959. *Birds of North Carolina*. Raleigh: North Carolina Department of Agriculture State Museum.

Pearson, T. G., and J. Burroughs, eds. 1917. *Birds of America*. Nature Lover's Library. New York: University Society Inc.

Pease, C. M., and J. A. Grzybowski. 1995. Assessing the consequences of brood parasitism and nest predation on seasonal fecundity in passerine birds. *Auk* 112: 343–363.

Peer, B. D., and E. K. Bollinger. 1997. Explanations for the infrequent cowbird parasitism on Common Grackles. *Condor* 99:151–161.

Penard, F. P., and A. P. Penard. 1910. *De vogels van Guyana*. Guyana: Martinus Nijhoff.

Pereyra, J. A. 1938. Aves de la zona ribería nordeste de la provincia de Buenos Aires. *Memorias del Jardin Zoologico, La Plata* 9:1–304.

Peterjohn, B. G., and J. R. Sauer. 1993. Temporal and geographic patterns in population trends of Brown-headed Cowbirds. Abstract 1. Austin: North American Workshop on the Ecology and Management of Cowbirds.

Petersen, A., and H. Young. 1950. A nesting study of the Bronzed Grackle. *Auk* 67: 466–475.

Petit, D. R., J. F. Lynch, R. L. Hutto, J. G. Blake, and R. B. Waide. 1993. Management and conservation of migratory landbirds overwintering in the neotropics. In *Status and management of neotropical migratory birds*, ed. D. M. Finch and P. W. Stangel. U.S.D.A. Forest Serv. Gen. Tech. Rep. RM-229, 70–92.

Petit, L. J. 1991. Adaptive tolerance of cowbird parasitism by Prothonotary Warblers: A consequence of nestsite limitation? *Anim. Behav.* 41:425–432.

Phillips, A., J. Marshall, and G. Monson. 1964. *The birds of Arizona*. Tucson: University of Arizona Press.

Phillips, R. S. 1951. Nest location, cowbird parasitism, and nesting success of the Indigo Bunting. *Wilson Bull.* 63:206–207.

Picman, J. 1986. Attempted nest parasitism of the Marsh Wren by a Brown-headed Cowbird. *Condor* 88:381–382.

———. 1989. Mechanism of increased puncture resistance of eggs of Brown-headed Cowbirds. *Auk* 106:577–583.

Pinxten, R., O. Honotte, M. Eens, R. F. Verheyen, A. A. Dhondt, and T. Burke. 1993. Extra-pair paternity and intraspecific brood parasitism in the European Starling, *Sturnus vulgaris*: Evidence from DNA fingerprinting. *Anim. Behav.* 45:795–809.

Piper, W. H. 1994. Courtship, copulation, nesting behavior and brood parasitism in the Venezuelan Stripe-backed Wren. *Condor* 96:654–671.

Pitelka, F. A., and E. J. Koestner. 1942. Breeding behavior of Bell's Vireo in Illinois. *Wilson Bull.* 54:97–106.

Poiani, A., and M. A. Elgar. 1994. Cooperative breeding in the Australian avifauna and brood parasitism by cuckoos (Cuculidae). *Anim. Behav.* 47:697–706.

Post, W. 1992. Dominance and mating success in male Boat-tailed Grackles. *Anim. Behav.* 44:917–929.

Post, W., A. Cruz, and D. B. McNair. 1993. The North American invasions pattern of the Shiny Cowbird. *J. Field Ornithol.* 64:32–41.

Post, W., T. K. Nakamura, and A. Cruz. 1990. Patterns of Shiny Cowbird parasitism in St. Lucia and southwestern Puerto Rico. *Condor* 92:461–469.

Post, W., and K. W. Post. 1987. Roosting behavior of the Yellow-shouldered Blackbird. *Florida Field Naturalist* 15:93–105.

Post, W., and J. W. Wiley. 1976. The Yellow-shouldered Blackbird: Present and future. *Am. Birds* 30(1):13–20.

———. 1977a. Reproductive interactions of the Shiny Cowbird and the Yellow-shouldered Blackbird. *Condor* 79:176–184.

———. 1977b. The Shiny Cowbird in the West Indies. *Condor* 79:119–121.

———. 1992. The head-down display in Shiny Cowbirds and its relation to dominance behavior. *Condor* 94:999–1002.

Potter, E. F., and G. T. Whitehurst. 1981. Cowbirds in the Carolinas. *Chat* 45:57–68.

Prather, J. W., and A. Cruz. 1995. Breeding biology of Florida Prairie Warblers and Cuban Yellow Warblers. *Wilson Bull.* 107:475–484.

Prescott, K. W. 1947. Unusual behavior of a cowbird and Scarlet Tanager at a Red-eyed Vireo nest. *Wilson Bull.* 59:210.

Probst, J. R., and J. Weinrich. 1993. Relating Kirtland's Warbler population to changing landscape composition and structure. *Landscape Ecology* 8:257–271.

Purrington, R. D. 1993. Central southern region. *Am. Birds* 47:1115–1118.

Putnam, L. S. 1949. The life history of the Cedar Waxwing. *Wilson Bull.* 61:141–182.

Raffaele, H. A. 1983. *A guide to the birds of Puerto Rico and the Virgin Islands.* San Juan, Puerto Rico: Fundo Educativo Interamericano.

Rahn, H., L. Curran-Everett, and D. T. Booth. 1988. Eggshell differences between parasitic and nonparasitic Icteridae. *Condor* 90:962–964.

Ramo, C., and B. Busto. 1981. La reproduccion de un ave parasita: el Tordo-mirlo (*Molothrus bonariensis*) en los Llanos de Apure (Venezuela). *Donana-Acta Vertebrada* 8:214–224.

Ranney, J. W., M. C. Bruner, and J. B. Levenson. 1981. The importance of edge in the structure and dynamics of forest islands. In *Forest island dynamics in man-dominated landscapes*, ed. R. L. Burgess and D. M. Sharpe. New York: Springer Verlag, 67–95.

Rappole, J. H., and M. V. McDonald. 1994. Cause and effect in population declines of migratory birds. *Auk* 111:652–660.

Ratcliffe, L., and R. Weisman. 1987. Phase order recognition by Brown-headed Cowbirds. *Anim. Behav.* 35:1260–1262.

Reed, R. A. 1968. Studies of the Diederik Cuckoo (*Chrysococcyx caprius*) in the Transvaal. *Ibis* 110:321–331.

Rees, E. C., and N. Hillgarth. 1984. The breeding biology of captive Black-headed Ducks and the behavior of their young. *Condor* 86:242–250.

Regosin, J. V. 1994. Scissor-tailed Flycatchers eject Brown-headed Cowbird eggs. *J. Field Ornithol.* 65:508–511.

Rensch, B. 1925. Verhalten von Singvogeln bei Aenderung des Geleges. *Ornith. Monat.* 33:169–173.

Rich, T., and S. I. Rothstein. 1985. Sage Thrashers reject cowbirds eggs. *Condor* 87: 561–562.

Riddiford, N. 1986. Why do cuckoos *(Cuculus canorus)* use so many species of hosts? *Bird Study* 33:1–5.

Ridgely, R. S., and J. A Gwynne Jr. 1989. *A guide to the birds of Panama.* 2d ed. Princeton: Princeton University Press.

Ridgely, R. S., and G. Tudor. 1989. *The birds of South America: The oscine passerines,* vol 1. Austin: University of Texas Press.

Robbins, C. S., J. R. Sauer, R. S. Greenberg, and S. Droege. 1989. Population declines in North American birds that migrate to the neotropics. *Proc. Natl. Acad. Sci.* 86: 7658–7662.

Robbins, C. S., J. R. Sauer, and B. G. Peterjohn. 1993. Population trends and management opportunities for neotropical migrants. In *Status and management of neotropical migratory birds,* ed. D. M. Finch and P. W. Stangel. U.S.D.A. Forest Serv. Gen. Tech. Rep. RM-229.

Robbins, M. B., and D. A. Easterla. 1981. Range expansion of the Bronzed Cowbird with the first Missouri Record. *Condor* 83:270–272.

Roberts, A. 1907. Remarks on the breeding habits of the Pin-tailed Widow-bird *(Vidua principalis). J. South African Ornithological Union* 3:9–11.

Robertson, R. J., and R. F. Norman. 1976. Behavioral defenses to brood parasitism by potential hosts of the Brown-headed Cowbird. *Condor* 78:166–173.

———. 1977. The function and evolution of aggressive host behavior towards the Brown-headed Cowbird *(Molothrus ater). Can. J. Zool.* 55:508–518.

Robinson, S. K. 1985. Coloniality in the Yellow-rumped Cacique as a defense against nest predators. *Auk* 102:506–519.

———. 1986. Benefits, costs, and determinants of dominance in a polygynous oriole. *Anim. Behav.* 34:241–255.

———. 1988. Foraging ecology and host relationships of Giant Cowbirds in southeastern Peru. *Wilson Bull.* 100:224–235.

———. 1992. Population dynamics of breeding Neotropical migrants in a fragmented Illinois landscape. In *Ecology and conservation of neotropical migrant landbirds,* ed. J. M. Hagan III and D. W. Johnson. Washington, D.C.: Smithsonian Institution Press, 408–418.

Robinson, S. K., J. A. Grzybowski, S. I. Rothstein, M. C. Brittingham, L. J. Petit, and F. R. Thompson. 1993. Management implications of cowbird parasitism on neotropical migrant songbirds. In *Status and management of neotropical migratory birds*, ed. D. M. Finch and P. W. Stangel. U.S.D.A. Forest Serv. Gen. Tech. Rep. RM-229, 93–102.

Roby, D. D., and N. H. Jackson. 1992. Breeding biology of captive Brown-headed Cowbirds. Abstract 34. Ames, Iowa: American Ornithologists' Union Meeting.

Rodrígues, D. 1918. Costumbres del <<Pato Picazo>>. *El Hornero* 1:185–187.

Rohwer, F. C., and M. G. Anderson. 1988. Female biased philopatry, monogamy, and the timing of pair formation in migratory waterfowl. *Curr. Ornithol.* 5:187–221.

Rohwer, F. C., and S. Freeman 1989. The distribution of conspecific nest parasitism in birds. *Can. J. Zool.* 67:239–253.

Rohwer, S., and C. D. Spaw. 1988. Evolutionary lag versus bill-size constraints: A comparative study of the acceptance of cowbird eggs by old hosts. *Evol. Ecol.* 2: 27–36.

Rohwer, S., C. D. Spaw, and E. Røskaft. 1989. Costs to Northern Orioles of puncture-ejecting parasitic cowbird eggs from their nests. *Auk* 106:734–738.

Romanoff, A. L., and A. J. Romanoff. 1949. *The avian egg.* New York: John Wiley & Sons, Inc.

Rosene, W. 1969. *The Bobwhite Quail.* New Brunswick: Rutgers University Press.

Røskaft E., G. H. Orians, and L. D. Beletsky. 1990. Why do Red-winged Blackbirds accept eggs of Brown-headed Cowbirds? *Evol. Ecol.* 4:35–42.

Røskaft, E., S. Rohwer, and C. D. Spaw. 1993. Cost of puncture ejection compared with costs of rearing cowbird chicks for Northern Orioles. *Ornis Scandinavica* 24: 28–32.

Roth, R. R., and R. K. Johnson. 1993. Long-term dynamics of a Wood Thrush population breeding in a forest fragment. *Auk* 110:37–48.

Rothstein, S. I. 1971a. A reanalysis of the interspecific invitation to preening display as performed by the Brown-headed Cowbird (*Molothrus ater*). *Am. Zool.* 11:638.

———. 1971b. Observation and experiment in the analysis of interaction between brood parasites and their hosts. *Am. Nat.* 105:71–74.

———. 1972a. Eggshell thickness and its variation in the Cedar Waxwing. *Wilson Bull.* 84:469–474.

———. 1972b. Territoriality and mating system in the parasitic Brown-headed Cowbird (*Molothrus ater*) as determined from captive birds. *Am. Zool.* 12:659.

———. 1973. Variation in the incidence of hatching failure in the Cedar Waxwing and other species. *Condor* 75:164–169.

———. 1974. Mechanisms of avian egg recognition: Possible learned and innate factors. *Auk* 91:796–807.

———. 1975a. An experimental and teleonomic investigation of avian brood parasitism. *Condor* 77:250–271.

———. 1975b. Evolutionary rates and host defenses against avian brood parasitism. *Am. Nat.* 109:161–176.

———. 1975c. Mechanisms of avian egg-recognition: Do birds know their own eggs? *Anim. Behav.* 23:268–278.

———. 1976a. Cowbird parasitism of the Cedar Waxwing and its evolutionary implications. *Auk* 93:498–509.

———. 1976b. Experiments on defenses Cedar Waxwings use against cowbird parasitism. *Auk* 93:675–691.

———. 1977a. Cowbird parasitism and egg recognition of the Northern Oriole. *Wilson Bull.* 89:21–32.

———. 1977b. The preening invitation or head-down display of parasitic cowbirds: I. Evidence for intraspecific occurrence. *Condor* 79:13–23.

———. 1978a. Geographic variation in the nestling coloration of parasitic cowbirds. *Auk* 95:152–160.

———. 1978b. Mechanisms of avian egg-recognition: Additional evidence for learned components. *Anim. Behav.* 26:671–677.

———. 1980. The preening invitation or head-down display of parasitic cowbirds: II. Experimental analyses and evidence for behavioural mimicry. *Behaviour* 75:148–184.

———. 1982a. Mechanisms of avian egg recognition: Which egg parameters elicit responses by rejecter species? *Behav. Ecol. Sociobiol.* 11:229–239.

———. 1982b. Successes and failures in avian egg and nestling recognition with comments on the utility of optimality reasoning. *Am. Zool.* 22:547–560.

———. 1986. A test of optimality: Egg recognition in the Eastern Phoebe. *Anim. Behav.* 34:1109–1119.

———. 1990. A model system for coevolution: Avian brood parasitism. *Ann. Rev. Ecol. Syst.* 21:481–508.

———. 1993. An experimental test of the Hamilton-Orians hypothesis for the origin of avian brood parasitism. *Condor* 95:1000–1005.

Rothstein, S. I., and R. C. Fleischer. 1987a. Brown-headed Cowbirds learn flight whistles after the juvenile period. *Auk* 104:512–516.

———. 1987b. Vocal dialects and their possible relation to honest status signalling in the Brown-headed Cowbird. *Condor* 89:1–23.

Rothstein, S. I., and A. O'Loghlen. Interpreting experimental studies of host aggression towards brood parasites. Unpubl. ms.

Rothstein, S. I., J. Verner, and E. Stevens. 1980. Range expansion and diurnal changes in dispersion of the Brown-headed Cowbird in the Sierra Nevada. *Auk* 97:253–267.

———. 1984. Radio-tracking confirms a unique diurnal pattern of spatial occurrence in the parasitic Brown-headed Cowbird. *Ecology* 65:77–88.

Rothstein, S. I., J. Verner, E. Stevens, and L. V. Ritter. 1987. Behavioral differences among sex and age classes of the Brown-headed Cowbird and their relation to the efficacy of a control program. *Wilson Bull.* 99:322–337.

Rothstein, S. I., D. A. Yokel, and R. C. Fleischer. 1986. Social dominance, mating and spacing systems, female fecundity, and vocal dialects in captive and free-ranging Brown-headed Cowbirds. *Curr. Ornithol.* 3:127–185.

———. 1988. The agonistic and sexual functions of vocalizations of male Brown-headed Cowbirds, *Molothrus ater*. *Anim. Behav.* 36:73–86.

Rowley, I. A confusion of cockatoos. 1991. *Natural History* (November):44–51.

Rowley, I., and G. Chapman. 1986. Cross-fostering, imprinting, and learning in two sympatric species of cockatoos. *Behaviour* 96:1–16.

Rowley, J. S. 1939. Breeding birds of Mono County, California. *Condor* 41:247–254.

———. 1962. Nesting of the birds of Morelos, Mexico. *Condor* 64:253–272.

Salvador, S. A. 1983. Parasitismo de cría del renegrido (*Molothrus bonariensis*) en Villa María, Córdoba, Argentina (Aves: Icteridae). *Historia Natural* 3:149–158.

———. 1984. Estudio de parasitismo de cría del Renegrido (*Molothrus bonariensis*) en Calandria (*Mimus saturninus*), en Villa María, Córdoba. *El Hornero* 12:141–149.

Sato, T. 1986. A brood parasitic catfish of mouthbrooding cichlid fishes in Lake Tanganyika. *Nature* 323:58–59.

Sauer, J. R., and S. Droege. 1992. Geographic patterns in population trends of neotropical migrants in North America. In *Ecology and conservation of neotropical migrant landbirds*, ed. J. M. Hagan III and D. W. Johnston. Washington, D.C.: Smithsonian Institution Press, 26–42.

Savage, D. 1897. Observations on the cowbird. *Iowa Ornithologist* 3:4–7.

Sayler, R. D. 1996. Behavioral interactions among brood parasites with precocial young: Canvasbacks and Redheads on the Delta Marsh. *Condor* 98:801–809.

Schäfer, E. 1957. Les Conotos: Etude comparative de *Psarocolius angustifrons* et *Psarocolius decumanus*. *Bonner Zool. Beit.*, Sonderheft.

Schönwetter, M. 1983. *Handbuch der Oologie*, ed. W. Meise. Berlin: Akademie Verlag.

Schram, B. 1994. An open solicitation for cowbird recipes. *Birding* 26:254–257.

Schrantz, F. C. 1943. Nest life of the Eastern Yellow Warbler. *Auk* 60:367–387.

Scott, D. M. 1963. Changes in the reproductive activity of the Brown-headed Cowbird within the breeding season. *Wilson Bull.* 75:123–129.

———. 1977. Cowbird parasitism on the Gray Catbird at London, Ontario. *Auk* 94:18–27.

———. 1978. Using sizes of unovulated follicles to estimate the laying rate of the Brown-headed Cowbird. *Can. J. Zool.* 56:2230–2234.

———. 1991. The time of day of egg laying by the Brown-headed Cowbird and other Icterines. *Can. J. Zool.* 69:2093–2099.

Scott, D. M., and C. D. Ankney. 1980. Fecundity of the Brown-headed Cowbird in southern Ontario. *Auk* 97:677–683.

———. 1983. The laying cycle of Brown-headed Cowbirds: Passerine chickens? *Auk* 100:583–592.

Scott, D. M., R. E. Lemon, and J. A. Darley. 1987. Relaying interval after nest failure in Gray Catbirds and Northern Cardinals. *Wilson Bull.* 99:708–712.

Scott, D. M., and A. L. A. Middleton. 1968. The annual testicular cycle of the Brown-headed Cowbird (*Molothrus ater*). *Can. J. Zool.* 46:77–87.

Scott, D. M., P. J. Weatherhead, and C. D. Ankney. 1992. Egg-eating by female Brown-headed Cowbirds. *Condor* 94:579–584.

Scott, P. E., and B. R. McKinney. 1994. Brown-headed Cowbird removes Blue-Gray Gnatcatcher nestlings. *J. Field Ornithol.* 65:363–364.

Scott, T. W., and J. M. Grumstrup-Scott. 1983. Why do Brown-headed Cowbirds perform the head-down display? *Auk* 100:139–148.

Sealy, S. G. 1992. Removal of Yellow Warbler eggs in association with cowbird parasitism. *Condor* 94:40–54.

———. 1994. Observed acts of egg destruction, egg removal, and predation on nests of passerine birds at Delta Marsh, Manitoba. *Can. Field-Nat.* 108:41–51.

———. 1995. Burial of cowbird eggs by parasitized Yellow Warblers: An empirical and experimental study. *Anim. Behav.* 49:877–889.

———. 1996. Evolution of host defenses against brood parasitism: Implications of puncture-ejection by a small passerine. *Auk* 113:346–355.

Sealy, S. G., and R. C. Bazin. 1995. Low frequency of observed cowbird parasitism on Eastern Kingbirds: Host rejection, effective nest defense, or parasite avoidance? *Behav. Ecol.* 6:140–145.

Sealy, S. G., K. A. Hobson, and J. V. Briskie. 1989. Responses of Yellow Warblers to experimental intraspecific brood parasitism. *J. Field Ornithol.* 60:224–229.

Sealy, S. G., and D. L. Neudorf. 1995. Male Northern Orioles eject cowbird eggs: Implications for the evolution of rejection behavior. *Condor* 97:369–375.

Sealy, S. G., D. L. Neudorf, and D. P. Hill. 1995. Rapid laying by Brown-headed Cowbirds *Molothrus ater* and other parasitic birds. *Ibis* 137:76–84.

Searcy, W. A. 1979. Morphological correlates of dominance in captive male Red-winged Blackbirds. *Condor* 81:417–420.

Sedgwick, J. A, and F. L. Knopf. 1988. A high incidence of Brown-headed Cowbird parasitism of Willow Flycatchers. *Condor* 90:253–256.

Seel, D. C. 1973. Egg-laying by the cuckoo. *Br. Birds* 66:528–535.

Selander, R. K. 1960. Failure of estrogen and prolactin treatment to induce brood patch formation in Brown-headed Cowbirds. *Condor* 62:65.

———. 1964. Behavior of captive South American cowbirds. *Auk* 81:394–402.

Selander, R. K., and D. R. Giller. 1960. First-year plumages of the Brown-headed Cowbird and Red-winged Blackbird. *Condor* 62:202–214.

Selander, R. K., and L. L. Kuich. 1963. Hormonal control and development of the incubation patch in Icterids, with notes on behavior of cowbirds. *Condor* 65:73–90.

Selander, R. K., and C. J. La Rue, Jr. 1961. Interspecific preening invitation display of parasitic cowbirds. *Auk* 78:473–504.

Seppa, J. 1969. The cuckoo's ability to find a nest where it can lay an egg. *Ornis Fennica* 46:78–79.

Shake, W. F., and J. P. Mattsson. 1975. Three years of cowbird control: An effort to save the Kirtland's Warbler. *Jack-Pine Warbler* 53:48–53.

Sharp, B., E. 1995. Brown-headed Cowbirds and grazing on national forests in the Pacific Northwest. *Northwestern Naturalist* 76:121.

Sheppard, J. M. 1996. Nestling Kentucky Warblers and cowbird attacked by Brown-headed Cowbird. *J. Field Ornithol.* 67:384–386.

Sherry, T. W., and R. T. Holmes. 1992. Population fluctuations in a long-distance neo-tropical migrant: Demographic evidence for the importance of breeding season events in the American Redstart. In *Ecology and conservation of neotropical migrant landbirds*, ed. J. M. Hagan III and D. W. Johnson. Washington, D.C.: Smithsonian Institution Press, 431–442.

———. 1993. Are populations of neotropical migrant birds limited in summer or winter? Implications for management. In *Status and management of neotropical migratory birds*, ed. D. M. Finch and P. W. Stangel.U.S.D.A. Forest Serv. Gen. Tech. Rep. RM-229, 47–57.

Sick, H. 1958. Notas biológicas sôbre o gaudério, *Molothrus bonariensis* (Gmelin) (Icteridae, Aves). *Rev. Bras. Biol.* 18:417–431.

Siegfried, W. R. 1976. Breeding biology and parasitism in the Ruddy Duck. *Wilson Bull.* 88:566–574.

Silverin, B. 1980. Effects of long-acting testosterone treatment of free-living Pied Flycatchers, *Ficedula hypoleuca*, during the breeding period. *Anim. Behav.* 28:823–840.

Simpkin, J. L., and A. A. Gubanich. 1991. Ash-throated Flycatchers (*Myiarchus cinerascens*) raise Mountain Bluebird (*Sialia currucoides*) young. *Condor* 93:461–462.

Skutch, A. F. 1949. Do tropical birds rear as many young as they can nourish? *Ibis* 91:430–455.

———. 1954. Life histories of Central American birds: I. Pacific Coast. *Avifauna* 31:1–448.

———. 1960. Life histories of Central American birds: II. Pacific Coast. *Avifauna* 34:1–593.

Slack, R. D. 1976. Nest guarding behavior by male Gray Catbirds. *Auk* 93:292–300.

Small, M. F., and M. L. Hunter. 1988. Forest fragmentation and avian predation in forested landscapes. *Oecologia* 76:62–64.

Smith, H. M. 1949. Irregularities in the egg laying behavior of the eastern cowbird. *J. Colorado-Wyoming Acad. Sci.* 4:60.

Smith, J. N. M. 1981. Cowbird parasitism, host fitness, and age of the host female in an island Song Sparrow population. *Condor* 83:152–161.

———. 1994. Cowbirds: Conservation villains or convenient scapegoats? *Birding* 26:257–259.

Smith, J. N. M., and P. Arcese. 1994. Brown-headed Cowbirds and an island population of Song Sparrows: A 16-year study. *Condor* 96:916–934.

Smith, J. N. M., P. Arcese, and I. G. McLean. 1984. Age, experience, and enemy recognition by wild Song Sparrows. *Behav. Ecol. Sociobiol.* 14:101–106.

Smith, J. P., and R. J. Atkins. 1979. Cowbird parasitism on Common Bushtit nest. *Wilson Bull.* 91:122–123.

Smith, N. G. 1968. The advantage of being parasitized. *Nature* 219:690–694.

———. 1979. Alternate responses by hosts to parasites which may be helpful or harmful. In *Host-Parasite interfaces*. New York: Academic Press, 7–15.

———. 1980. *Some evolutionary, ecological, and behavioural correlates of communal nesting by birds with wasps or bees*, vol 2. Berlin: XVII International Congress of Ornithology.

———. 1982. La vantaja de ser parasitado. In *Evolucion en los tropicos*, ed. G. A. de Alba and R. W. Rubinoff. Panama: Smithsonian Tropical Research Institute, 49–56.

———. 1983. *Zarhynchus wagleri* (Oropendola Cabelicastaña, Oropendola, Chestnut-headed Oropendola). In *Symposium on co-evolutional systems in birds*, ed. D. H. Jansen. Chicago: University of Chicago Press, 9:614–616.

Smith, P. W., and A. Sprunt IV. 1987. The Shiny Cowbird reaches the United States: Will the scourge of the Caribbean impact Florida's avifauna too? *Am. Birds* 41: 370–371.

Smith, S., and E. Hosking. 1955. *Birds fighting: Experimental studies of the aggressive displays of some birds*. London: Faber and Faber.

Smith, T. S. 1972. Cowbird parasitism of Western Kingbird and Baltimore Oriole nests. *Wilson Bull.* 84:497.

Snyder, L. L. 1957. Changes in the avifauna of Ontario. In *Changes in the Fauna of Ontario*, ed. F. A. Urquhart. Toronto: University of Toronto Press, 26–42.

Soler, M. 1990. Relationships between the Great Spotted Cuckoo *Clamator glandarius* and its corvid hosts in a recently colonized area. *Ornis Scandinavica* 21:212–223.

Soler, M., and A. P. Møller. 1990. Duration of sympatry and coevolution between the Great Spotted Cuckoo and its magpie host. *Nature* 343:748–750.

Southern, W. E., and L. K. Southern. 1980. A summary of the incidence of cowbird parasitism in northern Michigan from 1911–1978. *Jack-Pine Warbler* 58:77–84.

Spaw, C. D., and S. Rohwer. 1987. A comparative study of eggshell thickness in cowbirds and other passerines. *Condor* 89:307–318.

Spencer, K. G. 1953. *The lapwing in Britain*. London: A. Brown and Sons.

Sprunt, A., Jr. 1954. *Florida bird life*. New York: Coward-McCann, Inc.

Stacey, P. B., and W. D. Koenig, eds. 1990. *Cooperative breeding in birds*. Cambridge: Cambridge University Press.

Stedman, S. J. 1993. Central southern region. *Am. Birds* 47:266–269.

Stewart, B. G., W. A. Carter, and J. D. Tyler. 1988. Third known nest of the Slaty Vireo, *Vireo brevipennis* (Vireonidae), in Colima, Mexico. *Southwestern Naturalist* 33:252–253.

Stewart, J. R., Jr. 1976. Central southern region. *Am. Birds* 30:965–969.

Stewart, P. A. 1963. Abnormalities among Brown-headed Cowbirds trapped in Alabama. *Bird-Banding* 34:199–202.

Stewart, R. E. 1953. A life history study of the Yellow-throat. *Wilson Bull.* 65:99–115.

Stoner, E. A. 1919. Cowbird study in Iowa. *Oologist* 36:80–81.

Sturm, L. 1945. A study of the nesting activites of the American Redstart. *Auk* 62: 189–206.

Sugden, J. W. 1947. Exotic eggs in nests of California Gulls. *Condor* 49:93–96.

Sugden, L. G. 1980. Parasitism of Canvasback nests by Redheads. *J. Field Ornithol.* 51:361–364.

Sutton, G. M. 1928. The birds of Pymatung Swamp and Commeaut Lake, Crawford County, Pennsylvania. *Ann. Carnegie Mus.* 18:19–239.

Swynnerton, C. F. M. 1918. Rejection by birds of eggs unlike their own: With remarks on some of the cuckoo problems. *Ibis* 6:127–154.

Takasu, F., K. Kawasaki, H. Nakamura, J. E. Cohen, and N. Shigesada. 1993. Modeling the population dynamics of a cuckoo-host association and the evolution of host defenses. *Am. Nat.* 142:819–839.

Talent, L. G., G. L. Krapu, and R. L. Jarvis. 1981. Effects of Redhead nest parasitism on Mallards. *Wilson Bull.* 93:562–563.

Tate, J., Jr. 1967. Cowbird removes warbler nestling from nest. *Auk* 84:422.

Taylor, W. K. 1966. Additional records of Black-tailed Gnatcatchers parasitized by the dwarf Brown-headed Cowbird. *Am. Midl. Nat.* 76:242–243.

Teather, K. L., and R. J. Robertson. 1985. Female spacing patterns in Brown-headed Cowbirds. *Can. J. Zool.* 63:218–222.

———. 1986. Pair bonds and factors influencing the diversity of mating systems in Brown-headed Cowbirds. *Condor* 88:63–69.

Teather, K. L., and P. J. Weatherhead. 1995. The influence of age and access to females on dominance in captive male Brown-headed Cowbirds (*Molothrus ater*). *Can. J. Zool.* 73:1012–1018.

Temple, S. A. 1986. Predicting impacts of habitat fragmentation on forest birds: A comparison of two models. In *Wildlife 2000: Modeling habitat relationships of terrestrial vertebrates*, ed. J. Verner, M. L. Morrison, and C. J. Ralph. Madison: University of Wisconsin Press, 301–304.

Temple, S. A., and J. R. Cary. 1988. Modelling dynamics of habitat-interior bird populations in fragmented landscapes. *Conserv. Biol.* 2:340–347.

Terborgh, J. 1992. Why American songbirds are vanishing. *Sci. Am.* 166:98–104.

Terrill, L. M. 1961. Cowbird hosts in southern Quebec. *Can. Field-Nat.* 75:2–11.

Thompson, C. F., and B. M. Gottfried. 1976. How do cowbirds find and select nests to parasitize? *Wilson Bull.* 88:673–675.

———. 1981. Nest discovery and selection by Brown-headed Cowbirds. *Condor* 83:268–269.

Thompson, F. R. 1994. Temporal and spatial patterns of breeding Brown-headed Cowbirds in the midwestern United States. *Auk* 111:979–990.

Thurber, W. A., and A. Villeda. 1980. Notes on parasitism by Bronzed Cowbirds in El Salvador. *Wilson Bull.* 92:112–113.

Trail, P. W., and L. F. Baptista. 1993. The impact of Brown-headed Cowbird parasitism on populations of the Nuttall's White-crowned Sparrow. *Conserv. Biol.* 7:309–315.

Trine, C. L. 1993. Multiple parasitism by Brown-headed Cowbirds in Wood Thrush nests: Effects on fledgling production of parasite and hosts. Abstract 11. Austin: North American Workshop on the Ecology and Management of Cowbirds.

Trost, C. H., and C. L. Webb. 1986. Egg moving by two species of corvid. *Anim. Behav.* 34:294–295.

Twomey, A. C. 1948. California Gulls and exotic eggs. *Condor* 50:97–100.

U. S. Fish and Wildlife Service. 1986. North American Bird Banding Techniques, Part 6. Washington, D.C.: U.S. Fish and Wildlife Service, Department of Interior.

Uyehara, J. C., and P. M. Narins. 1995. Nest defense by Willow Flycatchers to brood-parasitic intruders. *Condor* 97:361–368.

Van Horne, B. 1983. Density as a misleading indicator of habitat quality. *J. Wildl. Manage.* 47:893–901.

Van Velzen, W. T. 1972. Distribution and abundance of the Brown-headed Cowbird. *Jack-Pine Warbler* 50:110–113.

Vehrencamp, S. L. 1977. Relative fecundity of parental effort in communally nesting anis, *Crotophaga sulcirostris. Science* 197:403–404.

Verner, J., and L. V. Ritter. 1983. Current status of the Brown-headed Cowbird in the Sierra National Forest. *Auk* 100:355–368.

Victoria, J. K. 1972. Clutch characteristics and egg discriminative ability of the African Weaverbird *(Ploceus cucullatus). Ibis* 114:367–376.

Visher, S. S. 1909. The capture of the Red-eyed Cowbird in Arizona. *Auk* 26:307.

Walkinshaw, L. H. 1949. Twenty-five eggs apparently laid by a cowbird. *Wilson Bull.* 61:82–85.

———. 1961. The effect of parasitism by the Brown-headed Cowbird on *Empidonax* flycatchers in Michigan. *Auk* 78:266–268.

———. 1972. Kirtland's Warbler: Endangered. *Am. Birds* 26:3–9.

Ward, D., A. K. Lindholm, and J. N. M. Smith. 1996. Multiple parasitism of the Red-winged Blackbird: Further experimental evidence of evolutionary lag in a common host of the Brown-headed Cowbird. *Auk* 113:408–413.

Warren, R. P., and R. A. Hinde. 1959. The effect of oestrogen and progesterone on the nest building of domesticated canaries. *Anim. Behav.* 7:209.

Weatherhead, P. J. 1989. Sex ratios, host-specific reproductive success, and impact of Brown-headed Cowbirds. *Auk* 106:358–366.

———. 1991. The adaptive value of thick-shelled eggs for Brown-headed Cowbirds. *Auk* 108:196–198.

Weatherhead, P. J., and G. F. Bennett. 1992. Ecology of parasitism of Brown-headed Cowbirds by Haematozoa. *Can. J. Zool.* 70:1–7.

Weatherhead, P. J., and K. L. Teather. 1987. The paradox of age-related dominance in Brown-headed Cowbirds *(Molothrus ater). Can J. Zool.* 65:2354–2357.

Webb, J. S., and D. K. Wetherbee. 1960. Southeastern breeding range of the Brown-headed Cowbird. *Bird Banding* 31:83–87

Webster, M. S. 1994. Interspecific brood parasitism of Montezuma Oropendolas by Giant Cowbirds: Parasitism or mutualism? *Condor* 96:794–798.

Weller, M. 1959. Parasitic egg-laying in the Redhead *(Aythya americana)* and other North American Anatidae. *Ecol. Monogr.* 29:333–365.

Weller, M. W. 1967a. Notes on plumages and weights of the Black-headed Duck, *Heteronetta atricapilla. Condor* 69:133–145.

———. 1967b. Notes on some marsh birds of Cape San Antonio, Argentina. *Ibis* 109: 391–411.

————. 1968. The breeding biology of the parasitic Black-headed Duck. *Living Bird* 7:169–207.

Welty, J. C. 1982. *The life of birds*. Philadelphia: Saunders College Publications.

West, M. J., A. P. King, and D. H. Eastzer. 1981. The cowbird: Reflections on development from an unlikely source. *Am. Sci.* 69:56–66.

West, R. L., and N. Wamer. 1993a. Florida region. *Am. Birds* 47:83–86.

————. 1993b. Florida region. *Am. Birds* 47:250–252.

Whitcomb, R. F., C. S. Robbins, J. F. Lynch, B. L. Whitcomb, M. K. Klimkiewicz, and D. Bystrak. 1981. Effects of forest fragmentation on avifauna of the eastern deciduous forests. In *Forest island dynamics in man-dominated landscapes*, ed. R. L. Burgess and D. M. Sharpe. New York: Springer-Verlag, 125–205.

Wiens, J. A. 1963. Aspects of cowbird parasitism in southern Oklahoma. *Wilson Bull.* 75:130–139.

Wigley, T. B., and J. M. Sweeney. 1993. Cooperative partnerships and the role of private landowners. In *Status and management of neotropical migratory birds*, ed. D. M. Finch and P. W. Stangel. U.S.D.A. Forest Serv. Gen. Tech. Rep. RM-229, 39–44.

Wilcove, D. S. 1985. Nest predation in forest tracts and the decline of migratory songbirds. *Ecology* 66:1211–1214.

Wiley, J. W. 1985. Shiny Cowbird parasitism in two avian communities in Puerto Rico. *Condor* 87:165–176.

————. 1988. Host selection of the Shiny Cowbird. *Condor* 90:289–303.

Wiley, J. W., W. Post, and A. Cruz. 1991. Conservation of the Yellow-shouldered Blackbird *Agelaius xanthomus*, an endangered West Indian species. *Biol. Conserv.* 55:119–138.

Wiley, R. H., and M. S. Wiley. 1980. Spacing and timing in the nesting ecology of a tropical blackbird: Comparison of populations in different environments. *Ecol. Monogr.* 50:153–178.

Williams, M. D. 1981. Discovery of the nest and eggs of Cinereous Finch (*Piezorhina cinerea*), a Peruvian endemic. *Auk* 98:187–189.

Williamson, S. C., J. K. Detling, J. L. Dodd, and M. I. Dyer. 1989. Experimental evaluation of the grazing optimization hypothesis. *J. Range Manage.* 42:149–152.

Willis, E. O., and E. Eisenmann. 1979. A revised list of birds of Barro Colorado Island, Panama. *Smithsonian Contrib. Zool.* 291:1–31.

Willis, E. O., and Y. Oniki. 1985. Bird specimens new for the state of São Paulo, Brazil. *Rev. Bras. Biol.* 45:105–108.

Wilson, A. S. 1923. Huevos de pato en un nido de Chimango. *El Hornero* 3:192.

Wilson, E. O., and E. O. Willis. 1975. Applied biogeography. In *Ecology and evolution of communities*, ed. M. L. Cody and J. M. Diamond. Cambridge: Belknap Press, 522–534.

Wittenberger, J. F., and R. L. Tilson. 1980. The evolution of monogamy: Hypotheses and evidence. *Ann. Rev. Ecol. Syst.* 11:197–232.

Wolf, L. 1987. Host-parasite interactions of Brown-headed Cowbirds and Dark-eyed Juncos in Virginia. *Wilson Bull.* 99:338–350.

Woodward, P. W. 1983. Behavioral ecology of fledgling Brown-headed Cowbirds and their hosts. *Condor* 85:151–163.

Woodworth, B. L. 1993. Ecology and behavior of the Shiny Cowbird (*Molothrus bonariensis*) in a dry subtropical forest. Abstract 21. Austin: North American Workshop on the Ecology and Management of Cowbirds.

Wunderle, J. M., and K. H. Pollock. 1985. The Bananaquit-wasp nesting association and a random choice model. *Ornithol. Monogr.* 36:595–603.

Wyllie, I. 1981. *The cuckoo.* New York: Universe Books.

Yahner, R. H., and C. A. DeLong. 1992. Avian predation and parasitism on artificial nests and eggs in two fragmented landscapes. *Wilson Bull.* 104:162–168.

Yahner, R. H., and D. P. Scott. 1988. Effects of forest fragmentation on depredation of artificial nests. *J. Wildl. Manage.* 52:158–161.

Yezerinac, S. M., and K. W. Dufour. 1994. On testing the Hamilton-Orians hypothesis for the origin of brood parasitism. *Condor* 96:1115–1116.

Yokel, D. A. 1986. Monogamy and brood parasitism: An unlikely pair. *Anim. Behav.* 34:1348–1358.

———. 1989a. Intrasexual aggression and the mating behavior of Brown-headed Cowbirds: Their relation to population densities and sex ratios. *Condor* 91:43–51.

———. 1989b. Payoff asymmetries in contests among male Brown-headed Cowbirds. *Behav. Ecol. Sociobiol.* 24:209–216.

Yokel, D. A., and S. I. Rothstein. 1991. The basis for female choice in an avian brood parasite. *Behav. Ecol. Sociobiol.* 29:39–45.

Yom Tov, Y. 1980. Intraspecific nest parasitism in birds. *Biol. Rev.* 55:93–108.

Yom Tov, Y., G. M. Dunnet, and A. Anderson. 1974. Intraspecific nest parasitism in the starling *Sturnus vulgaris*. *Ibis* 116:87–90.

Young, C. G. 1929. A contribution to the ornithology of the coastland of British Guiana: Part III. *Ibis* 12 ser., 5:221–261.

Young, H. 1963. Breeding success of the cowbird. *Wilson Bull.* 75:115–122.

Zimmerman, J. L. 1983. Cowbird parasitism of Dickcissels in different habitats and at different nest densities. *Wilson Bull.* 95:7–22.

Index

About the Author

Dr. Catherine Ortega received her Bachelor of Arts, Summa Cum Laude, and doctoral degree from the Department of Environmental, Population, and Organismic Biology at the University of Colorado, Boulder, in 1987 and 1991, respectively, under the guidance of Dr. Alexander Cruz. Her interest in cowbirds began in 1983, when she worked with Shiny Cowbirds in Puerto Rico. In 1984, she began working with Brown-headed Cowbirds and their relationship with Red-winged Blackbirds and Yellow-headed Blackbirds in cattail marshes near Boulder, Colorado. Among several awards she won for this study was the Creative Works Award from the University of Colorado. Dr. Ortega is an assistant professor in the Department of Biology at Fort Lewis College in Durango, Colorado. Currently, her research interests focus on the management of cowbirds and predators. In particular, she is interested in the effects of cowbirds and predators on avian communities, and how grazing and different grazing regimes affect the nesting success of migratory songbirds in western landscapes. Together with Dr. Joseph Ortega, she has been conducting one of the longest-term demographic studies of Brown-headed Cowbirds, which began in 1992, in southwestern Colorado. Her recent scientific publications include papers on topics such as improved methods of aging male Brown-headed Cowbirds, effects of researcher activity on the nesting success of birds, and predation patterns.